Nanostructured Light-Emitters

Nanostructured Light-Emitters

Editor

Hieu Pham Trung Nguyen

MDPI • Basel • Beijing • Wuhan • Barcelona • Belgrade • Manchester • Tokyo • Cluj • Tianjin

Editor
Hieu Pham Trung Nguyen
New Jersey Institute of Technology
USA

Editorial Office
MDPI
St. Alban-Anlage 66
4052 Basel, Switzerland

This is a reprint of articles from the Special Issue published online in the open access journal *Micromachines* (ISSN 2072-666X) (available at: https://www.mdpi.com/journal/micromachines/special_issues/Nanostructured_Light_Emitters).

For citation purposes, cite each article independently as indicated on the article page online and as indicated below:

LastName, A.A.; LastName, B.B.; LastName, C.C. Article Title. *Journal Name* **Year**, *Article Number*, Page Range.

ISBN 978-3-03936-904-1 (Hbk)
ISBN 978-3-03936-905-8 (PDF)

Contents

About the Editor

Hieu Pham Trung Nguyen is an Associate Professor of Electrical Engineering at New Jersey Institute of Technology (NJIT), and the Director of the Nano-optoelectronic Materials and Devices Laboratory. He received the Ph.D. degree in Electrical Engineering from McGill University, Canada in 2012. He has authored/coauthored more than 160 publications including journal papers and conference presentations. His current research interests focus on developing high-performance nanophotonic devices using nanostructures and their applications in a variety of areas, including light emitting diodes, lasers, photodiodes, and solar cells. He is a recipient of the 2020 National Science Foundation CAREER Award, the 2019 Saul K Fenster Innovation in Engineering Education Ward, the SPIE scholarship in Optics and Photonics 2012, the Best Student Paper Award at the IEEE Photonics Conference 2011, and the Outstanding Student Paper Award at the 28th North American Molecular Beam Epitaxy Conference, 2011.

micromachines

MDPI

Editorial

Editorial of Special Issue "Nanostructured Light-Emitters"

Hieu P. T. Nguyen

Department of Electrical and Computer Engineering, New Jersey Institute of Technology, Newark, NJ 07102, USA; hieu.p.nguyen@njit.edu; Tel.: +1-973-596-3523

Received: 10 June 2020; Accepted: 20 June 2020; Published: 21 June 2020

Significant progress has been made in the development of nanophotonic devices and the use of nanostructured materials for optoelectronic devices, including light-emitting diodes (LEDs) and laser diodes, has recently attracted tremendous attention due to the fact of their unique geometry. Nanostructures in small dimensions, comprising nanowires, nanotubes, nanoparticles, etc., can be perfectly integrated into a variety of technological platforms, offering novel physical and chemical properties for high-performance light-emitting devices. Exploring new nanostructured light-emitters and their emerging practical applications is necessary and has attracted considerable investigations from industry and academia.

This Special Issue contains 3 review papers and 11 research articles, presenting the most recent advances in the device design, simulation, synthesis, characterization and application of such novel nanostructures. The research areas on nanophotonics reported in this issue are broad, covering from ultraviolet (UV) to visible and infrared wavelength regimes. The authors presented their novel results on different types of devices that include inorganic III-V based LEDs, organic LEDs (OLEDs), perovskite LEDs (PeLEDs), and other light emitters/absorbers.

Wei et al. [1] presented a comprehensive review on β-Ga_2O_3 with details in material properties, growth approach, doping, and its application in vertical structure LEDs. Study on β-Ga_2O_3 and its function as the substrate for vertical structure LEDs have been intensively investigated. Some of the key advantages of β-Ga_2O_3 are high *n*-type conductivity, low lattice mismatch with III-nitrides, and high transparency (>80%) in blue and UV region, enabling β-Ga_2O_3 as a highly potential substrates for high-performance light-emitters.

In another review paper, Zhou et al. [2] summarized recent studies on the interface/surface properties of low-dimensional materials and structures. Importantly, the performance of nanostructured LEDs can be significantly improved by engineering their surface/interface characteristics with a focus on the surface/interface purification, quantum dots-emitting layer, surface ligands, and optimization of device architecture.

The recent progress made in AlGaN nanowires for UV LEDs is reported in the third review paper, entitled "AlGaN nanowires for UV LEDs: Recent progress, challenges, and prospects" [3]. In this paper, Zhao et al. [3] discussed recent developments, the prospects, and the general challenges of AlGaN nanowire UV LEDs. Moreover, AlGaN nanowire UV LEDs on Si, foreign substrates as well as patterned growth approaches with the related device performance are presented.

Enhancing light output power of III-nitride LEDs has been an emerging topic toward the achievement of high efficiency LEDs. In this Special Issue, several approaches have been reported for the enhanced light extraction efficiency (LEE) of III-nitride LEDs. Lei et al. [4] demonstrated that the light output power of the InGaN/GaN LEDs was increased proximately 1.52 times by employing liquid-phase-deposited silver nanoparticle embedded ZnO (LPD-Ag NP/ZnO) thin-film on the top surface of the LEDs. The LPD-Ag NP/ZnO layer served as a window layer in InGaN/GaN LEDs, providing effective surface texture and localized surface plasmon coupling effect for the enhanced light output power of the related InGaN/GaN LEDs. Choi et al. [5]. reported their study on the enhanced photon emission efficiency of AlGaN based UV LEDs by engineering the surface plasmon effect of Pt nanoparticles. They demonstrated that the emission intensity of the surface plasmon-based UV LEDs could be increased 70% higher than that of UV LEDs without using this approach. In another paper, Wang et al. [6] investigated, both theoretically and experimentally, the enhancement in LEE and modulation bandwidth of flip-chip GaN LEDs using photonic crystal. This study shows a promising approach for achieving high-frequency visible light communication. Moreover, Lan et al. [7] theoretically studied the impact of the thin-film flip-chip structure on the enhanced LEE of red/green/blue LEDs. They engineered systematically the impact of the substrate thickness, encapsulation, surface texture, microstructures between the substrate and epilayer, the size, cutting shape, and angle on the performance of the related LEDs.

A novel design of narrow linewidth InGaN/GaN laser diode using distributed Bragg reflector (DBR) is reported in this special issue. Xie et al. [8] demonstrated that the spectral linewidth could be achieved at as narrow was 0.45 nm, using a narrow-band DBR filtering approach.

This Special Issue includes two papers reporting current trends in micro-LEDs (μLEDs) that include nanowire and flip-chip structures. Bui et al. [9] reported that full-color μLEDs with stable emission can be varied from blue to red were successfully demonstrated using nanowire heterostructures grown by molecular beam epitaxy. The μLEDs exhibit excellent optical and electrical properties, showing their promises for the next generation of high-resolution micro-display applications. In another study, Lan et al. reported their study on the LEE of red/green/blue μLEDs using Monte Carlo ray tracing method [7]. Different types of substrates were investigated in this study. They claimed that the LEEs of the μLEDs showed a sharp rise when the chip-size reduced from 30 to 10 μm. This study delivers important approaches for developing high efficiency μLEDs for practical applications.

Studies on OLEDs and PeLEDs are also reported in this Special Issue with two novel structures. Liu et al. [10] studied the effect of charge transport materials on blue OLEDs using a host-dopant system with four different hole transport layers and two different electron transport layers. They reported that the light intensity and lifetime of the OLEDs can be optimized by engineering the charge transport material properties. Moreover, a high brightness emission with a maximum luminance of 3640 cd/m^2 at 50 mA/cm^2 was recorded for their blue OLEDs. In another paper, Zhou et al. [11] demonstrated high efficiency PeLEDs using a three-step spin-coated $CH_3NH_3PbBr_3$ emitter and a PEDOT:PSS/MoO_3-ammonia composite hole transport layer. They claimed that these approached enable high efficiency PeLEDs that could reach a maximum luminance of 5044 cd/m^2 and maximum current efficiency of 3.12 cd/A, offering highly promising approach for generating high performance PeLEDs.

Presented in the paper "The Composition-Dependent Photoluminescence Properties of Non-Stoichiometric $Zn_xAg_yInS_{1.5+x+0.5y}$ Nanocrystals", Feng et al. [12] reported an effective method to synthesize high-quality $Zn_xAg_yInS_{1.5+x+0.5y}$ nanocrystals using a hot facile injection approach. The nanocrystals exhibit tunable photoluminescence from 450–700 nm by varying the Ag composition accordingly. Moreover, this study shows the first investigation of the synthesis of $Zn_xAg_yInS_{1.5+x+0.5y}$ nanocrystal with Zn of 25% and tunable emission wavelength with high quantum yield of 35%. Dang et al. presented a special design of a broadband absorber employing epsilon-near-zero mode with plasmonic metasurfaces [13]. The proposed absorber exhibits a wide range absorption bandwidth of mid-infrared irradiation wavelengths which is perfectly suitable for several applications in detection, sensing, and imaging.

In the paper "Large-Scale Fabrication of Photonic Nanojet Array via Template-Assisted Self-Assembly", Zhang et al. [14] demonstrated an advanced manufacturing method to fabricate large-scale homogenized photonic nanojet array with defined pattern and spacing. A template-assisted self-assembly approach was introduced in this study. Importantly, high patterning efficiency of 95% was demonstrated, enabling highly promising applications in super-resolution imaging, subwavelength-resolution nanopatterning, nano objects trapping and detection technology.

We would like to take this opportunity to thank all authors for their valuable contributions to this Special Issue. We would also like to acknowledge all reviewers of this Special Issue for their time and comments.

Funding: This research received no external funding.

Conflicts of Interest: The author declare no conflict of interest.

References

1. Li, W.; Zhang, X.; Meng, R.; Yan, J.; Wang, J.; Li, J.; Wei, T. Epitaxy of III-Nitrides on β-Ga_2O_3 and Its Vertical Structure LEDs'. *Micromachines* **2019**, *10*, 322. [CrossRef] [PubMed]
2. Cao, L.; Liu, X.; Guo, Z.; Zhou, L. Surface/Interface Engineering for Constructing Advanced Nanostructured Light-Emitting Diodes with Improved Performance: A Brief Review. *Micromachines* **2019**, *10*, 821. [CrossRef] [PubMed]
3. Zhao, S.; Lu, J.; Hai, X.; Yin, X. AlGaN Nanowires for Ultraviolet Light-Emitting: Recent Progress, Challenges, and Prospects. *Micromachines* **2020**, *11*, 125. [CrossRef] [PubMed]
4. Lei, P.-H.; Yang, C.-D.; Huang, P.-C.; Yeh, S.-J. Enhancement of Light Extraction Efficiency for InGaN/GaN Light-Emitting Diodes Using Silver Nanoparticle Embedded ZnO Thin Films. *Micromachines* **2019**, *10*, 239. [CrossRef] [PubMed]
5. Choi, H.-J.; Kim, S.; Chu, E.-K.; Noh, B.-R.; Lee, W.-S.; Kwon, S.-H.; Oh, S.; Kim, K.-K. Enhanced Photon Emission Efficiency Using Surface Plasmon Effect of Pt Nanoparticles in Ultra-Violet Emitter. *Micromachines* **2019**, *10*, 528. [CrossRef] [PubMed]
6. Wang, H.; Zhong, M.; Tan, L.; Shi, W.; Zhou, Q. Study on Modulation Bandwidth and Light Extraction Efficiency of Flip-Chip Light-Emitting Diode with Photonic Crystals. *Micromachines* **2019**, *10*, 767. [CrossRef] [PubMed]
7. Lan, S.; Wan, H.; Zhao, J.; Zhou, S. Light Extraction Analysis of AlGaInP Based Red and GaN Based Blue/Green Flip-Chip Micro-LEDs Using the Monte Carlo Ray Tracing Method. *Micromachines* **2019**, *10*, 860. [CrossRef] [PubMed]
8. Xie, W.; Li, J.; Liao, M.; Deng, Z.; Wang, W.; Sun, S. Narrow Linewidth Distributed Bragg Reflectors Based on InGaN/GaN Laser. *Micromachines* **2019**, *10*, 529. [CrossRef] [PubMed]
9. Bui, H.Q.T.; Velpula, R.T.; Jain, B.; Aref, O.H.; Nguyen, H.-D.; Lenka, T.R.; Nguyen, H.P.T. Full-Color InGaN/AlGaN Nanowire Micro Light-Emitting Diodes Grown by Molecular Beam Epitaxy: A Promising Candidate for Next Generation Micro Displays. *Micromachines* **2019**, *10*, 492. [CrossRef] [PubMed]
10. Liu, N.; Mei, S.; Sun, D.; Shi, W.; Feng, J.; Zhou, Y.; Mei, F.; Xu, J.; Jiang, Y.; Cao, X. Effects of Charge Transport Materials on Blue Fluorescent Organic Light-Emitting Diodes with a Host-Dopant System. *Micromachines* **2019**, *10*, 344. [CrossRef] [PubMed]
11. Zhou, Y.; Mei, S.; Sun, D.; Liu, N.; Shi, W.; Feng, J.; Mei, F.; Xu, J.; Jiang, Y.; Cao, X. Improved Efficiency of Perovskite Light-Emitting Diodes Using a Three-Step Spin-Coated $CH_3NH_3PbBr_3$ Emitter and a PEDOT:PSS/MoO_3-Ammonia Composite Hole Transport Layer. *Micromachines* **2019**, *10*, 459. [CrossRef] [PubMed]
12. Feng, J.; Yang, X.; Li, R.; Yang, X.; Feng, G. The Composition-Dependent Photoluminescence Properties of Non-Stoichiometric $Zn_xAg_yInS_{1.5+x+0.5y}$ Nanocrystals. *Micromachines* **2019**, *10*, 439. [CrossRef] [PubMed]
13. Dang, P.T.; Le, K.Q.; Lee, J.-H.; Nguyen, T.K. A Designed Broadband Absorber Based on ENZ Mode Incorporating Plasmonic Metasurfaces. *Micromachines* **2019**, *10*, 673. [CrossRef] [PubMed]
14. Zhang, P.; Chen, X.; Yang, H. Large-Scale Fabrication of Photonic Nanojet Array via Template-Assisted Self-Assembly. *Micromachines* **2020**, *11*, 473. [CrossRef]

micromachines

MDPI

Review

AlGaN Nanowires for Ultraviolet Light-Emitting: Recent Progress, Challenges, and Prospects

Songrui Zhao *, Jiaying Lu, Xu Hai and Xue Yin

Department of Electrical and Computer Engineering, McGill University, 3480 University Street, Montreal, QC H3A 0E9, Canada; jiaying.lu@mail.mcgill.ca (J.L.); xu.hai@mail.mcgill.ca (X.H.); xue.yin@mail.mcgill.ca (X.Y.)

* Correspondence: songrui.zhao@mcgill.ca; Tel.: +1-514-398-3244

Received: 21 December 2019; Accepted: 22 January 2020; Published: 23 January 2020

Abstract: In this paper, we discuss the recent progress made in aluminum gallium nitride (AlGaN) nanowire ultraviolet (UV) light-emitting diodes (LEDs). The AlGaN nanowires used for such LED devices are mainly grown by molecular beam epitaxy (MBE) and metalorganic chemical vapor deposition (MOCVD); and various foreign substrates/templates have been investigated. Devices on Si so far exhibit the best performance, whereas devices on metal and graphene have also been investigated to mitigate various limitations of Si substrate, e.g., the UV light absorption. Moreover, patterned growth techniques have also been developed to grow AlGaN nanowire UV LED structures, in order to address issues with the spontaneously formed nanowires. Furthermore, to reduce the quantum confined Stark effect (QCSE), nonpolar AlGaN nanowire UV LEDs exploiting the nonpolar nanowire sidewalls have been demonstrated. With these recent developments, the prospects, together with the general challenges of AlGaN nanowire UV LEDs, are discussed in the end.

Keywords: compound semiconductor; nanostructure; ultraviolet; light-emitting diode (LED); molecular beam epitaxy; GaN; AlN

1. Introduction

Compared with bulk materials, low-dimensional materials such as nanowires can have different electrical and optical properties, such as the strong confinement of charge carriers and photons associated with the reduced dimensions. Motivated by exploring novel electrical and optical properties at low dimensions as well as new material platforms for future generation electronic and photonic devices, tremendous efforts have been devoted in the past two decades to the study of semiconductor nanowires and their device applications; and remarkable progresses have been made in applying various semiconductor nanowires to light-emitting devices, solar energy conversion devices, transistors, and biosensors [1–17].

Among various semiconductor nanowires, aluminum gallium nitride (AlGaN) nanowires, due to their direct and tunable bandgap energies from ~3.4 eV to 6 eV (corresponding to ~207–364 nm), are of particular interest for mid-deep ultraviolet (UV) light-emitting diodes (LEDs) and lasers [18]. Semiconductor UV light-emitting technologies are positioned to replace conventional UV light-emitting technologies, which are predominantly relying on mercury lamps, for a wide range of applications, such as water disinfection, curing, sensing, to name just a few [18–21].

Besides the suitable bandgap energies, there are a few other important reasons to investigate AlGaN nanowire structures for UV light-emitting: (1) In nanowire structures, due to the large surface to the bulk volume ratio, the lattice strain, due to the lattice mismatches between AlGaN and commonly used substrates as well as between GaN and AlN, can be better accommodated compared with planar counterparts, promising a better material quality [22]; (2) In nanowire structures, the doping concentration can be higher than that in planar counterparts due to the enhanced surface doping [23–26], which could be highly beneficial for the electrically injected light-emitting devices; and (3) The possibility of having devices on different substrates, including flexible substrates, through an *in-situ* integration [27–29].

The past few years have witnessed the rapid development of AlGaN nanowire UV LEDs, as well as lasing under the direct electrical injection [30–35]. For example, AlGaN nanowire deep UV LEDs with milli- to sub-milli-watt light output power have been demonstrated [36,37]. These devices are made viable due to the use of high-quality AlGaN nanowires grown by molecular beam epitaxy (MBE), the improved p-type doping with the use of nanowire structures, the use of tunnel junction to improve the carrier injection (in particular the hole injection), and the presence of high Al content AlGaN passivation shell that can confine the charge carriers in the nanowire bulk region [36,37]. In this review paper, we discuss the recent progress made in AlGaN nanowire UV LEDs.

This paper is organized as follows: Section 2 presents a brief overview of various AlGaN nanowire synthesis techniques, with a focus on the bottom-up approach. It is noted, though, AlGaN nanowires can also be obtained through the top-down etching, e.g., Ref. [38]. As the majority device studies are on Si substrate, following Section 2 we first discuss the recent progress of devices on Si in Section 3, focusing on AlGaN nanowire tunnel junction deep UV LEDs, which show the state of the art performance for large-area devices on Si; and such devices also have much better performance compared with large-area devices on other foreign substrates. Given various limitations of Si substrate, such as the UV light absorption, in Section 4 we discuss some possible solutions using other foreign substrates/templates, including metal and graphene; and we again highlight the best performance achieved so far. We then move on to AlGaN nanowire UV LEDs with nanowires grown on patterned substrates in Section 5, aiming to show some possible solutions to issues related to the spontaneously formed nanowires. In Section 6, we discuss nonpolar AlGaN nanowire quantum well UV LEDs, exploiting the nonpolar sidewalls of the wurtzite nanowire structure. This represents an alternative approach to reduce the quantum confined Stark effect (QCSE) in AlGaN quantum well UV LEDs, in addition to the use of challenging nonpolar/semipolar substrates [39]. The prospects, together with the general challenges of AlGaN nanowire UV LEDs, are discussed in Section 7.

2. A Brief Overview of Synthesis Techniques

A wide range of techniques have been explored to synthesize AlGaN ternary nanowires (including AlN nanowires). The detailed growth studies can be found in a number of review papers [15,30,32,34]. In what follows, we briefly discuss the major synthesis techniques for AlGaN ternary nanowires.

2.1. CVD and PVD

Chemical vapor deposition (CVD) and Physical vapor deposition (PVD) typically involve precursors and carrier gases, with or without catalysts. The growth is generally described by the vapor–liquid–solid (VLS) mechanism [40–43]. Using these techniques, AlGaN nanowires with a wide range of Al contents (from 0 to 100%) have been achieved. Nonetheless, AlGaN nanowires synthesized by these techniques typically emit light in the near UV and/or visible spectral ranges due to defects, making device development challenging.

2.2. MBE and MOCVD

Hitherto, large-area AlGaN nanowire UV LEDs are mainly fabricated using AlGaN nanowires grown by epitaxy tools, including MBE and metalorganic chemical vapor deposition (MOCVD, also called metalorganic vapor phase epitaxy, MOVPE); and the shortest wavelength with AlGaN ternary nanowires is 236 nm [44], whereas 207 nm emission has been achieved using AlN nanowires [25,45]. The early efforts of growing AlGaN nanowires using such large-scale epitaxy tools can be dated back to around 2000, when AlGaN nanowires with low Al contents were first investigated by MBE [46,47]. These early efforts were followed by tremendous efforts from a large number of groups who have been working on the epitaxial growth of AlGaN nanowires (primarily by MBE) [25,28,48–54]. In these studies, the AlGaN nanowires are typically spontaneously formed on 2-inch or 3-inch Si substrates under the nitrogen rich conditions, with the help of GaN nanowire template. These substrate sizes are mainly limited by reactor design, and in principle there are no fundamental limitations to scale up the growth to larger substrate sizes. The growth is generally understood through a diffusion-driven, self-organized mechanism [55,56]. Due to different chemical potentials on the nanowire top surface and the sidewall, the impinged atoms diffuse at the substrate surface and then migrate to the nanowire top, promoting a spontaneous vertical growth. Furthermore, for GaN nanowires grown by MBE, lattice registration (a requirement for the growth of epi-layers) is not needed [57], which enables the formation of an AlGaN nanowire segment on a wide range of substrates [27–29].

2.3. Selective Area Growth

To further improve the nanowire uniformity, AlGaN nanowires on patterned substrates have also been demonstrated [58–63]. In such a growth process, a mask layer is typically required; and the nanowire nucleation site is determined by the opening, due to the different chemistries of the impinged adatoms on the surface of the substrate and the surface of the mask material. Using such a technique, highly uniform AlGaN nanowires, across a wide range of Al contents, have been reported [60]. Alternatively, such a selective area epitaxy can also be achieved using etched GaN nanopillars [64]. It is noted, though, that in general, selective area growth can be achieved with various growth techniques, including CVD/PVD, MOCVD/MOVPE, and MBE.

3. AlGaN Nanowire UV LEDs on Si

Si has been playing a dominant role in modern information and communication technologies, and it is thus of great interest in integrating light sources with Si technologies and/or on Si substrates. Further, given the low cost of Si substrate, the majority of studies of group-III nitride nanowire UV LEDs are on Si substrate. These nanowire LED structures are primarily grown by MBE (through a spontaneous formation process, as afore-discussed), and predominantly with AlGaN ternary nanowires [25,36,37,51,65–69]. The relatively longer history of investigating the MBE growth of AlGaN nanowires on Si, compared with the growth on other foreign substrates, has also made AlGaN nanowire UV LEDs on Si of better performance compared with devices on other foreign substrates, albeit with various limitations of using Si substrate (see Section 4). In this section, we focus on the recent advances of AlGaN nanowire deep UV LEDs on Si.

3.1. Basic Device Structure

Figure 1a shows the layer-by-layer structure of an individual AlGaN nanowire that is used to form the large-area AlGaN nanowire deep UV LEDs. Figure 1b shows the SEM image of the AlGaN nanowires at a large scale. It is seen that a relatively uniform nanowire height and top-surface diameter can be achieved even if the nanowires are spontaneously formed. The device fabrication process involves photolithography and metallization [36,37]. The typical device size varies from 300 μm × 300 μm to 1 mm × 1 mm. Comparing the light emission intensity under optical pumping and electrical injection for device structures with and without the tunnel junction, a drastic improvement of light intensity under electrical injection (by more than two orders of magnitude) is measured, whereas a similar intensity is measured under optical pumping; this indicates that the improvement of light intensity under electrical injection is due to the improved carrier injection (i.e., the injection of charge carriers into the active region) [37].

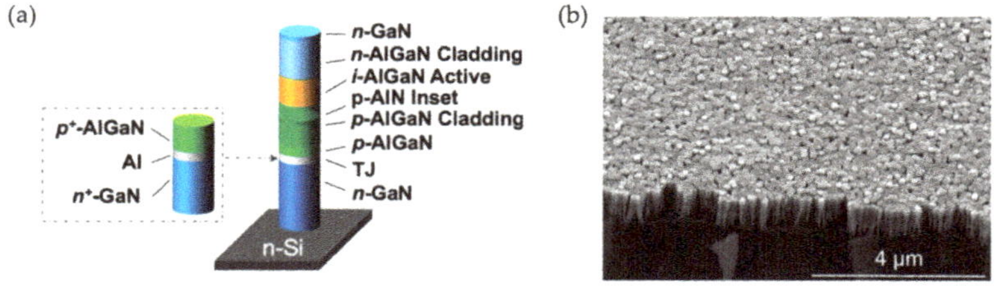

Figure 1. (**a**) Schematic of an individual AlGaN nanowire used for the large-area AlGaN nanowire tunnel junction deep ultraviolet light-emitting diodes (UV LEDs). (**b**) SEM image of the AlGaN nanowires at a large scale [70].

It is noted, though, tunnel junctions involving large bandgap thin films have remained challenging to realize. The success of having the GaN-based nanowire tunnel junction is due to the enhanced dopant incorporation in nanowire structures [23,25,26,71,72].

3.2. Electrical Properties

The detailed I-V characteristics of such AlGaN nanowire tunnel junction deep UV LEDs have also been investigated [70]. It is found that the impurity band conduction, associated with the heavily p-doped AlGaN cladding layer, plays an important role in the electrical properties. First, the deviation from the low injection regime of the diode occurs at low injection currents. As shown in the inset of Figure 2, the deviation from the low injection regime, marked by the dashed line, occurs at a relatively low injection current (~0.1 mA). This is because the impurity band conduction is typically associated with low carrier mobility [25,45,73,74], which immediately leads to a large difference between the electron mobility and the hole mobility; and this large mobility difference can lead to the deviation from the low injection regime at low injection currents [75]. Secondly, I-V characteristics are nearly temperature-independent under high injections. As shown by Figure 2, the I-V curves at different temperatures show a similar slope at a forward voltage of around 10 V. This is because, under high injections, the bottleneck for conduction is the p-AlGaN cladding layer and the conduction of the p-AlGaN cladding layer is dominated by the impurity band condition, which is associated with small active energies for electrical conduction [25,73,74].

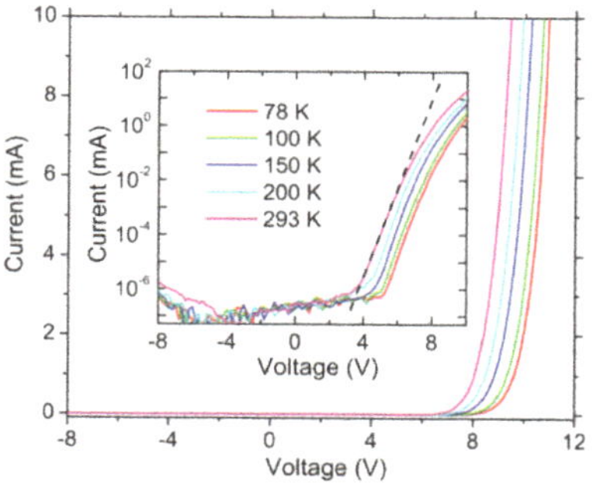

Figure 2. I-V characteristics of AlGaN nanowire tunnel junction deep UV LEDs. Device size: 1 mm × 1 mm [70]. The dashed line is a guide for the eye.

3.3. *Light-Emitting Properties*

3.3.1. Electroluminescence Spectra

In general, for large-area AlGaN nanowire deep UV LEDs, besides the near band-edge emission peak, additional emission components have been observed [36,37,70]. This can be seen from the electroluminescence (EL) spectra of devices emitting at 242 and 274 nm in the semi-log scale (inset of Figure 3a,b; the EL spectra in the linear scale is shown in Figure 3a,b). The emission at around 320 and 380 nm for both samples could be attributed to radiative recombinations from the p-AlGaN and p-GaN layers, respectively [36,37,70], whereas the emission component at around 300 nm for both samples could be related to the localized states due to the compositional fluctuations in AlGaN nanowires grown in the nitrogen rich conditions [70].

The emission component at around 480 nm for the 242 nm emitting device is not discussed previously. Here, we suggest that it is likely related to the Al vacancy (V_{Al}^{3-}), as in the previously reported unintentionally and/or *n*-type doped AlGaN thin films and/or thin-film quantum wells [76–78]. This explanation is further supported by the absence of this emission component and/or the negligible contribution of this emission component to the entire EL spectrum (Figure 3b) for the device emitting at 274 nm, as the formation energy of V_{Al}^{3-} has been suggested to increase as the Al content decreases by first principle calculations, becoming unfavorable [79–84]. Furthermore, as V_{Al}^{3-} exists in the unintentionally and/or n-type doped AlGaN, it further suggests that V_{Al}^{3-} presumably exists in the active region. It is also worthy of noting that, a similar defect luminescence has also been observed in AlGaN thin-film quantum wells that show more than 80% IQE [78].

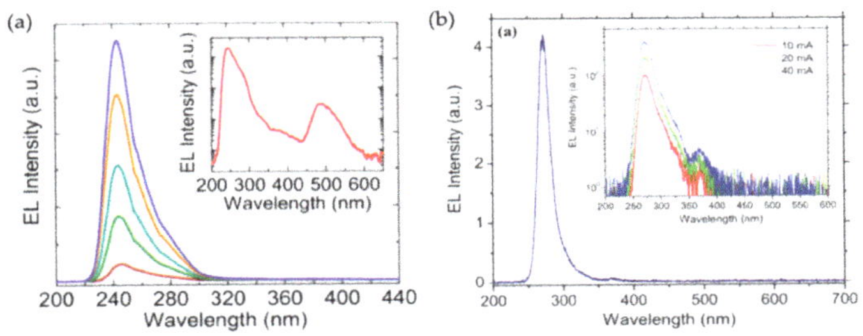

Figure 3. (**a**) EL spectra of AlGaN nanowire tunnel junction deep UV LEDs emitting at 242 nm under injection currents varying from 2 to 60 mA. Device size: 0.5 mm × 0.5 mm. Inset: the EL spectrum in the semi-log scale under an injection current of 20 mA. (**b**) EL spectra of AlGaN nanowire tunnel junction deep UV LEDs emitting at 274 nm under an injection current of 20 mA. Device size: 1 mm × 1 mm. Inset: EL spectra in the semi-log scale under different injection currents [37,70].

3.3.2. Light Output Power

The light output power of devices operating at 242 and 274 nm has also been investigated in detail [36,37]. The light output power vs. the injection current under continuous-wave (CW) and pulse operations for a device emitting at 242 nm is shown in Figure 4a. It is seen that under the CW biasing, a maximum power of 0.06 mW is measured; and under the pulsed biasing a maximum power of 0.38 mW is measured, largely due to the minimization of Joule heating under the pulsed biasing. A maximum external quantum efficiency (EQE) is further derived to be ~0.012%. For devices operating at around 274 nm (Figure 4b), a maximum light output power of 8 mW is measured, with a maximum EQE of 0.4 %. These EQE numbers are within the range of typical AlGaN thin-film quantum well deep UV LEDs, i.e., ~0.04–0.2% for devices operating at around 240 nm and ~0.1–20% for devices operating at around 275 nm [19]. It is noted that the performance of the AlGaN nanowire deep UV LEDs is evaluated by measuring the light output power from the device top surface without any packaging. The use of a relatively thick top-contact metal layer (~ 20 nm) also blocks the light emission severely.

Other losses could be attributed to the light absorption by the Si substrate and the light trapping effect in the spontaneously formed nanowires [85–88].

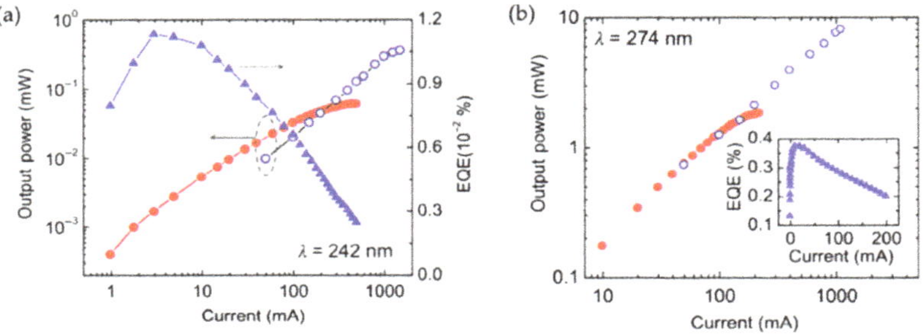

Figure 4. (**a**,**b**) Light output power and EQE vs. the injection current for AlGaN nanowire tunnel junction deep UV LEDs emitting at 242 and 274 nm, respectively. Device size: 1 mm × 1 mm [32]. Open symbols represent devices under the pulse operation, whereas filled symbols denote devices under the continuous-wave (CW) operation.

3.3.3. Efficiency Droop

The efficiency droop has been further analyzed for devices operating at 274 nm [70]. For such AlGaN nanowire deep UV LEDs, the detailed analysis suggests that the efficiency droop occurs at a current density in the range of 0.3–3 A/cm^2 [70]. As a large Shockley–Read-Hall (SRH) rate can overshadow the efficiency droop [89], it is thus noted that the efficiency droop onset current density for such devices could occur at an even lower current density. Further given the relatively thick active region (around 40 nm), it is thus suggested that the Auger process might not be a dominant reason for the efficiency droop. In addition, given the bifurcation current density under the CW operation and pulse operation for the light output power vs. the injection current is much higher than the efficiency droop onset current density, Joule heating is not likely playing a major role in the efficiency droop. Further detailed analysis suggests that the efficiency droop is largely due to the poor hole mobility, fundamentally associated with the impurity band conduction in highly p-doped AlGaN alloys [70]. This is also consistent with the observation that the efficiency droop occurs in the high injection regime (Figure 2). A similar efficiency droop mechanism might be applied to devices emitting at 242 nm.

4. AlGaN Nanowire UV LEDs on Other Foreign Substrates

Despite the progress made for devices on Si substrate, the limitations of using Si substrate are also obvious, e.g., the strong light absorption in the UV spectral range, the spontaneously formed SiN_x that might be a barrier for the electrical charge transport [90]. This motivates the studies of AlGaN nanowire UV LEDs on other foreign substrates, including metal and graphene. In this section, we discuss the recent development of AlGaN nanowire UV LEDs on these substrates.

4.1. Metal Foils and Metal-coated Substrates

Over the past few years, various metal foils (e.g., Ti, Ta) [27,91] and metal-coated substrates (e.g., Al, Pt, Ti, Mo) [28,29,92–97] have been investigated for the growth of AlGaN nanowire UV LED structures, motivated by the excellent physical properties of metals, including thermal and electrical conductivity, light reflection, as well as flexibility. In addition, by coating a metal layer to Si substrate one can also reduce the formation of SiN_x. These LED structures are primarily grown by MBE. So far, the shortest emission wavelength of AlGaN nanowire UV LEDs with the use of a metal layer is at 288 nm [29]. Such AlGaN nanowire deep UV LEDs use a simple AlGaN p-i-n axial junction and are fabricated on Al-coated Si substrate for a better UV light reflection compared with Ti (Figure 5a).

The EL spectra under different injection currents are shown in the inset of Figure 5b. A maximum EQE of ~0.04% is reported from such AlGaN nanowire deep UV LEDs (Figure 5b).

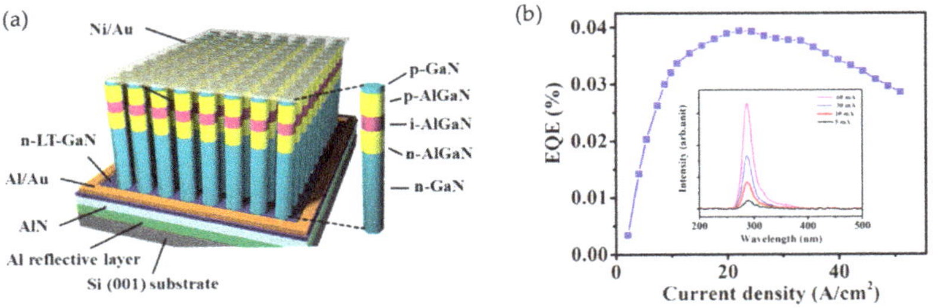

Figure 5. (**a**) Schematic of AlGaN nanowire deep UV LEDs on Al-coated Si substrate. (**b**) EQE vs. the injection current, with the inset showing the EL spectra under different injection currents (from 5 mA to 60 mA) [29].

It is seen that, despite the progress made for devices on metal foils and/or metal-coated substrates, the device performance has remained inferior to that of devices on Si. One major issue is delamination, which leads to non-uniform nanowires. The metal delamination is proposed to be associated with the different thermal expansion coefficients of the underlying substrate and the metal layer on top [94]. Another issue is associated with the crystalline form of the commonly used metal foils, i.e., polycrystalline. In general, the orientation of nanowires is strongly correlated to the microstructure of the underlying substrate; and due to the different grains associated with the polycrystalline metal foils, the nanowires tend to be tilted with respect to the c-axis (the growth direction) at a large scale, leading to the coalescence of nanowires and the metal deposition on the nanowire sidewalls that causes electrical current leakage paths and thus deteriorates the device performance [27]. Solutions using metallic glasses (amorphous) and nanocrystalline metal films have been investigated; and an improved nanowire uniformity has been achieved, due to the reduced grain size [92]. Addressing these issues could lead to a further improvement of the device performance on metal. It is also noted that, although with the above issues, a further improvement of the device performance might also be obtained by using tunnel junctions as in devices on Si (Section 3).

4.2. Graphene

AlGaN nanowire UV LEDs on the graphene-coated glass substrate have been demonstrated recently [98]. The UV light is emitted through the substrate. This is made possible due to the use of UV light transparent glass substrates and the graphene bottom electrode that is not only transparent to the UV light but also possesses a low sheet resistance.

In this work, GaN/AlGaN double heterostructures are used for light-emitting. Schematically shown in Figure 6a, such AlGaN nanowire LED structures are grown on a double-layer graphene coated amorphous silica glass by MBE. The device fabrication and the emission schematic are shown in Figure 6b–d. The electrical current injection is realized through the bottom metal contact to graphene and the top metal contact to p-type GaN.

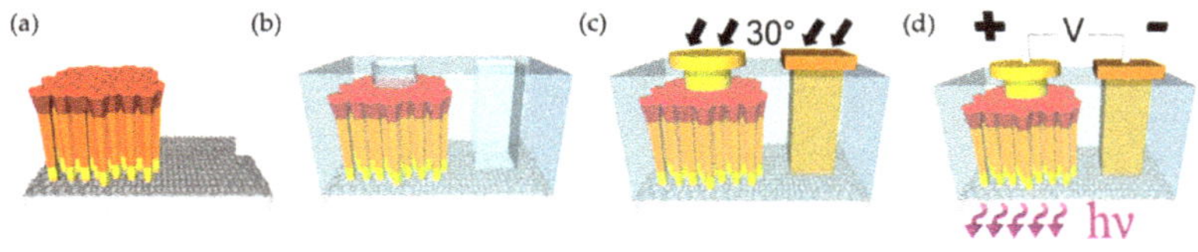

Figure 6. (**a**) Schematic of AlGaN nanowire UV LED structures grown on the graphene-coated glass substrate. (**b**) Openings in polymer for metal contacts to AlGaN nanowires and graphene. (**c**) Schematic of metal contacts deposition at a tilting angle of 30°. Au is used for the contact to graphene. (**d**) Schematic of light-emitting through the substrate [98].

The I-V characteristics of such AlGaN nanowire UV LEDs are shown in Figure 7a. It is noted that the operation voltage is quite high, which is ascribed to the drastically increased sheet resistance of the double-layer graphene after the MBE growth. Protecting the graphene layer during the MBE growth of AlGaN nanowires represents a challenge for devices on graphene. Also shown in Figure 7a is the light output power vs. the injection current for light emission around 365 nm (the emission spectra are shown in the inset of Figure 7a); and it is seen that under an injection current of 1 mA (the circular p-metal diameter is 150 µm), a light output power of ~250 nW is measured. The maximum EQE is further derived to be ~0.01% (Figure 7b).

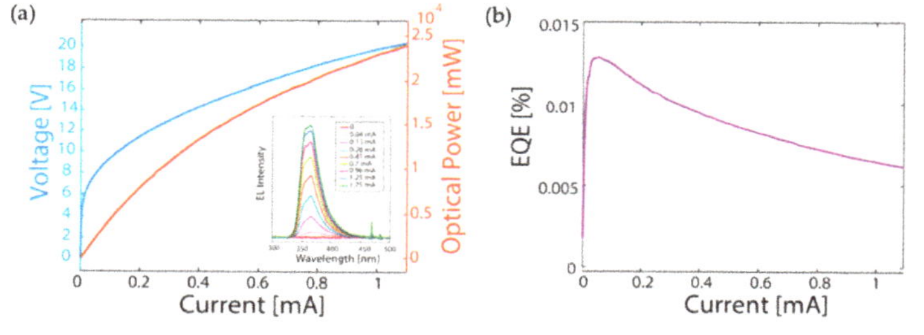

Figure 7. (**a**) I-V characteristics and the light output power vs. the injection current for AlGaN nanowire UV LEDs on graphene. Inset: EL spectra under different injection currents (from 0 mA to 1.75 mA). (**b**) EQE vs. the injection current [98].

5. AlGaN Nanowire UV LEDs on Patterned Substrates

For the above-discussed AlGaN nanowire UV LEDs, the nanowires are spontaneously formed. In order to further improve the device performance, it is necessary to control the nanowire size and spacing. In the past, AlGaN nanowire UV LEDs on patterned substrates using both GaN and AlGaN active regions have been reported [58,60,61]. In this section, we focus on the recent demonstration of AlGaN nanowire deep UV LEDs on patterned GaN-on-sapphire templates [60,61]. The SEM image of AlGaN nanowires on such a patterned template is shown in Figure 8a. It is seen that nanowires with highly identical size and spacing are obtained. Figure 8b shows the schematic of the fabricated device. The electrical injection is realized through the top metal contact to the p-type contact layer and the bottom metal contact to the n-type GaN template [60]. The EL emission spectra under different injection currents are shown in Figure 8c. The two peaks at around 275 and 260 nm correspond to the emission from the active region and the cladding layer, respectively [60]. For such AlGaN nanowire UV LEDs, a maximum power density of 1 W/cm^2 is measured under an injection current density of 250 A/cm^2 (Figure 8d).

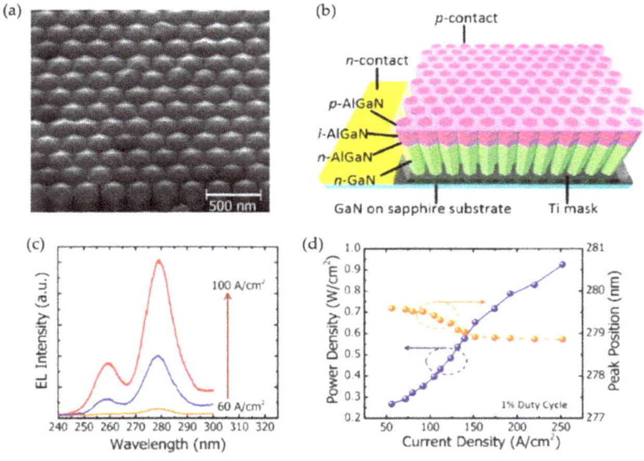

Figure 8. (**a**) SEM image of AlGaN nanowires grown on a patterned GaN-on-sapphire template [61]. (**b**) Schematic of AlGaN nanowire UV LEDs on such a patterned template. [Reprinted/Adapted] with permission from Ref. [60] The Optical Society. (**c**,**d**) EL spectra and the light output power vs. the injection current density, respectively. Also shown in (**d**) is the peak position vs. the injection current density. Device size: 50 μm × 50 μm. [Reprinted/Adapted] with permission from Ref. [60] The Optical Society.

Using regular AlGaN nanowire arrays, a high light extraction efficiency of around 70% has also been suggested for an emission wavelength of 280 nm, through exploiting photonic bandgap effects [99]. This could greatly mitigate the light trapping effect in the spontaneously formed nanowires [85–88]. The precise control on the nanowire size and spacing could enable a rational design-to-realization of AlGaN nanowires UV LEDs; and it also opens a door to engineer the generation and propagation properties of deep UV photons. Moreover, the high nanowire uniformity could lead to highly uniform AlGaN passivation shells in each individual nanowire, improving the charge carrier confinement. This is in contrary to the case with the spontaneously formed nanowires (Section 3). In addition, through controlling the lateral growth rate, a coalesced nanowire top can also be achieved; this can largely mitigate the challenge to make the top metal contact [61].

6. Nonpolar AlGaN Nanowire UV LEDs

Due to the presence of the large electrical polarization fields in c-plane group-III nitrides, strong QCSE is present in quantum well LEDs based on c-plane group-III nitrides. Making devices on nonpolar/semipolar planes promises the reduced electrical polarization fields and thus the improved device performance [39,100–103]. Nonetheless, having high quality epitaxy-ready nonpolar/semipolar substrates/templates has remained challenging [39]. This makes using nanowire structures appealing: For group-III nitride nanowires grown either by MOCVD or MBE, the sidewalls are naturally nonpolar, due to the wurtzite structure.

In this context, different approaches have been investigated to obtain nonpolar light-emitting regions using AlGaN nanowires. Coulon et al. have investigated the overgrowth of the nonpolar AlN/AlGaN/AlN single quantum well on the top-down etched AlN core using MOCVD, and cathodoluminescence experiments indicate a deep UV emission around 229 nm [104]. Brubaker et al. have investigated the growth of the core-shell AlGaN/GaN p-i-n UV LED structures by MBE, and demonstrated a 5× higher light-emitting intensity compared with the axial p-i-n junctions. In this work, the AlGaN shell is realized by reducing the substrate temperature [105].

Recently, single AlGaN nanowire UV LEDs using lateral quantum wells have also been reported [106]. Such single nanowire LED structures are grown by MOCVD on Si substrate. The lateral quantum wells are achieved by reducing the substrate temperature. The schematic of the device structure is shown in Figure 9a. For the device fabrication, the as-grown AlGaN nanowires are dispersed to SiO_2-coated Si substrate, followed by a focused ion beam (FIB) etching to expose the n-GaN nanowire core. The metal contacts are realized by standard e-beam lithography and metallization processes. The electrodes are annealed at 550 °C for 2 min in N_2 environment. Ti/Au (20/50 nm) and Ni/Au (20/50 nm) are used for n- and p-contacts, respectively. The schematic of the single nanowire LED and the SEM image of a fabricated device are shown in Figure 9b,c, respectively.

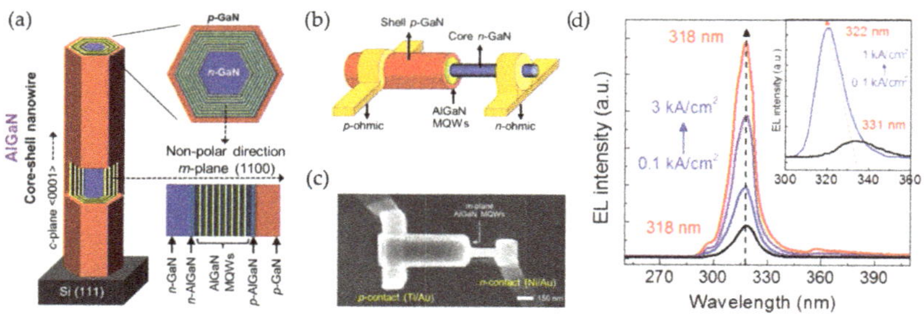

Figure 9. (**a**) Schematic of the structure of single AlGaN nanowire UV LEDs using lateral nonpolar quantum wells. (**b**) Schematic and (**c**) SEM image of the fabricated single AlGaN nanowire UV LED with lateral nonpolar quantum wells. (**d**) EL spectra under different injection currents. Inset: EL spectra of an axial nanowire UV LED for a comparison purpose (see the main text) [106].

The EL spectra under different injection currents are shown in Figure 9d. The inset of Figure 9d shows the EL spectra measured from an axial single nanowire LED for a comparison purpose. It is seen that compared with the strong blueshift in the axial single nanowire LED, the single nanowire LED using lateral nonpolar quantum wells shows a negligible blueshift, suggesting a greatly reduced QCSE. Compared with the EL spectral linewidth of devices using ensemble nanowires (~18 nm) [36,37], the single nanowire LED using lateral nonpolar quantum wells possesses a narrower linewidth (~11 nm), allowing the examination of more detailed spectral characteristics. The EQE of such nonpolar single nanowire LEDs is further derived to be ~3% [106].

7. Conclusions and Prospects

In this paper, we have reviewed the recent progress made in AlGaN nanowire UV LEDs. Using the spontaneously formed AlGaN nanowire structures on Si, devices with optical performance comparable to conventional AlGaN quantum well UV LEDs have been achieved. Furthermore, devices on other foreign substrates have been investigated to mitigate various limitations of Si substrate. Moreover, a selective area epitaxy technique has been used to fabricate AlGaN nanowire UV LEDs, to overcome issues with the spontaneously formed nanowires. In addition, nonpolar AlGaN quantum well UV LEDs, exploiting the natural nonpolar facets (sidewalls) of wurtzite AlGaN nanowires, have been demonstrated, representing another approach to reduce the QCSE in AlGaN quantum wells.

Further improving the performance of AlGaN nanowire UV LEDs requires solving the challenges related to nanowires, including surface defects/states, light extraction, and fabrication (in particular for the top metal contact). Nonetheless, as discussed in this paper, there are solutions readily available for these challenges. For example, Al-rich AlGaN shells can passivate the nanowire surface (and thus surface effects can be minimized); and the uniformity issue of the AlGaN shells with the spontaneously formed nanowires can be addressed by using nanowires grown by selective area epitaxy, which can give highly uniform nanowires. Moreover, with the use of transparent substrate/template, light can be extracted from the backside; and thus, in principle the light blocking by contacts is not a concern for nanowire devices. It is also noted that the light trapping effect in the spontaneously formed nanowires can also be greatly alleviated by using nanowires grown by selective area epitaxy. In addition, the coalescence of the nanowire top, in a controlled manner, has been shown by selective area epitaxy, which can greatly address the top metal contact issue.

AlGaN quantum well UV LEDs have been investigated in the past two decades and significant progress has been made [20]. However, the development of AlGaN quantum well deep UV LEDs is hindered by two grand challenges. The first one arises from the lack of suitable substrates, which leads to large dislocation densities. Bulk AlN substrate emerges in recent years [107], nonetheless, the high price, the small substrate size, and the deep UV light absorption due to impurities of bulk AlN hold back its applications [20]. Another challenge is p-type doping. Fundamentally, to activate a p-type dopant (Mg) in high Al content AlGaN alloys, large activation energies in the range of 400–600 meV are required at room temperature, which is a major challenge for electrically injected light-emitting devices with AlGaN alloys [108,109]. If a high Mg concentration can be achieved, it is possible to reduce the activation energy for p-type conduction, e.g., through the strong doping induced band fluctuations or utilizing the impurity band. These effects have been observed in GaN [110]. For AlGaN, however, due to the elevated growth temperature compared with that of GaN (it is noted that in order to have high-quality AlGaN alloys high substrate temperatures are required), it is difficult to achieve high Mg doping concentrations, largely due to the high Mg desorption rate at high substrate temperatures. The low Mg doping concentration in high Al content AlGaN alloys limits the free holes available at room temperature. In addition, the compensation effect from n-type defect donors also limits the free hole concentration in high Al content AlGaN alloys at room temperature. For example, for the end compound AlN, the free hole concentration is only ~10^{11} cm^{-3} [109,111].

Nanowire structures, on the other hand, could be a viable approach to solve the above two grand challenges for AlGaN deep UV LEDs: the large surface-to-bulk volume ratio can relax the lattice strain efficiently to the nanowire surface, so that the bulk region can be dislocation free; the large surface-to-bulk volume ratio can also make the dopant incorporation more efficient in nanowires. For example, free hole concentrations on the order of 10^{17} cm^{-3} have been reported from AlN nanowires [73]. This leads to a drastically improved electrical performance comparing AlN nanowire LEDs to AlN thin film LEDs [25,45,109].

These key advantages of AlGaN nanowires, compared with AlGaN thin films, together with the recent progress made in AlGaN nanowire UV LEDs, could make AlGaN nanowires an alternative path for semiconductor UV LEDs; and the further improved performance can be expected in the near future with addressing the above issues for nanowire devices. In addition, the much-reduced dislocation density and the drastically improved p-type doping in AlGaN nanowires could also render them as a promising path for semiconductor deep UV lasers [85–88].

Author Contributions: S.Z. conceived the idea and wrote the first draft. All authors contributed to improving the draft, with S.Z. throughout all sections, J.L. to Sections 1 and 4., X.H. to Section 3, and X.Y. to Section 7. All authors have read and agreed to the published version of the manuscript.

Funding: This work is supported by Natural Sciences and Engineering Research Council of Canada (NSERC) RGPIN-2019-04726, NSERC DGECR-2019-00494, and McGill University.

Conflicts of Interest: The authors declare no conflict of interest.

References

1. Hu, J.; Odom, T.W.; Lieber, C.M. Chemistry and Physics in One Dimension: Synthesis and Properties of Nanowires and Nanotubes. *Acc. Chem. Res.* **1999**, *32*, 435–445. [CrossRef]
2. Duan, X.; Lieber, C.M. Laser-Assisted Catalytic Growth of Single Crystal GaN Nanowires. *J. Am. Chem. Soc.* **2000**, *122*, 188–189. [CrossRef]
3. Cui, Y.; Lieber, C.M. Functional nanoscale electronic devices assembled using silicon nanowire building blocks. *Science* **2001**, *291*, 851–853. [CrossRef] [PubMed]

4. Huang, M.H.; Mao, S.; Feick, H.; Yan, H.; Wu, Y.; Kind, H.; Weber, E.; Russo, R.; Yang, P. Room-temperature ultraviolet nanowire nanolasers. *Science* **2001**, *292*, 1897–1899. [CrossRef] [PubMed]
5. Johnson, J.C.; Choi, H.J.; Knutsen, K.P.; Schaller, R.D.; Yang, P.; Saykally, R.J. Single gallium nitride nanowire lasers. *Nat. Mater.* **2002**, *1*, 106–110. [CrossRef]
6. Duan, X.; Niu, C.; Sahi, V.; Chen, J.; Parce, J.W.; Empedocles, S.; Goldman, J.L. High-performance thin-film transistors using semiconductor nanowires and nanoribbons. *Nature* **2003**, *425*, 274–278. [CrossRef]
7. Lu, W.; Lieber, C.M. Semiconductor nanowires. *J. Phys. D Appl. Phys.* **2006**, *39*, R387. [CrossRef]
8. Hochbaum, A.I.; Yang, P. Semiconductor Nanowires for Energy Conversion. *Chem. Rev.* **2010**, *110*, 527. [CrossRef]
9. Yang, P.; Yan, R.; Fardy, M. Semiconductor nanowire: what's next? *Nano Lett.* **2010**, *10*, 1529–1536. [CrossRef]
10. Zhao, S.; Nguyen, H.P.T.; Kibria, M.G.; Mi, Z. III-Nitride nanowire optoelectronics. *Prog. Quan. Electron.* **2015**, *44*, 14–68. [CrossRef]
11. Eaton, S.W.; Fu, A.; Wong, A.B.; Ning, C.-Z.; Yang, P. Semiconductor nanowire lasers. *Nat. Rev. Mater.* **2016**, *1*, 16028. [CrossRef]
12. Zhang, A.; Zheng, G.; Lieber, C.M. *Nanowires: Building Blocks for Nanoscience and Nanotechnology*; Springer: Berlin/Heidelberg, Germany, 2016.
13. Ishikawa, F.; Buyanova, I.A. *Novel Compound Semiconductor Nanowires*; Pan Standford Publishing Pte. Ltd.: Singapore, 2017.
14. Fan, S.; Zhao, S.; Chowdhury, F.A.; Wang, R.; Mi, Z. Molecular Beam Epitaxial Growth of III-Nitride Nanowire Heterostructures and Emerging Device Applications. In *Handbook of GaN Semiconuctor Materials and Devices*; Bi, W.W., Kuo, H.H., Ku, P., Shen, B., Eds.; CRC Press: Boca Raton, FL, USA, 2017.
15. Zhao, C.; Alfaraj, N.; Chandra Subedi, R.; Liang, J.W.; Alatawi, A.A.; Alhamoud, A.A.; Ebaid, M.; Alias, M.S.; Ng, T.K.; Ooi, B.S. III-nitride nanowires on unconventional substrates: From materials to optoelectronic device applications. *Prog. Quan. Electron.* **2018**, *61*, 1–31. [CrossRef]
16. Quan, L.N.; Kang, J.; Ning, C.Z.; Yang, P. Nanowires for Photonics. *Chem. Rev.* **2019**, *119*, 9153–9169. [CrossRef] [PubMed]
17. Zhao, S.; Wang, R.; Chu, S.; Mi, Z. Molecular Beam Epitaxy of III-Nitride Nanowires: Emerging Applications From Deep-Ultraviolet Light Emitters and Micro-LEDs to Artificial Photosynthesis. *IEEE Nanotechnol. Mag.* **2019**, *13*, 6–16. [CrossRef]
18. Moustakas, T.D.; Paiella, R. Optoelectronic device physics and technology of nitride semiconductors from the UV to the terahertz. *Rep. Prog. Phys.* **2017**, *80*, 106501. [CrossRef]
19. Kneissl, M.; Seong, T.-Y.; Han, J.; Amano, H. The emergence and prospects of deep-ultraviolet light-emitting diode technologies. *Nat. Photonics* **2019**, *13*, 233–244. [CrossRef]
20. Kneissl, M.; Rass, J. *III-Nitride Ultraviolet Emitters*; Springer: Berlin/Heidelberg, Germany, 2016.
21. Hirayama, H.; Maeda, N.; Fujikawa, S.; Toyoda, S.; Kamata, N. Recent progress and future prospects of AlGaN-based high-efficiency deep-ultraviolet light-emitting diodes. *Jpn. J. Appl. Phys.* **2014**, *53*, 100209. [CrossRef]
22. Glas, F. Critical dimensions for the plastic relaxation of strained axial heterostructures in free-standing nanowires. *Phys. Rev. B* **2006**, *74*, 121302. [CrossRef]
23. Xie, P.; Hu, Y.; Fang, Y.; Huang, J.; Lieber, C.M. Diameter-dependent dopant location in silicon and germanium nanowires. *Proc. Natl. Acad. Sci. USA* **2009**, *106*, 15254–15258. [CrossRef]
24. Fang, Z.; Robin, E.; Rozas-Jimenez, E.; Cros, A.; Donatini, F.; Mollard, N.; Pernot, J.; Daudin, B. Si Donor Incorporation in GaN Nanowires. *Nano Lett.* **2015**, *15*, 6794–6801. [CrossRef]
25. Zhao, S.; Connie, A.T.; Dastjerdi, M.H.; Kong, X.H.; Wang, Q.; Djavid, M.; Sadaf, S.; Liu, X.D.; Shih, I.; Guo, H.; et al. Aluminum nitride nanowire light-emitting diodes: Breaking the fundamental bottleneck of deep ultraviolet light sources. *Sci. Rep.* **2015**, *5*, 8332. [CrossRef] [PubMed]
26. Zhao, S.; Le, B.H.; Liu, D.P.; Liu, X.D.; Kibria, M.G.; Szkopek, T.; Guo, H.; Mi, Z. p-Type InN Nanowires. *Nano Lett.* **2013**, *13*, 5509–5513. [CrossRef]
27. May, B.J.; Sarwar, A.T.M.G.; Myers, R.C. Nanowire LEDs grown directly on flexible metal foil. *Appl. Phys. Lett.* **2016**, *108*, 141103. [CrossRef]
28. Sun, H.; Shakfa, M.K.; Muhammed, M.M.; Janjua, B.; Li, K.-H.; Lin, R.; Ng, T.K.; Roqan, I.S.; Ooi, B.S.; Li, X. Surface-Passivated AlGaN Nanowires for Enhanced Luminescence of Ultraviolet Light-emitting Diodes. *ACS Photonics* **2017**, *5*, 964–970. [CrossRef]

29. Wu, Y.; Wang, Y.; Sun, K.; Mi, Z. Molecular beam epitaxy and characterization of AlGaN nanowire ultraviolet light-emitting diodes on Al coated Si (0 0 1) substrate. *J. Cryst. Growth* **2019**, *507*, 65–69. [CrossRef]
30. Zhao, S.; Mi, Z. *Chapter Five-Al (Ga) N Nanowire Deep Ultraviolet Optoelectronics*; Elsevier: Amsterdam, The Netherlands, 2017; Volume 96, pp. 167–199.
31. Zhao, S.; Mi, Z. AlGaN Nanowires: Path to Electrically Injected Semiconductor Deep Ultraviolet Lasers. *IEEE J. Quan. Electron.* **2018**, *54*, 1–9. [CrossRef]
32. Zhao, S.; Mi, Z. Molecular Beam Epitaxy of Al(Ga)N Nanowire Heterostructures and Their Application in Ultraviolet Optoelectronics. In *Molecular Beam Epitaxy*; Henini, M., Ed.; Elsevier: Amsterdam, The Netherlands, 2018.
33. Alias, M.S.; Tangi, M.; Holguin-Lerma, J.A.; Stegenburgs, E.; Alatawi, A.A.; Ashry, I.; Subedi, R.C.; Priante, D.; Shakfa, M.K.; Ng, T.K.; et al. Review of nanophotonics approaches using nanostructures and nanofabrication for III-nitrides ultraviolet-photonic devices. *J. Nanophotonics* **2018**, *12*, 043508. [CrossRef]
34. Ooi, B.S.; Ng, T.K.; Alyamani, A.Y.; Albadri, A.M.; Alaskar, Y.; Al-Maghrabi, L.; Zhao, C.; Prabaswara, A.; Kang, C.H.; Liu, G.; et al. Unleashing the potential of molecular beam epitaxy grown AlGaN-based ultraviolet-spectrum nanowires devices. *J. Nanophotonics* **2018**, *12*, 043511.
35. Liu, X.; Mashooq, K.; Laleyan, D.A.; Reid, E.T.; Mi, Z. AlGaN nanocrystals: Building blocks for efficient ultraviolet optoelectronics. *Photonics Res.* **2019**, *7*, B12–B23. [CrossRef]
36. Sadaf, S.M.; Zhao, S.; Wu, Y.; Ra, Y.H.; Liu, X.; Vanka, S.; Mi, Z. An AlGaN Core-Shell Tunnel Junction Nanowire Light-Emitting Diode Operating in the Ultraviolet-C Band. *Nano Lett.* **2017**, *17*, 1212–1218. [CrossRef]
37. Zhao, S.; Sadaf, S.M.; Vanka, S.; Wang, Y.; Rashid, R.; Mi, Z. Sub-milliwatt AlGaN nanowire tunnel junction deep ultraviolet light-emitting diodes on silicon operating at 242 nm. *Appl. Phys. Lett.* **2016**, *109*, 201106. [CrossRef]
38. Zhang, L.; Guo, Y.; Yan, J.; Wu, Q.; Lu, Y.; Wu, Z.; Gu, W.; Wei, X.; Wang, J.; Li, J. Deep ultraviolet light-emitting diodes based on a well-ordered AlGaN nanorod array. *Photonics Res.* **2019**, *7*, B66–B72. [CrossRef]
39. Monavarian, M.; Rashidi, A.; Feezell, D. A Decade of Nonpolar and Semipolar III-Nitrides: A Review of Successes and Challenges. *Phys. Status Solidi (A)* **2018**, *216*, 1800628. [CrossRef]
40. Wang, H.; Northwood, D.O.; Han, J.; Du, S. Combustion synthesis of AlN whiskers. *J. Mater. Sci.* **2006**, *41*, 1697–1703. [CrossRef]
41. Yong, K.-T.; Yu, S.F. AlN nanowires: Synthesis, physical properties, and nanoelectronics applications. *J. Mater. Sci.* **2012**, *47*, 5341–5360.
42. He, C.; Wu, Q.; Wang, X.; Zhang, Y.; Yang, L.; Liu, N.; Zhao, Y.; Lu, Y.; Hu, Z. Growth and Characterization of Ternary AlGaN Alloy Nanocones across the Entire Composition Range. *ACS Nano* **2011**, *5*, 1291–1296. [CrossRef]
43. Chen, F.; Ji, X.; Lu, Z.; Shen, Y.; Zhang, Q. Structural and Raman properties of compositionally tunable AlxGa1−xN ($0.66 \leq x \leq 1$) nanowires. *Mater. Sci. Eng. B* **2014**, *183*, 24–28. [CrossRef]
44. Zhao, S.; Woo, S.Y.; Sadaf, S.M.; Wu, Y.; Pofelski, A.; Laleyan, D.A.; Rashid, R.T.; Wang, Y.; Botton, G.A.; Mi, Z. Molecular beam epitaxy growth of Al-rich AlGaN nanowires for deep ultraviolet optoelectronics. *APL Mater.* **2016**, *4*, 086115. [CrossRef]
45. Zhao, S.; Djavid, M.; Mi, Z. Surface emitting, high efficiency near-vacuum ultraviolet light source with aluminum nitride nanowires monolithically grown on silicon. *Nano Lett.* **2015**, *15*, 7006–7009. [CrossRef]
46. Calleja, E.; Sánchez-Garcıa, M.; Sanchez, F.; Calle, F.; Naranjo, F.; Munoz, E.; Molina, S.; Sanchez, A.; Pacheco, F.; Garcıa, R. Growth of III-nitrides on Si (111) by molecular beam epitaxy doping, optical, and electrical properties. *J. Cryst. Growth* **1999**, *201*, 296–317. [CrossRef]
47. Ristic, J.; Sanchez-Garcia, M.A.; Calleja, E.; Sanchez-Paramo, J.; Calleja, J.M.; Jahn, U.; Ploog, K.H. AlGaN Nanocolumns Grown by Molecular Beam Epitaxy: Optical and Structural Characterization. *Phys. Status Solidi (A)* **2002**, *192*, 60–66. [CrossRef]
48. Holmes, M.J.; Choi, K.; Kako, S.; Arita, M.; Arakawa, Y. Room-temperature triggered single photon emission from a III-nitride site-controlled nanowire quantum dot. *Nano Lett.* **2014**, *14*, 982–986. [CrossRef] [PubMed]
49. Songmuang, R.; Ben, T.; Daudin, B.; Gonzalez, D.; Monroy, E. Identification of III-N nanowire growth kinetics via a marker technique. *Nanotechnology* **2010**, *21*, 295605. [CrossRef] [PubMed]
50. Bertness, K.A.; Roshko, A.; Sanford, N.A.; Barker, J.M.; Davydov, A.V. Spontaneously grown GaN and AlGaN nanowires. *J. Cryst. Growth* **2006**, *287*, 522–527. [CrossRef]

51. Carnevale, S.D.; Kent, T.F.; Phillips, P.J.; Sarwar, A.T.; Selcu, C.; Klie, R.F.; Myers, R.C. Mixed Polarity in Polarization-Induced pn Junction Nanowire Light-emitting Diodes. *Nano Lett.* **2013**, *13*, 3029–3035. [CrossRef] [PubMed]
52. Wang, Q.; Zhao, S.; Connie, A.T.; Shih, I.; Mi, Z.; Gonzalez, T.; Andrews, M.P.; Du, X.Z.; Lin, J.Y.; Jiang, H.X. Optical properties of strain-free AlN nanowires grown by molecular beam epitaxy on Si substrates. *Appl. Phys. Lett.* **2014**, *104*, 223107. [CrossRef]
53. Himwas, C.; den Hertog, M.; Dang, L.S.; Monroy, E.; Songmuang, R. Alloy inhomogeneity and carrier localization in AlGaN sections and AlGaN/AlN nanodisks in nanowires with 240–350 nm emission. *Appl. Phys. Lett.* **2014**, *105*, 241908. [CrossRef]
54. Park, M.-D.; Min, J.-W.; Lee, J.-Y.; Hwang, H.-Y.; Kim, C.; Kang, S.; Kang, C.-M.; Park, J.-H.; Jho, Y.-D.; Lee, D.-S. Effects of nitrogen flow rate on the morphology and composition of AlGaN nanowires grown by plasma-assisted molecular beam epitaxy. *J. Cryst. Growth* **2019**, *528*, 125233. [CrossRef]
55. Bertness, K.; Roshko, A.; Mansfield, L.; Harvey, T.; Sanford, N. Mechanism for spontaneous growth of GaN nanowires with molecular beam epitaxy. *J. Cryst. Growth* **2008**, *310*, 3154–3158. [CrossRef]
56. Ristić, J.; Calleja, E.; Fernández-Garrido, S.; Cerutti, L.; Trampert, A.; Jahn, U.; Ploog, K.H. On the mechanisms of spontaneous growth of III-nitride nanocolumns by plasma-assisted molecular beam epitaxy. *J. Cryst. Growth* **2008**, *310*, 4035–4045. [CrossRef]
57. Zhao, S.; Kibria, M.G.; Wang, Q.; Nguyen, H.P.T.; Mi, Z. Growth of large-scale vertically aligned GaN nanowires and their heterostructures with high uniformity on SiO_x by catalyst-free molecular beam epitaxy. *Nanoscale* **2013**, *5*, 5283. [CrossRef] [PubMed]
58. Sekiguchi, H.; Kato, K.; Tanaka, J.; Kikuchi, A.; Kishino, K. Ultraviolet GaN-based nanocolumn light-emitting diodes grown on n-(111) Si substrates by rf-plasma-assisted molecular beam epitaxy. *Phys. Status Solidi (A)* **2008**, *205*, 1067–1069. [CrossRef]
59. Kishino, K.; Sekiguchi, H.; Kikuchi, A. Improved Ti-mask selective-area growth (SAG) by rf-plasma-assisted molecular beam epitaxy demonstrating extremely uniform GaN nanocolumn arrays. *J. Cryst. Growth* **2009**, *311*, 2063–2068. [CrossRef]
60. Liu, X.; Le, B.H.; Woo, S.Y.; Zhao, S.; Pofelski, A.; Botton, G.A.; Mi, Z. Selective area epitaxy of AlGaN nanowire arrays across nearly the entire compositional range for deep ultraviolet photonics. *Opt. Express* **2017**, *25*, 30494. [CrossRef]
61. Le, B.H.; Zhao, S.; Liu, X.; Woo, S.Y.; Botton, G.A.; Mi, Z. Controlled Coalescence of AlGaN Nanowire Arrays: An Architecture for Nearly Dislocation-Free Planar Ultraviolet Photonic Device Applications. *Adv. Mater.* **2016**, *28*, 8446–8454. [CrossRef]
62. Bertness, K.A.; Sanders, A.W.; Rourke, D.M.; Harvey, T.E.; Roshko, A.; Schlager, J.B.; Sanford, N.A. Controlled Nucleation of GaN Nanowires Grown with Molecular Beam Epitaxy. *Adv. Funct. Mater.* **2010**, *20*, 2911–2915. [CrossRef]
63. Gacevic, Z.; Gomez Sanchez, D.; Calleja, E. Formation mechanisms of GaN nanowires grown by selective area growth homoepitaxy. *Nano Lett.* **2015**, *15*, 1117–1121. [CrossRef]
64. Yamano, K.; Kishino, K.; Sekiguchi, H.; Oto, T.; Wakahara, A.; Kawakami, Y. Novel selective area growth (SAG) method for regularly arranged AlGaN nanocolumns using nanotemplates. *J. Cryst. Growth* **2015**, *425*, 316–321. [CrossRef]
65. Sun, H.; Priante, D.; Min, J.-W.; Subedi, R.C.; Shakfa, M.K.; Ren, Z.; Li, K.-H.; Lin, R.; Zhao, C.; Ng, T.K.; et al. Graded-Index Separate Confinement Heterostructure AlGaN Nanowires: Toward Ultraviolet Laser Diodes Implementation. *ACS Photonics* **2018**, *5*, 3305–3314. [CrossRef]
66. Sarwar, A.T.M.G.; May, B.J.; Deitz, J.I.; Grassman, T.J.; McComb, D.W.; Myers, R.C. Tunnel junction enhanced nanowire ultraviolet light-emitting diodes. *Appl. Phys. Lett.* **2015**, *107*, 101103. [CrossRef]
67. Kent, T.F.; Carnevale, S.D.; Sarwar, A.T.; Phillips, P.J.; Klie, R.F.; Myers, R.C. Deep ultraviolet emitting polarization induced nanowire light-emitting diodes with AlxGa(1)-xN active regions. *Nanotechnology* **2014**, *25*, 455201. [CrossRef] [PubMed]
68. Mi, Z.; Zhao, S.; Woo, S.Y.; Bugnet, M.; Djavid, M.; Liu, X.; Kang, J.; Kong, X.; Ji, W.; Guo, H.; et al. Molecular beam epitaxial growth and characterization of Al(Ga)N nanowire deep ultraviolet light-emitting diodes and lasers. *J. Phys. D Appl. Phys.* **2016**, *49*, 364006. [CrossRef]

69. Velpula, R.T.; Philip, M.R.; Jain, B.; Nguyen, H.D.; Wang, R.; Nguyen, H.P.T. Epitaxial Growth and Characterization of AlInN Based Core-Shell Nanowire Light-Emitting Diodes Operating in the Ultraviolet Spectrum. 2019. Available online: https://arxiv.org/abs/1908.02364 (accessed on 31 July 2019).
70. Hai, X.; Rashid, R.T.; Sadaf, S.M.; Mi, Z.; Zhao, S. Effect of low hole mobility on the efficiency droop of AlGaN nanowire deep ultraviolet light-emitting diodes. *Appl. Phys. Lett.* **2019**, *114*, 101104. [CrossRef]
71. Zhao, S.; Fathololoumi, S.; Bevan, K.H.; Liu, D.P.; Kibria, M.G.; Li, Q.; Wang, G.T.; Guo, H.; Mi, Z. Tuning the surface charge properties of epitaxial InN nanowires. *Nano Lett.* **2012**, *12*, 2877–2882. [CrossRef]
72. Zhao, S.; Mi, Z. Recent Advances on p-Type III-Nitride Nanowires by Molecular Beam Epitaxy. *Crystals* **2017**, *7*, 268. [CrossRef]
73. Tran, N.H.; Le, B.H.; Zhao, S.; Mi, Z. On the mechanism of highly efficient p-type conduction of Mg-doped ultra-wide-bandgap AlN nanostructures. *Appl. Phys. Lett.* **2017**, *110*, 032102. [CrossRef]
74. Connie, A.T.; Zhao, S.; Sadaf, S.M.; Shih, I.; Mi, Z.; Du, X.; Lin, J.; Jiang, H. Optical and electrical properties of Mg-doped AlN nanowires grown by molecular beam epitaxy. *Appl. Phys. Lett.* **2015**, *106*, 213105. [CrossRef]
75. Meyaard, D.S.; Lin, G.-B.; Shan, Q.; Cho, J.; Fred Schubert, E.; Shim, H.; Kim, M.-H.; Sone, C. Asymmetry of carrier transport leading to efficiency droop in GaInN based light-emitting diodes. *Appl. Phys. Lett.* **2011**, *99*, 251115. [CrossRef]
76. Nam, K.B.; Nakarmi, M.L.; Lin, J.Y.; Jiang, H.X. Deep impurity transitions involving cation vacancies and complexes in AlGaN alloys. *Appl. Phys. Lett.* **2005**, *86*, 222108. [CrossRef]
77. Nepal, N.; Nakarmi, M.L.; Lin, J.Y.; Jiang, H.X. Photoluminescence studies of impurity transitions in AlGaN alloys. *Appl. Phys. Lett.* **2006**, *89*, 092107. [CrossRef]
78. Bryan, Z.; Bryan, I.; Xie, J.; Mita, S.; Sitar, Z.; Collazo, R. High internal quantum efficiency in AlGaN multiple quantum wells grown on bulk AlN substrates. *Appl. Phys. Lett.* **2015**, *106*, 142107. [CrossRef]
79. Reshchikov, M.A.; Morkoç, H. Luminescence properties of defects in GaN. *J. Appl. Phys.* **2005**, *97*, 061301. [CrossRef]
80. Limpijumnong, S.; Van de Walle, C. Diffusivity of native defects in GaN. *Phys. Rev. B* **2004**, *69*, 035207. [CrossRef]
81. Stampfl, C.; Van de Walle, C.G. Theoretical investigation of native defects, impurities, and complexes in aluminum nitride. *Phys. Rev. B* **2002**, *65*, 155212. [CrossRef]
82. Neugebauer, J.; Van de Walle, C.G. Atomic geometry and electronic structure of native defects in GaN. *Phys. Rev. B* **1994**, *50*, 8067–8070. [CrossRef] [PubMed]
83. Mattila, T.; Nieminen, R.M. Point-defect complexes and broadband luminescence in GaN and AlN. *Phys. Rev. B* **1997**, *55*, 9571–9576. [CrossRef]
84. Mattila, T.; Nieminen, R.M. Ab initio study of oxygen point defects in GaAs, GaN, and AlN. *Phys. Rev. B* **1996**, *54*, 16676. [CrossRef] [PubMed]
85. Zhao, S.; Liu, X.; Wu, Y.; Mi, Z. An electrically pumped 239 nm AlGaN nanowire laser operating at room temperature. *Appl. Phys. Lett.* **2016**, *109*, 191106. [CrossRef]
86. Zhao, S.; Woo, S.Y.; Bugnet, M.; Liu, X.; Kang, J.; Botton, G.A.; Mi, Z. Three-Dimensional Quantum Confinement of Charge Carriers in Self-Organized AlGaN Nanowires: A Viable Route to Electrically Injected Deep Ultraviolet Lasers. *Nano Lett.* **2015**, *15*, 7801–7807. [CrossRef]
87. Zhao, S.; Liu, X.; Woo, S.Y.; Kang, J.; Botton, G.A.; Mi, Z. An electrically injected AlGaN nanowire laser operating in the ultraviolet-C band. *Appl. Phys. Lett.* **2015**, *107*, 043101. [CrossRef]
88. Li, K.H.; Liu, X.; Wang, Q.; Zhao, S.; Mi, Z. Ultralow-threshold electrically injected AlGaN nanowire ultraviolet lasers on Si operating at low temperature. *Nat. Nanotechnol.* **2015**, *10*, 140–144. [CrossRef]
89. Kim, M.H.; Schubert, M.F.; Dai, Q.; Kim, J.K.; Schubert, E.F.; Piprek, J.; Park, Y. Origin of efficiency droop in GaN-based light-emitting diodes. *Appl. Phys. Lett.* **2007**, *91*, 183507. [CrossRef]
90. Sun, H.; Li, X. Recent Advances on III-Nitride Nanowire Light Emitters on Foreign Substrates–Toward Flexible Photonics. *Phys. Status Solidi (A)* **2018**, *216*, 1800420. [CrossRef]
91. Calabrese, G.; Corfdir, P.; Gao, G.; Pfüller, C.; Trampert, A.; Brandt, O.; Geelhaar, L.; Fernández-Garrido, S. Molecular beam epitaxy of single crystalline GaN nanowires on a flexible Ti foil. *Appl. Phys. Lett.* **2016**, *108*, 202101. [CrossRef]
92. May, B.J.; Hettiaratchy, E.C.; Selcu, C.; Wang, B.; Esser, B.D.; McComb, D.W.; Myers, R.C. Enhanced uniformity of III-nitride nanowire arrays on bulk metallic glass and nanocrystalline substrates. *J. Vac. Sci. Technol. B* **2019**, *37*, 031212. [CrossRef]

93. Wolz, M.; Hauswald, C.; Flissikowski, T.; Gotschke, T.; Fernandez-Garrido, S.; Brandt, O.; Grahn, H.T.; Geelhaar, L.; Riechert, H. Epitaxial Growth of GaN Nanowires with High Structural Perfection on a Metallic TiN Film. *Nano Lett.* **2015**, *15*, 3743–3747. [CrossRef] [PubMed]
94. Sarwar, A.T.; Carnevale, S.D.; Yang, F.; Kent, T.F.; Jamison, J.J.; McComb, D.W.; Myers, R.C. Semiconductor Nanowire Light-Emitting Diodes Grown on Metal: A Direction Toward Large-Scale Fabrication of Nanowire Devices. *Small* **2015**, *11*, 5402–5408. [CrossRef] [PubMed]
95. Janjua, B.; Sun, H.; Zhao, C.; Anjum, D.H.; Priante, D.; Alhamoud, A.A.; Wu, F.; Li, X.; Albadri, A.M.; Alyamani, A.Y.; et al. Droop-free $Al_xGa_{1-x}N$/AlyGa1-yN quantum-disks-in-nanowires ultraviolet LED emitting at 337 nm on metal/silicon substrates. *Opt. Express* **2017**, *25*, 1381. [CrossRef]
96. Priante, D.; Janjua, B.; Prabaswara, A.; Subedi, R.C.; Elafandy, R.T.; Lopatin, S.; Anjum, D.H.; Zhao, C.; Ng, T.K.; Ooi, B.S. Highly uniform ultraviolet-A quantum-confined AlGaN nanowire LEDs on metal/silicon with a TaN interlayer. *Opt. Mater. Express* **2017**, *7*, 4214. [CrossRef]
97. Rozhavskaya, M.M.; Lundin, W.V.; Lundina, E.Y.; Sakharov, A.V.; Troshkov, S.I.; Smirnov, A.N.; Davydov, V.Y. Synthesis of GaN nano- and microwire crystals induced by a titanium nanolayer. *Tech. Phys. Lett.* **2014**, *40*, 372–374. [CrossRef]
98. Hoiaas, I.M.; Liudi Mulyo, A.; Vullum, P.E.; Kim, D.C.; Ahtapodov, L.; Fimland, B.O.; Kishino, K.; Weman, H. GaN/AlGaN Nanocolumn Ultraviolet Light-Emitting Diode Using Double-Layer Graphene as Substrate and Transparent Electrode. *Nano Lett.* **2019**, *19*, 1649–1658. [CrossRef]
99. Djavid, M.; Mi, Z. Enhancing the light extraction efficiency of AlGaN deep ultraviolet light-emitting diodes by using nanowire structures. *Appl. Phys. Lett.* **2016**, *108*, 051102. [CrossRef]
100. Zhao, Y.; Tanaka, S.; Pan, C.-C.; Fujito, K.; Feezell, D.; Speck, J.S.; DenBaars, S.P.; Nakamura, S. High-Power Blue-Violet Semipolar ($20\bar{2}\bar{1}$) InGaN/GaN Light-Emitting Diodes with Low Efficiency Droop at 200 A/cm^{2}. *Appl. Phys. Express* **2011**, *4*, 082104. [CrossRef]
101. Holder, C.; Speck, J.S.; DenBaars, S.P.; Nakamura, S.; Feezell, D. Demonstration of Nonpolar GaN-Based Vertical-Cavity Surface-Emitting Lasers. *Appl. Phys. Exp.* **2012**, *5*, 092104. [CrossRef]
102. Zhao, Y.; Fu, H.; Wang, G.T.; Nakamura, S. Toward ultimate efficiency: Progress and prospects on planar and 3D nanostructured nonpolar and semipolar InGaN light-emitting diodes. *Adv. Opt. Photonics* **2018**, *10*, 246–308. [CrossRef]
103. Poyiatzis, N.; Athanasiou, M.; Bai, J.; Gong, Y.; Wang, T. Monolithically integrated white light LEDs on (11–22) semi-polar GaN templates. *Sci. Rep.* **2019**, *9*, 1383. [CrossRef]
104. Coulon, P.M.; Kusch, G.; Martin, R.W.; Shields, P.A. Deep UV Emission from Highly Ordered AlGaN/AlN Core-Shell Nanorods. *ACS Appl. Mater. Interfaces* **2018**, *10*, 33441–33449. [CrossRef]
105. Brubaker, M.D.; Genter, K.L.; Roshko, A.; Blanchard, P.T.; Spann, B.T.; Harvey, T.E.; Bertness, K.A. UV LEDs based on p-i-n core-shell AlGaN/GaN nanowire heterostructures grown by N-polar selective area epitaxy. *Nanotechnology* **2019**, *30*, 234001. [CrossRef]
106. Ra, Y.H.; Kang, S.; Lee, C.R. Ultraviolet Light-Emitting Diode Using Nonpolar AlGaN Core–Shell Nanowire Heterostructures. *Adv. Opt. Mater.* **2018**, *6*, 1701391. [CrossRef]
107. Herro, Z.G.; Zhuang, D.; Schlesser, R.; Sitar, Z. Growth of AlN single crystalline boules. *J. Cryst. Growth* **2010**, *312*, 2519–2521. [CrossRef]
108. Nakarmi, M.L.; Nepal, N.; Lin, J.Y.; Jiang, H.X. Photoluminescence studies of impurity transitions in Mg-doped AlGaN alloys. *Appl. Phys. Lett.* **2009**, *94*, 091903. [CrossRef]
109. Taniyasu, Y.; Kasu, M.; Makimoto, T. An aluminium nitride light-emitting diode with a wavelength of 210 nanometres. *Nature* **2006**, *441*, 325–328. [CrossRef] [PubMed]
110. Gunning, B.; Lowder, J.; Moseley, M.; Alan Doolittle, W. Negligible carrier freeze-out facilitated by impurity band conduction in highly p-type GaN. *Appl. Phys. Lett.* **2012**, *101*, 082106. [CrossRef]
111. Taniyasu, Y.; Kasu, M. Surface 210 nm light emission from an AlN p–n junction light-emitting diode enhanced by A-plane growth orientation. *Appl. Phys. Lett.* **2010**, *96*, 221110. [CrossRef]

micromachines

MDPI

Review

Surface/Interface Engineering for Constructing Advanced Nanostructured Light-Emitting Diodes with Improved Performance: A Brief Review

Lianzhen Cao [1,2], Xia Liu [1,2,*], Zhen Guo [2,3,4,*] and Lianqun Zhou [2,5,*]

1 Department of Physics and Optoelectronic Engineering, Weifang University, Weifang 261061, China; lianzhencao@wfu.edu.cn
2 CASKey Lab of Bio-Medical Diagnostics, Suzhou Institute of Biomedical Engineering and Technology, Chinese Academy of Sciences, Suzhou 215163, China
3 Shandong Guo Ke Medical Technology Development Co., Ltd., Jinan 25001, China
4 Zhongke Mass Spectrometry (Tianjin) Medical Technology Co., Ltd. Tianjin 300399, China
5 Jihua Laboratory, Foshan 528200, China
* Correspondence: liuxia@wfu.edu.cn (X.L.); guozhen@sibet.ac.cn (Z.G.); zhoulq@sibet.ac.cn (L.Z.); Tel.: +86-536-878-5376 (X.L.)

Received: 1 November 2019; Accepted: 25 November 2019; Published: 27 November 2019

Abstract: With the rise of nanoscience and nanotechnologies, especially the continuous deepening of research on low-dimensional materials and structures, various kinds of light-emitting devices based on nanometer-structured materials are gradually becoming the natural candidates for the next generation of advanced optoelectronic devices with improved performance through engineering their interface/surface properties. As dimensions of light-emitting devices are scaled down to the nanoscale, the plentitude of their surface/interface properties is one of the key factors for their dominating device performance. In this paper, firstly, the generation, classification, and influence of surface/interface states on nanometer optical devices will be given theoretically. Secondly, the relationship between the surface/interface properties and light-emitting diode device performance will be investigated, and the related physical mechanisms will be revealed by introducing classic examples. Especially, how to improve the performance of light-emitting diodes by using factors such as the surface/interface purification, quantum dots (QDs)-emitting layer, surface ligands, optimization of device architecture, and so on will be summarized. Finally, we explore the main influencing actors of research breakthroughs related to the surface/interface properties on the current and future applications for nanostructured light-emitting devices.

Keywords: nanostructured materials; surface/interface properties; nanostructured light-emitting devices; physical mechanism; surface/interface modification; surface/interface control

1. Introduction

The area of nanostructure materials refers to a new system that is composed or assembled according to certain rules based on the material units at the nanoscale (usually below 100 nm). According to the spatial dimension of the nanometer scale, nanostructured materials can be divided into zero-dimensional (0D) nanometer materials (such as nanoparticles, artificial atoms, clusters, etc.), one-dimensional (1D) nanomaterials (such as nanowires, fine rice filaments, nanorods, nanotubes, and nanofibers, etc.), and two-dimensional (2D) nanomaterials (such as nanoribbons, nanoscale disks, superlattices, multilayer membranes, etc.). As nanomaterials are in line with development trends related to the miniaturization and miniaturization of future devices, they will be regarded as the basic component unit of various components in the future once discovered; furthermore, they have been widely considered already [1–10].

In addition, the quantum confinement, size, and surface/interace effects of nanometer materials make them different from similar component bulk materials in mechanics, calorifics, optics, electricity, magnetism, and so on. Many of the physical properties such as optical properties can be controlled and optimized through the appropriate method. So, nanomaterials in both science and engineering fields have been widely studied and used. In the early research on nanomaterials, people usually started from the synthesis of nanomaterials and studied the changes of material properties by adjusting the composition and structure of materials. At the same time, the continuous development of micro/nanoprocessing technology promotes the research of nanodevices based on nanomaterials and helps people to better find the relationship between material properties and device performance, as well as deeply understand the novel physical phenomena in the nanometer world.

Nowadays, photoelectric devices, which can be broadly divided into three categories, play an indispensable role in people's daily lives. First, devices are used as light sources to convert incoming electrical energy into light radiation, such as light-emitting diodes (LEDs) and laser diodes (LDs). The second is an optical detector, which converts weak light signals into the electrical signals of devices such as photomultiplier tubes, photodiodes, and phototransistors. The third category is photovoltaic devices, or solar cells, which convert light into electricity. From the classification of photoelectric devices, we can see that the interaction between photons and electrons in materials plays an important role in the process of energy conversion from electrons to photons or photons to electrons, and is an important determinant of device performance. Compared with the traditional optoelectronic devices with bulk materials, nanostructured optoelectronic devices, such as LEDs, have attracted more and more attention due to their high quantum efficiency, flexibility, and stability [11–15], and some nanostructure LEDs have been applied commercially. For example, the industrialization of organic light-emitting diode (OLED) technology has been realized, and it is gradually applied in the commercial field, such as mobile phones, digital cameras, and televisions [11–15].

Similar to traditional light-emitting devices, the energy conversion between photons and electrons or between electrons and photons of nanostructured light-emitting devices usually involves the generation of carriers, recombination, separation, collection, and other dynamic processes, all of which are closely related to the surface/interface of devices. When the size of the device is scaled down to the nanometer level, the proportion of the surface/interface in the device increases sharply. Thus, the surface/interface properties of nanostructured materials are an important factor affecting the performance of the photoelectric devices [16–20]. So, how to explore the surface/interface properties of nanostructured light-emitting devices for constructing advanced light-emitting devices with improved performance has become a hot pursuit for the researchers. We need to comprehensively study the surface/interface of nanostructured light-emitting devices from the microscopic perspective in depth and understand the correlation between the properties of nanostructured devices and the physical mechanism of the surface/interface in the materials and devices. Here, in this paper, how the surface/interface properties dominating the performance of the built photoelectric devices will be illustrated theoretically and summarized experimentally.

2. Surface/Interface Properties

The surface is broadly defined as all the atomic layers that do not have the three-dimensional periodicity of bulk materials. The presence of the nanosurface will destroy the periodicity of the three-dimensional infinite lattice in the direction of the vertical surface. Therefore, the potential energy of the lattice electrons does not have translational symmetry in the direction of the vertical surface. If the Schrodinger equation is solved directly by using the theory of quantum mechanics, some new eigenvalues will appear in the Hamiltonian eigenvalue spectrum. This means that there will be a new electronic energy state (energy level), which is defined as a surface state, as shown in Figure 1. There are two kinds of surface states; one is called an intrinsic surface state, which is the electronic state on the surface without foreign impurities. The other is called the induced surface state, which is due to the presence of impurities, adsorbed atoms, and other imperfections on the surface. According to the different action with the electron, the surface state also can be divided into a donor state and acceptor state. The donor state is a neutral surface state occupied by an electron. A negatively charged surface state is occupied by an electron. Surface states have three very important characteristics: first, they can be the effective generation and composite center of a few carriers in the material, which determines the surface composite characteristics; second, the surface state scatters most carriers, reduces the surface mobility, and affects the surface conductance; third, the electric field generated on the vertical material surface causes the surface electric field effect [21–23].

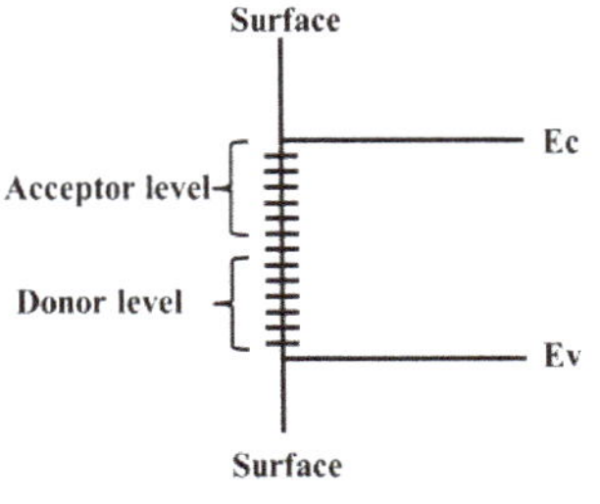

Figure 1. Schematic diagram of surface level.

The interface refers to the interface between different substance phases or different kinds of substances. Similar to surface states, interface states are introduced at the interface state and occur at the interface where two different substances are in contact or at the interface of a heterogeneous junction due to the interruption of a periodic lattice constant. These interface states can also be introduced by lattice mismatch and interface roughness. In addition, the interface state can also be generated because the thermal expansion of the two materials does not match. The interface state is a local electronic state that cannot propagate in the material. Interface states are generally divided into a donor and recipient. Regardless of the position of the energy level in the forbidden band, if the energy level is neutral when occupied by electrons and positively charged after releasing electrons, it is called the donor interface state. If the energy level is neutral when it is empty and the electron is negatively charged, it is called the acceptor interface state. The interface state is another key factor to determine the performance of nanostructured devices [24–26].

As mentioned above, theoretically studying the influence of surface interface on nanomaterials and devices is of great significance to improve the performance of nanostructured devices. Fortunately, with the development of material characterization technology, there are many means and instruments to characterize the properties of surface interfaces, which can be used to validate and promote theories and help scientists to understand their impact on nanomaterials and devices. The analysis of surface interface properties mainly includes surface composition, surface morphology, surface structure, and surface energy state. The main characterization methods used in the experiment and the feedback information are shown in Table 1 [27–31]. Through these characterization methods, we can have a deep and comprehensive understanding of the surface or interface properties of nanomaterials and devices, which provide beneficial help for the further utilization of the surface/interface properties. Figure 2 contains three-dimensional images of the surface/interface topography using the confocal microscope with far-field configuration.

Table 1. Surface/interface characterization of nanomaterials and devices [1].

Analysis Method	Acquired Information
Auger electron spectrum (AES) X-ray photoelecgon spectroscopy (XPS) Secondary ion mass spectrometry (SIMS) Ion scattering spectroscopy (ISS)	Surface/interface composition
Transmission electron microscopy (TEM) Scanning electron microscope (SEM) Scanning tunneling microscopy (STM) Atomic force microscope (AFM)	Surface/interface morphology
Low-energy electron diffraction (LEED) Field ion microscope (FIM) High-resolution transmission electron Microscopy (HRTEM)	Surface/interface structure
X-ray photoelecgon spectroscopy (XPS) Ultraviolet photoelectron spectroscopy (UPS) Electron energy loss spectroscopy (EELS) Ion neutralization spectroscopy (INS) Deep level transient spectrum (DLTS) Pulsed I–V measurement Conductance	Surface/interface energy state

[1] References [27–31].

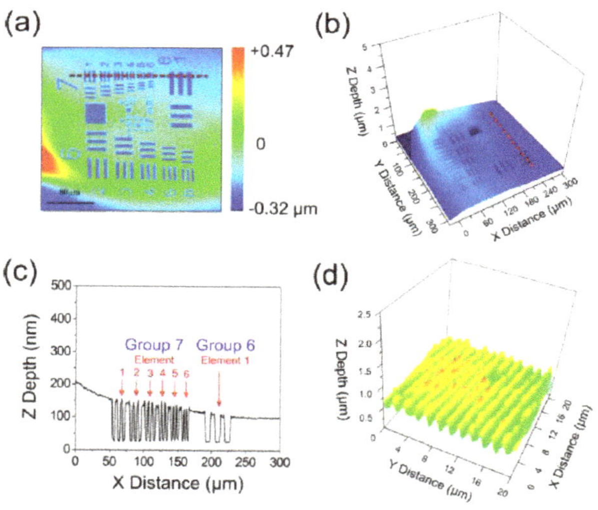

Figure 2. Three-dimensional (3D) surface profiles of metallic film. (**a**) Top view. (**b**) Perspective view. (**c**) Cross-sectional view. (**d**) Perspective view of a 600 line/mm diffraction grating. Reproduced with permission from reference [27].

3. Influence of Surface and Interface on LED and Optimization Method

In the last few years, tremendous progress has been achieved in increasing the efficiencies, stabilities, and lifetime of light-emitting devices [32]. In this section, the application of nanostructured materials in LED devices will be introduced according to the dimension of materials, especially the influence of surface/interface properties on materials and devices, and the optimization mechanism will be summarized.

3.1. *Influence of Surface and Interface of 0D Nanomaterials on LED and Optimization Method*

Quantum dot-based LEDs (QD-LEDs) have attracted considerable attention owing to their high color purity, thermal stability, and size-dependent emission wavelength tenability, making them suitable candidates for next-generation solid-state lighting [13]. As noted in the introduction, QDs have a very large surface-to-volume ratio due to their small size. The surface/interface composition and structure of QDs can significantly affect the photoluminescence (PL) emission, charge injection, and charge transport. Thus, the surface/interface engineering of QDs plays a vital role in the realization of high performance QD-LEDs. In order to reduce the surface/interface effects and improve the device performance, some advances related to surface purification, different functional layers, the design of nanostructures, and different device architects have been made in the engineering of nanostructures and surfaces of QDs. The zero-dimensional material name, material structure, and optical properties of LED devices are summarized in Table 2.

3.1.1. Surface Purification Methods

Inevitably, the synthesis of QDs will introduce many different type impurities, including metal carboxylate precursors, free ligands, and non-crystalline side products, which significantly diminish the stability, radiation efficiency, and potential applications of QDs in the field of nano-optoelectronic devices. Therefore, a surface purification for QDs is necessary for the improvement of the optical properties of QDs and device performance. Yang et al. fabricated the colloidal QDs from solution by the addition of a non-solvent [33]. The most common solvent and non-solvent pair was toluene and methanol, although other combinations including trichloromethane or hexanes as the solvent and methanol or acetone as the non-solvent have often been used as purification processes, as reported previously [34]. In addition, the hexane–methanol extraction process can also be used to purify the surface of QDs [35]. However, the non-solvent and hexane–methanol extraction approaches did not completely remove raw materials from the solution of colloidal QDs. Yang et al. developed a new purification scheme in which trichloromethane was selected as the solvent additive to enhance the solubility of the cadmium sources, while acetonitrile was selected as the solvent additive to precipitate the QDs, which was more effective than methanol [36]. Therefore, the improved purification scheme can effectively remove residual impurities in colloidal QD solution, including metal carboxylic acid precursors, non-volatile solvents, and non-crystalline by-products. Therefore, this benign purification process provides a good foundation for the surface engineering of quantum dots, and provides more opportunities for the photoelectric application of quantum dots. To solve the lifetime issue, Cao et al. tackle the hole barrier issue directly by tailoring the band-energy levels of QDs to reduce the injection barriers and improve device performance. A high-quality colloidal QD usually comprises an inorganic semiconductor core and a semiconductor shell with a wider energy bandgap to passivate dangling bonds of core surface and confine electron and hole wavefunctions for good luminescent properties and reliability [37].

3.1.2. QDs Emitting Layer Method

The QDs emitting layer method, especially involving perovskite QDs, has significant potential and has been extensively studied for light-emitting devices. Organo-inorganic lead halide perovskite is a direct bandgap semiconductor material with many excellent properties and is attractive in electroluminescence (EL) applications [38–42]. Many high-efficiency perovskite light-emitting diodes, such as the green LED with peak external quantum efficiency (EQE) $\approx 8\%$ [43] and the near-infrared LED with peak EQE of 11.7% [41], have the nanolevel structure of perovskite materials as their photon emission core. Kovalenko et al. synthesized monodisperse $CsPbX_3$ perovskite quantum dots (X=Cl, Br and I, or Cl/Br and Br/I in the mixed halide system) [44,45]. These all-inorganic perovskite quantum dots cover the entire visible region with very high luminescence stability and photoluminescence quantum yields (PLQYs), and the gamut covers 140% of the NTSC standard. In addition, compared with organic–inorganic hybrid perovskite materials, all-inorganic perovskite QDs have better environmental stability [40,46,47].

3.1.3. Surface Ligand Method

The developments of surface coordination chemistry allow facile ligand-displacement reactions which enable the rational design of surface ligands for QDs used in LEDs [48]. Shen et al. report high-efficiency blue-violet QD-LEDs by using high quantum yield ZnCdS/ZnS graded core–shell QDs with proper surface ligands. Such ligand exchange results in an even greater increase in hole injection into the QD layer, thus improving the overall charge balance in the LEDs and yielding a 70% increase in quantum efficiency [11]. Next year, Zhong et al. developed an in situ ion exchange method to improve the performance of QD-LED devices [49]. The in situ ligand exchange process is shown in Figure 3. The results show that this method is very effective and the photoluminescence (PL) quantum yields are almost unchanged after the ligand exchange process. As a result, significant device performance improvements have been shown. Li et al. demonstrate a highly efficient solution-processed $CsPbBr_3$ QD-LED through balancing surface passivation and carrier injection via ligand density control [50]. Compared to surface ligand exchange, the control of ligand density on QD surfaces is a more proper strategy to promote the performance of $CsPbX_3$ QLEDs. In general, the solubility of QDs will decrease with the decrease of ligand size. On the other hand, the spatial separation between quantum dots caused by surface ligands will affect the colloidal stability of quantum dot solution [51]. In 2016, Peng et al. used "entropy ligands" [52,53] to improve the performance of nanocrystalline optoelectronic devices. Pan et al. have shown that the charge transfer characteristics of QD films can be regulated by using specially designed polymer ligands with colloids and perovskite groups instead of insulating ligands [54,55]. Therefore, a QD/polymer hybrid can be used as an important candidate material for the emitting layer of QLEDs [56]. Brown et al. demonstrated that the ligand-induced generation of surface dipoles is an effective way to control the absolute energy levels of QD films [57]. It is found that the strength of the surface dipole induced by the ligands can be regulated by the chemical binding group and dipole moment of the ligands, and the size of the surface energy level can be controlled. This method enables fine-tailoring of the band-energy calibration of PbS quantum dots, thus improving the optical performance of the device [58]. Yang et al. used this method to make QLEDs. Combined with the size control of QDs, the bandgap and band position of a PbS QD film can be fine-tuned by the quantum effect, so that the QD film can be used as electron transport layers (ETLs), hole transporting layers(HTLs), and the transmitting layer of LEDs [59].

3.1.4. Core/Shell Interface Structure Method

Although surface purification is an important way to reduce impurities on the surface of QDs, the purification process may also bring about electron lone pairs or vacancies, which serve as surface traps by capturing excitons, thereby leading to non-radiative recombination. QDs often suffer from surface-related trap states, which act as non-radiative de-excitation channels for photogenerated charge carriers, thus decreasing their PL QDs. The photochemical stability, thermal stability, and photochemical stability of the quantum dot can be improved by covering the epitaxial layer and optimizing the growth parameters of the epitaxial layer, and the non-radiative recombination can be effectively reduced. Therefore, core–shell structure QDs are widely used to improve the performance of QLEDs. Bawendi et al. obtained high-quality CdSe/CdS core/shell QDs by using octanethiol and cadmium oleate as precursors and maintaining an appropriate growth rate at 310 °C. The obtained core/shell structure of the QDs has high uniformity; PLQYs can reach 97%, and the full width at half maximum (FWHM) of the PL peak is only 67.1 meV [60]. Pal et al. studied the optical properties of CdSe/CdS core/shell quantum dot films with different shell thicknesses and compared them with the luminescence properties of corresponding quantum dot solutions. The results show that the luminescence properties of core/shell materials are obviously improved, and the redshift effect of the PL spectrum is gradually weakened with the increase of the thickness of the CdS shell [61]. Li et al. showed that ZnCdSe-based core/shell QDs featuring a ZnS shell with 10 monolayers offered the highest external quantum efficiency of ~17%, which could compare favorably with the highest efficiency of green QLEDs with traditional multilayered structures [62]. The epitaxy of a gradient alloy shell on a core crystal can synthesize quantum dots with gradient band structures. Klimov et al. have shown that "smoothing" the shape of the constraint potential through the interfacial alloying of the core–shell interface can effectively inhibit an auger recombination in CdSe/CdS QDs [63,64]. Peng et al. prepared a series of small, scintillation-free and bleach-resistant high-quality CdSe/$Cd_xZn_{1-x}S$ and CdSe and $CdSe_yS_{1-y}$/$Cd_xZn_{1-x}S$ core/shell QDs. Quantum dots with a continuously adjustable luminous range and high PLQYs have important potential applications in LEDs [65].

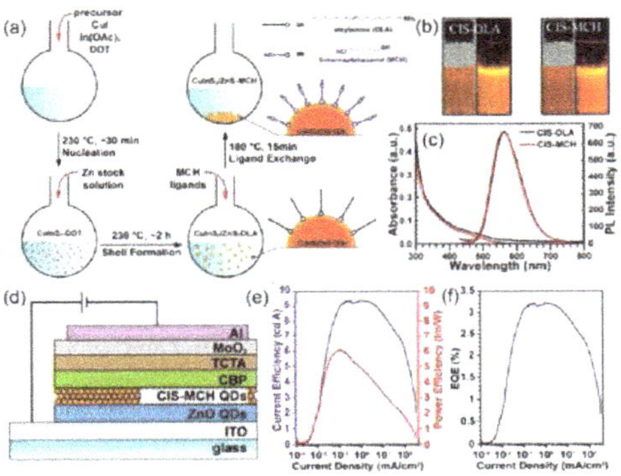

Figure 3. Schematic diagram of the in situ ligand exchange process and the relevant material and device measurement results. (**a**) The ligand exchange process and surface structure of the QDs. (**b**) The photos of QDs in solution before and after the ligand exchange process. (**c**) Absorption and photoluminescence (PL) results. (**d**) The device structure diagram of a quantum dot-based light-emitting diodes (QD-LED). (**e**,**f**) The power efficiency and current efficiency as a function of the current density. Reproduced with permission from reference [49].

Table 2. The summaries of zero-dimensional material name, material structure, and optical properties of light-emitting diode (LED) devices. QD: quantum dots.

Materials	Structures	Spectral Range of LEDs	Ref.
$CsPbBr_3$	Nanocrystals	Green	[34]
CdSe/ZnS	QDs	Blue, Green, Red	[36]
CdSe/$Cd_{1-x}Zn_xSe$	QDs	Red	[37]
$CsPbBr_3$	QDs	Deep-red	[38]
$CH_3NH_3Pb_{3-x}Cl_x$	Films	Near-infrared	[39]
$CsPbX_3$	QDs	Red, Orange, Green, Blue	[40]
$NFPI_7$	QWs	Near-Infrared	[41]
$CH_3NH_3PbI_3$	Quasi-2D	Near-Infrared	[42]
$CH_3NH_3PbBr_3$	QDs to Nanograins	Green	[43]
$CsPbX_3$ (X = Cl, Br, I)	Nanocrystals	Blue, Green	[44]
$CsPbX_3$ (X = Cl, Br, I)	Nanocrystals	Blue to Red	[45]
$CsPbBr_3$	QDs	Blue to Red	[46]
$CsPbX_3$	QDs	Blue to Red	[47]
$CuInS_2$	QDs	Green	[49]
$CsPbX_3$	QDs	Green	[50]
$CsPbI_3$	Nanocrystals	Red	[55]
Colloidal	QDs	Near-Infrared	[59]
CdSe/CdS	Nanocrystals	~	[60]
CdSe/CdS	QDs	Red	[61]

3.1.5. Optimization of LED Device Interface Architecture Methods

Device architecture using different charge transport layers (CTLs) and interfacial engineering is another important method to improve the performance of luminescent devices. In a typical QD-LED, apart from the QD-emitting layer, the charge-injection layers (CILs) and CTLs also significantly contribute to the overall device performance. The design of these layers can be made to favor the balance of the carrier injection, charge transport, and radiative recombination of excitons in the QD-emitting layer. Therefore, interfacial engineering between the QD-emitting layer and the CTL plays a critical role in enhancing the device performance of the QD-LEDs. Kim et al. used an all-solution processed method to fabricate highly efficient green QD-LEDs with an inverted architecture [66]. An interfacial polymeric surface modifier of polyethylenimine ethoxylated (PEIE) is inserted between a QD-emitting layer and a hole transport layer. At the same time, a MoO_x hole injection layer is solution deposited on top of the hole transport layer. Among the inverted QLEDs with varied PEIE thicknesses, the device with an optimal PEIE thickness of 15.5 nm shows record maximum efficiency values of 65.3 cd/A in current efficiency and 15.6% in external quantum efficiency (EQE). The all-solution processed fabrication of inverted QLEDs is further implemented on a flexible platform by developing a high-performing transparent conducting composite film of ZnO nanoparticles overcoated on Ag nanowires. The resulting flexible inverted device possesses 35.1 cd/A in current efficiency and 8.4% in EQE, which are also the highest efficiency values ever reported in flexible QLEDs. Apart from engineering the compositions, the size, structure, and shape control of QDs may provide additional benefits regarding the accessibility in band structure engineering and enhance the out-coupling efficiency in QD-LEDs. Nam et al. synthesized double-heterojunction nanorods consisting of two offset and staggered bandgaps, which offered independent control over the electron- and hole-injection processes in devices [12]. The out-coupling efficiency was significantly enhanced due to the nanorods and was assembled parallel to the substrates. More importantly, the anisotropic shape introduces a transition dipole along the rod axis. Liu et al. used the balanced charge-injection process to enhance the external quantum efficiency of nonblinking blue QD-LEDs [67]. Using nonblinking ZnCdSe/ZnS/ZnS QDs as the emissive layer, highly efficient blue QD-LEDs were prepared. The charge-injection balance within the QD active layer was improved by introducing a nonconductive layer of poly-(methyl methacrylate) (PMMA) between the electron transport layer (ETL) and the QD layer, where the PMMA layer takes the role of coordinator to impede excessive electron flux. The optimized LED device shows excellent performance such as a maximum luminance of 14,100 cd/m^2, current efficiency of 11.8 cd/A, and external quantum efficiency (EQE) of 16.2%.

3.2. *Influence of Surface and Interface of 1D Nanomaterials on LED and Optimization Method*

1D nanomaterials possessing natural structures that can act as resonant cavities are ideal platforms to realize laser diodes and light-emitting diodes. Various nanostructures such as nanowires, nanotubes, and nanobelts have been synthesized to fabricate optoelectronic devices. In order to reduce the surface/interface effects and improve the device performance, some advances such as surface or interface structure design, interface control, and interface modification and modulation have been made in the engineering of 1D nanomaterial. The one-dimensional material name, material structure, and optical properties of LED devices are summarized in Table 3.

3.2.1. Surface or Interface Structure Design Methods

Similar to 0D nanomaterials, material and device structure design is a common method to suppress the surface states of 1D nanomaterials. For example, in order to reduce the reflection and improve the transmission of light, nanostructure arrays have been developed as effective antireflective surfaces [68]. By controlling the surface wetting properties of a polydimethylsiloxane release template, Liu et al. was able to pattern a random AgNWs network with uniform conducting property [69]. Modifying the core–shell structure is another method to overcome the total internal reflection at the semiconductor and air interface, improve the escape probability of the light, and consequently increase the light extraction efficiency [70,71], as shown in Figure 4. Zhang, Yao, and Zheng's team used the high directionality of waveguide mode transmission and the efficient energy transfer of localized surface plasmon (LSP) resonances to increase the spontaneous emission rate of LEDs, respectively [72–74].

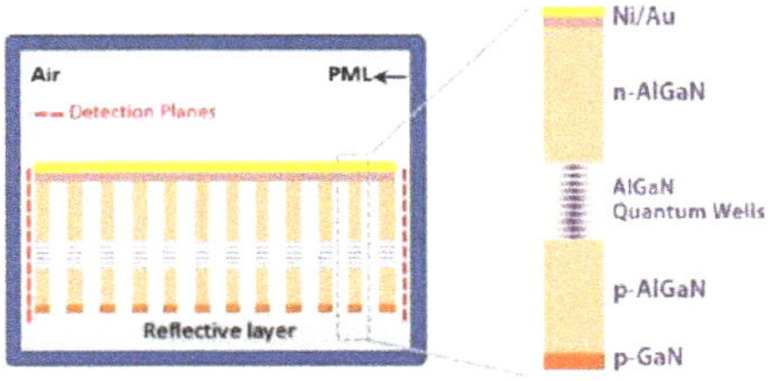

Figure 4. Schematic illustration of P-GaN/P-AlGaN/AlGaN core–shell UV LED structure. [71].

3.2.2. Interface Control 1D Nanomaterial Methods

ZnO has inspired considerable attention to develop ultraviolet (UV) LEDs and LDs due to its wide direct bandgap of 3.37 eV and high exciton binding energy up to 60 meV. However, in the process of emission of 1D ZnO nanometer materials, it is difficult to avoid an extra emission from interface states. Hence, the interface design and optimization are necessary to realize efficient UV EL. So, both a high quality active layer and good interface at the p-i-n junction are critical factors to realize pure UV LED. You et al. successfully prepared the vertically aligned ZnO NRs and showed high crystal and optical quality. Nanostructured LED arrays were constructed by directly bonding ZnO NRs onto on AlN-coated p-GaN wafer. This simple and feasible method can effectively suppress the interface defects induced by the buffer layer formation [75,76].

3.2.3. Interface Modification and Modulation 1D Nanomaterial Methods

As we all know, total internal reflection (TIR) will occur at the epitaxial layers/substrate interface and substrate/air interface because of the large difference in refraction indices. The surface/interface modification and modulation are effective methods to avoid the TIR. Guo et.al reported AlGaN-based 282-nm LEDs grown on nanopatterned sapphire substrates (NPSS), exhibiting 98% better performance relative to those grown on flat sapphire substrates [77]. The AlN epitaxial lateral overgrowth on pattered substrates or templates can not only improve the crystal quality of the overgrown epitaxial layers but also form embedded air voids in the AlN layer. The effective refraction index around the interface is thereby between the AlN layer and the substrates. The internal quantum efficiency (IQE) enhancement is estimated to be 60%, so the light-extraction efficiency (LEE) enhancement would be more than 20%. Lee et al. fabricated DUV LEDs on NP-AlN/sapphire templates, with air surrounding the AlN nanorods. Light emitted from the multiple quantum wells can propagate vertically by passing through the embedded nanostructures, and thus the TIR is avoided [78,79].

3.2.4. Core/Shell Structure Methods

Due to the large surface-to-volume ratios of the 1D nanomaterials, the surface plays a key role in the optical properties of 1D nanometer material LEDs [80–90]. Due to the existence of surface states, a Fermi-level pinning effect will occur, and the resulting transverse electric field effect and related surface non-radiative recombination will be very unfavorable to one-dimensional nanostructured-led devices [80–82]. Optimizing the growth of core–shell structures can reduce non-radiative recombination and improve the photo-, electrical- and photochemical stability of 1D nanometer materials. Among them, InGaN/GaN and InGaN/AlGaN core–shell nanowires have attracted extensive attention due to their important applications in variable wavelength and UV LED devices. Ledig et al. characterized in detail the structural, optical, and electrical properties of InGaN/GaN core–shell structure LEDs [83].

It turns out that the 3D core shell structure design of the LED based on GaN has many advantages over conventional planar LED counterparts. In addition, in 2016, Müller et al. prepared the InGaN/GaN core–shell nanowires using the selective area metal organic vapor phase epitaxy method and studied the effects of InGaN thickness on cathodoluminescence spectroscopy [84]. The following year, InGaN/AlGaN and AlGaN core–shell tunnel junction nanowire LEDs were prepared by Philip's research group and Sadaf's research group, respectively [85,86]. The results show that the GaN-based nanowires LED can realize relatively high internal quantum efficiency from the deep green to red wavelength range. In 2018, Sim et al. designed highly efficient white LEDs using the 3D InGaN/GaN structure [87]. It can be proved by the experimental results that white LEDs based on dodecagonal ring structures are a platform enabling a high-efficiency warm white light-emitting source. Recently, high-performance InGaN/GaN and InGaN/AlGaN nanowire heterostructure LEDs were prepared by different research groups [88–94]. A schematic diagram of a nanowire LED with an InGaN/AlGaN core–shell heterostructure is shown in Figure 5. The results show that by controlling and optimizing the core shell structure, various surface defects and surface states can be effectively reduced and suppressed, and a high-quality crystal structure can be achieved, which can significantly improve the performance of LED devices.

Table 3. The summaries of one-dimensional material name, material structure, and optical properties of LED devices.

Materials	Structures	Spectral Range of LEDs	Ref.
Ag	Nanowires	Green	[69]
InGaN/GaN	Nanostructure	~	[70]
AlGaN	Nanowires	Deep ultraviolet	[71]
ZnO/MgZnO	Nanorods	Ultraviolet	[72]
ZnO/GaN	Nanorods	Ultraviolet	[73]
CdS	Nanowires	Red	[74]
ZnO	Nanowires	Ultraviolet	[75]
AlGaN	Nanowires	Ultraviolet	[77]
AlGaN	Nanoscale	Deep ultraviolet	[78]
InGaN/GaN/AlGaN	Nanowires	White	[80]
GaN	Nanowires	White	[81]
InGaN/AlGaN	Nanowires	Full color	[82]
InGaN/GaN	Nanorods	Green	[84]
InGaN/AlGaN	Nanowires	Green, Yellow	[85]
AlGaN	Nanowires	Ultraviolet	[86]
InGaN/AlGaN	Nanowires	Full color	[89]
InGaN/AlGaN	Quantum disks	~	[90]

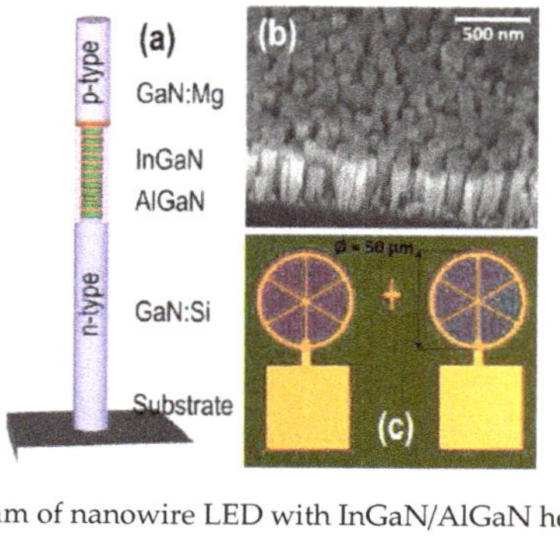

Figure 5. (**a**) Schematic diagram of nanowire LED with InGaN/AlGaN heterostructures. (**b**) The SEM image of InGaN/AlGaN nanowires. (**c**) The optical image and electrode pads of LEDs. Reproduced with permission from reference [89].

3.3. *Influence of Surface and Interface of 2D Nanomaterials on LED and Optimization Method*

The application of 2D materials in LEDs mainly involves the synthesis of various thin films and the preparation of devices with heterogeneous LED structures [95–98]. Similar to zero and one-dimensional nanomaterials, it is also very important to quantify the influence of the 2D surface/interface on LED devices and to adopt effective methods to reduce the useless surface/interface recombination mechanism [99].

3.3.1. Surface Modification and Interface Engineering Method

Carbon nanotubes and graphene are used for surface/interface modification or as the protecting layer to optimize LED device performance [100–102]. Due to the high refractive index, the light extraction efficiency of GaAs-based LED devices is limited. Nanoscale surface modification is an effective method to enhance the output light power of GaAs-based LED devices. Jin et al. report a simple method for nanostructure fabrication using super-aligned multiwalled carbon nanotube (SACNT) thin films as etching masks for top–down etching processes [100]. The morphology of the carbon nanotube (CNT) networks can be transferred to the substrate GaAs material at the macro scale. With this method, the nanostructured SACNT network morphology significantly increases the optical output power of GaAs devices in comparison with planar GaAs LED devices. Graphene is a useful material for conducting electrodes in LED device applications [101,102]. Seo et al. study the impact of the graphene quality on the performance of a hybrid electrode of graphene on AgNWs in GaN-based UV LED. The hybrid electrode using two-step graphene showed good ambient stability with stable sheet resistance over time. The UV LED using this TCE offered a low forward voltage, an increase in the EL intensity, and a reduction of efficiency droop. In addition, surface or interface passivation and the modulation of LED devices based on amine and perovskite materials have been studied [103–106]. Yang et al. made a green LED based on a quasi-2D perovskite composition and phase with surface passivation [104]. The surface passivation is realized through coating molecules of trioctylphosphine oxide on the surface of the perovskite thin film. The measured results of optimized LED based on quasi-2D perovskite reach a current efficiency of 62.4 cd A^{-1} and an external quantum efficiency of 14.36%, as shown in Figure 6. All the results show that the surface-sensitive characterization is expected to help to reveal the role that the interface plays in various devices and to identify specific strategies to regulate and tune the properties of the surface/interface, leading to enhanced device performance.

3.3.2. Interface Structure Design Methods

For the 2D materials, core–shell, 3D pixel configuration, back-end-of-line material, and device structures are designed to optimize the device performance [107–113]. For example, Zhang et al. insert a 4-nm Si_3N_4 layer between the ZnSe core and the CdS shell of p-ZnSe/n-CdS core–shell heterojunctions to passivate the interface defects and reduce the recombination and the saturation current [107]. Liu et al. study the band states at the crystallized interface between GaN and SiNx and the influence of interface roughness on the material, as shown in Figure 7 [109]. Zhang et al. designed a three-dimensional reflective concave structure coated with a high refractive index material and achieved an increase in OLED display pixels by embedding the OLED into the three-dimensional reflective concave structure. This structure allows the coupling region of the light to be defined so that as much of the coupling interior is emitted to the filled region and then redirected. The optical simulation results show that if the optimized structure and highly transparent top electrode material are adopted, the efficiency of light extraction can be improved by $\approx$ 80%. [110]. Bulling and Venter use two methods to improve light extraction efficiency: first, the design of an improved back-end-of-line (BEOL) light directing structure, and second, the use of surface texturing. The design of an optimized pipe-like BEOL light directing structure has resulted in a 1.35-factor improvement in luminance and a 1.38-factor improvement in light extraction efficiency over the previously designed parabolic BEOL light directing structure; furthermore, it has also resulted in an improved BEOL light-directing structure for improved light extraction efficiency. In addition, the directionality of the light emission radiation pattern has also significantly improved. Once the internal TIR is reduced and the light radiation propagation is improved, surface texturing techniques can be used to further improve the light extraction efficiency [108]. Recently, Lei et al. used the surface texture and LSP coupling effect to enhance the light extraction efficiency for InGaN/GaN LEDs [113].

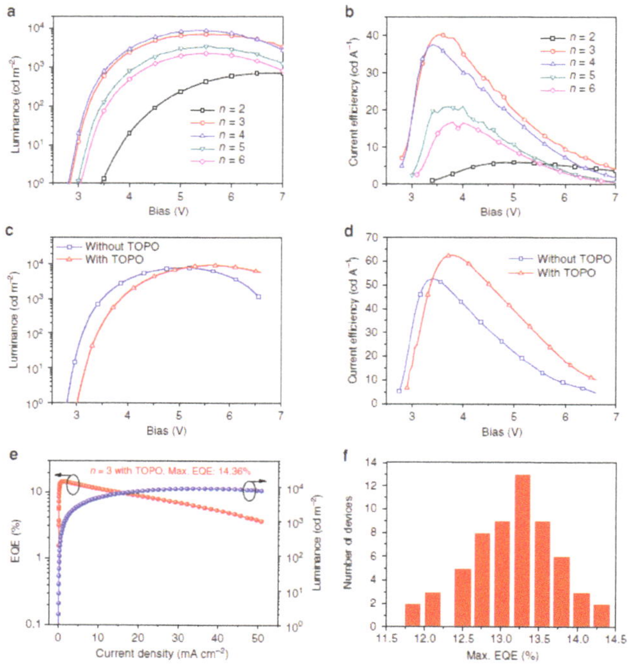

Figure 6. Device performance of perovskite LEDs with different compositions and surface passivation. (**a**) Luminance-voltage (L-V) curves with different compositions. (**b**) Current efficiency voltage (CE-V) curves with different compositions. (**c**) L-V curves with and without trioctylphosphine oxide (TOPO) layer. (**d**) CE-V curves without TOPO layer. (**e**) EQE with TOPO layer. (**f**) Maximum EQEs measured from 60 devices. Reproduced with permission from reference [104].

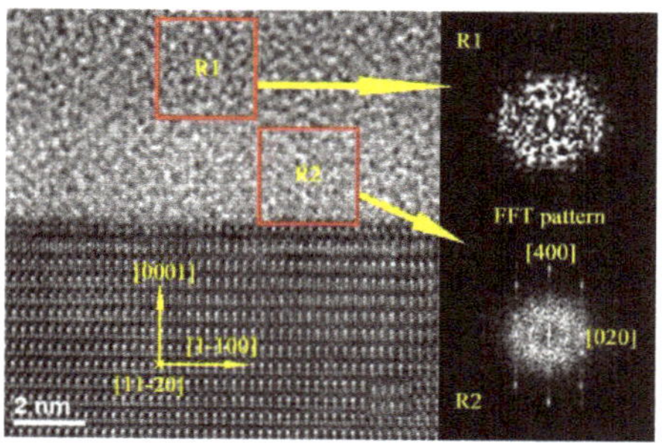

Figure 7. The high-resolution transmission electron microscope images across the SiN_x/GaN interface. Reproduced with permission from reference [109].

4. Conclusions and Perspective

In summary, we introduce the effects of surface/interface properties of nanostructured materials on light-emitting diodes, taking some II–VI, III–V, IV, and perovskite nanomaterials as representatives. According to the review, we can see that the surface/interface properties of nanostructured materials are an important factor affecting the performance of the device, which involves the generation of carriers, recombination, separation, collection, and other dynamic processes. Most important of all, as shown in the paper, through surface or interface engineering such as surface purification, surface ligands, or introducing the core/shell structure and others in constructing optoelectric devices, the performance could be improved remarkably. Therefore, continuing to study the surface/interface of nanostructured light-emitting devices from the microscopic perspective and understand the correlation between the properties of nanostructured devices in depth is crucial. Even though this progress in the improvement of the device performance of LEDs is very encouraging, there are several shortcomings and challenges that stand in the way of commercialization. The following aspects are critical to improving the device performance and accelerating the commercialization of LEDs.

1. Reducing Surface/Interface Recombination

As we all know, when the atomic lattice is abruptly broken at a surface/interface, unsatisfied dangling bonds (or foreign bonds) introduce electronic energy levels inside of the bandgap that enhance electron–hole non-radiative recombination at the surface/interface by acting as stepping stones for charge carrier transitions between the conduction and valence bands. However, achieving these values in operational device architecture has remained elusive because contacting the nanomaterials with extracting contacts generally induces new, non-radiative loss pathways at the surface, resulting in a decrease in the PLQE and PL lifetime.

2. The Choice of the Appropriate Surface Ligands

The surface ligand methods play an important role in improving the properties of nanomaterials and devices. On the one hand, surface ligands can effectively combine with surface atoms to passivate surface defects and reduce the surface states of materials. On the other hand, the intrinsic insulation of the surface ligands will affect the effective charge injection and transmission characteristics of the emitting layer, and reduce the efficiency and performance of the nanoluminescent devices. Therefore, balanced exciton recombination and charge injection/transport are key to the effective use of the method of surface ligand in nanostructured LEDs.

3. In-Depth Understanding of the Interactions of Nanomaterials with CTLs

It is generally considered that the exciton quenching of nanomaterials is critical to the performance of nanostructure LEDs. In nanostructure LEDs, nanomaterials can acquire a net charge due to the interactions with the CTLs, and the excess charges in the charging nanomaterials can lead to exciton quenching by non-radiative energy transfer to the CTLs or the defects within the CTLs films, thus diminishing the device efficiency. Therefore, it is necessary to develop effective characterization techniques to study the interactions of the nanomaterials with different CTLs in order to gain a more comprehensive understanding of the mechanisms.

4. Reduce Internal/Reflection TIR Effect

As mentioned earlier, TIR occurs at the epitaxial layers/substrate interface and substrate/air interface because of the large difference in refraction indices. A large amount of photons are trapped inside the LED structure and finally absorbed after multiple internal reflections. Disturbing the TIR at the interfaces would be beneficial to achieve high-efficient LEDs. Therefore, how to use highly reflective techniques and surface/interface modification to mitigate TIR's influence on nanostructured light-emitting devices is still a difficult task.

Author Contributions: Conceptualization, X.L. and Z.G. writing—original draft preparation, L.C.; writing—review and editing, X.L. and Z.G.; project administration, L.C., Z.G. and L.Z.

Funding: This research was funded by the Natural Science Foundation of China (nos. 11404246, 51675517, and 61874133), STS Projects of the Chinese Academy of Sciences (No. KFJ-STS-SCYD-217, KFJ-STS-ZDTP-061), Natural Science Foundation of Shandong Province (No. ZR2018LA014 and ZR2019QEE038), Key Research and

Development Plan of Shandong Province (2019GGX101073) and Higher School Science and Technology Plan of Shandong Province (J17KA188), Key Research and Development Program of Jiangsu Province (No. BE2018080), Tianjin Science and Technology Project (No. 19YFYSQY00040), and Jihua Laboratory (No. X190181TD190).

Acknowledgments: The authors thank Wang Cong and Chen Disheng for their meaningful discussion.

Conflicts of Interest: The author declares no conflict of interest.

References

1. Jie, J.S.; Zhang, W.J.; Bello, I.; Lee, C.S.; Lee, S.T. One-dimensional II-VI nanostructures: Synthesis, properties and optoelectronic applications. *Nano Today* **2010**, *5*, 313–336. [CrossRef]
2. Panda, D.; Tseng, T.Y. One-dimensional ZnO nanostructures: Fabrication, optoelectronic properties, and device applications. *J. Mater. Sci.* **2013**, *48*, 6849–6877. [CrossRef]
3. Pascal, A.; Peter, R.; Jens, W.T.; Jean-Charles, R.; Iwan, M. Semiconductor nanostructures towards electronic and optoelectronic device applications. *Phys. Status Solidi C* **2014**, *11*, 193–194.
4. Burak, G.; Hilmi, V.D. Near-field energy transfer using nanoemitters for optoelectronics. *Adv. Funct. Mater.* **2016**, *26*, 8158–8177.
5. Sudaraka, M.; Sarath, D.G.; Mark, I.S.; Malin, P. Generalized superradiant assembly for nanophotonic thermal emitters. *Phys. Rev. B* **2018**, *97*, 125406.
6. He, X.; Htoon, H.; Doorn, S.K.; Pernice, W.H.P.; Pyatkov, F.; Krupke, P.; Jeantet, A.; Chassagneux, Y.; Voisin, C. Carbon nanotubes as emerging quantum-light sources. *Nat. Mater.* **2018**, *17*, 663–670. [CrossRef]
7. Gu, Q.Y.; Chen, J.N. Carbon-nanotube-based nano-emitters: A review. *J. Lumin.* **2018**, *200*, 181–188. [CrossRef]
8. Liang, F.X.; Gao, Y.; Xie, C.; Tong, X.W.; Li, Z.J.; Luo, L.B. Recent advances in the fabrication of graphene-ZnO heterojunctions for optoelectronic device applications. *J. Mater. Chem. C* **2018**, *6*, 3815–3833. [CrossRef]
9. Xie, C.; Wang, Y.; Zhang, Z.X.; Wang, D.; Luo, L.B. Graphene/semiconductor hybrid heterostructures for optoelectronic device applications. *Nano Today* **2018**, *19*, 41–83. [CrossRef]
10. Arko, G.; Caroline, M.; Yuriy, Z.; Jana, Z.; Malte, C.G. Infrared organic light-emitting diodes with carbon nanotube emitters. *Adv. Mater.* **2018**, *30*, 1706711.
11. Shen, H.; Cao, W.; Shewmon, N.T.; Yang, C.; Li, L.S.; Xue, J. High-efficiency, low turn-on voltage blue-violet quantum-dot-based light-emitting diodes. *Nano Lett.* **2015**, *15*, 1211–1216. [CrossRef] [PubMed]
12. Nam, S.; Oh, N.; Zhai, Y.; Shim, M. High efficiency and optical anisotropy in double-heterojunction nanorod light-emitting diodes. *ACS Nano* **2015**, *9*, 878–885. [CrossRef] [PubMed]
13. Zhang, H.; Chen, S.; Sun, X.W. Efficient red/green/blue tandem quantum-dot light-emitting diodes with external quantum efficiency exceeding. *ACS Nano* **2018**, *12*, 697–704. [CrossRef] [PubMed]
14. Wang, W.H.; Qi, L.M. Light management with patterned micro- and nanostructure arrays for photocatalysis, photovoltaics, and optoelectronic and optical devices. *Adv. Funct. Mater.* **2019**, *29*, 1807275. [CrossRef]
15. Zhang, Y.B.; Zhang, F.J.; Wang, H.Z.; Wang, L.; Wang, F.F.; Lin, Q.L.; Shen, H.B.; Li, L.S. High-efficiency CdSe/CdS nanorod-based red light-emitting diodes. *Opt. Express* **2019**, *27*, 7935–7944. [CrossRef]
16. Allan, G.; Delerue, C.; Lannoo, M. Nature of luminescent surface states of semiconductor nanocrystallites. *Phys. Rev. Lett.* **1996**, *76*, 2961–2964. [CrossRef]
17. He, G.H.; Jiang, M.M.; Li, B.H.; Zhang, Z.H.; Zhao, H.F.; Shan, C.X.; Shen, D.Z. Sb-doped ZnO microwires: Emitting filament and homojunction light-emitting diodes. *J. Mater. Chem. C* **2017**, *5*, 10938–10946. [CrossRef]
18. Zhao, L.F.; Yeh, Y.W.; Tran, N.L.; Wu, F.; Xiao, Z.G.; Kerner, R.A.; Lin, Y.H.L.; Scholes, G.D.; Yao, N.; Rand, B.P. In situ preparation of metal halide perovskite nanocrystal thin films for improved light-emitting devices. *ACS Nano* **2017**, *11*, 3957–3964. [CrossRef]
19. Ding, M.; Guo, Z.; Zhou, L.Q.; Fang, X.; Zhang, L.L.; Zeng, L.Y.; Xie, L.N.; Zhao, H.B. One-dimensional zinc oxide nanomaterials for application in high-performance advanced optoelectronic devices. *Crystals* **2018**, *8*, 223. [CrossRef]
20. Zhu, L.L.; Tang, X.Y.; Wang, J.C.; Hou, Y. Modeling phonon thermal conductivity in spatially confined GaN nanofilms under stress fields and phonon surface scattering. *AIP Adv.* **2019**, *9*, 015024. [CrossRef]
21. Shalish, I.; Temkin, H.; Narayanamurti, V. Size-dependent surface luminescence in ZnO nanowires. *Phys. Rev. B* **2004**, *69*, 245401. [CrossRef]

22. Wang, X.F.; Zhao, F.L.; Xie, P.B.; Deng, S.Z.; Xu, N.S.; Wang, H.Z. Surface emission characteristics of ZnO nanoparticles. *Chem. Phys. Lett.* **2006**, *423*, 361–365. [CrossRef]
23. Wen, X.M.; Chen, W.J.; Yang, J.F.; Ou, Q.D.; Yang, T.S.; Zhou, C.H.; Lin, H.; Wang, Z.Y.; Zhang, Y.P.; Conibeer, G.J.; et al. The role of surface recombination in halide perovskite nanoplatelets. *ACS Appl. Mater. Interfaces* **2018**, *10*, 31586–31593. [CrossRef] [PubMed]
24. Zhang, Z.F.; Qian, Q.K.; Li, B.K.; Chen, K.J. Interface engineering of monolayer MoS_2/GaN hybrid heterostructure: Modified band alignment for photocatalytic water splitting application by nitridation treatment. *ACS Appl. Mater. Interfaces* **2018**, *10*, 17419–17426. [CrossRef]
25. Mishra, K.C.; Piquette, A. The limits of transmittance of an interface and their effects on light extraction. *ECS J. Solid State Sci. Technol.* **2018**, *9*, 3016–3022. [CrossRef]
26. Ciccullo, F.; Calzolari, A.; Bader, K.; Neugebauer, P.; Gallagher, N.M.; Rajca, A.; Slageren, J.V.; Casu, M.B. Interfacing a potential purely organic molecular quantum bit with a real-life surface. *ACS Appl. Mater. Interfaces* **2019**, *11*, 1571–1578. [CrossRef]
27. Zhao, G.Y.; Hsu, C.H.; Wang, Y.H.; Chan, M.C. Chromatic confocal microscopy to rapidly reveal nanoscale surface/interface topography by position-sensitive detection. *Appl. Phys. Lett.* **2018**, *113*, 083106. [CrossRef]
28. Angelika, M.; Dieter, I.; Thomas, B.; Mirjam, R.; Christof, P.; Herwig, B.; Holger, B.; Christina, S.; Jürgen, P. Analysis of organic multilayer structures using a combined grazing incidence X-ray fluorescence/X-ray reflectometry approach. *Spectrochim. Acta Part B* **2018**, *148*, 188–192.
29. Singh, A.; Mathur, A.; Pal, D.; Sengupta, A.; Singh, R.; Chattopadhyay, S. Structure and morphology of atomic layer deposition grown ZnO thin film/nanostructure on polymeric template. *Mater. Today Proc.* **2019**, *18*, 1517–1523. [CrossRef]
30. Daniela, C.; Maria, A.F. Electronic transitions in low dimensional semiconductor structures measured by surface photovoltage spectroscopy. *Mater. Sci. Semicon. Proc.* **2019**, *92*, 28–38.
31. Park, J.H.; Sanne, A.; Guo, Y.Z.; Amani, M.; Zhang, K.H.; Movva, H.C.P.; Robinson, J.A.; Javey, A.; Robertson, J.; Banerjee, S.K.; et al. Defect passivation of transition metal dichalcogenides via a charge transfer van der Waals interface. *Sci. Adv.* **2017**, *3*, 1701661. [CrossRef]
32. Chen, F.; Guan, Z.Y.; Tang, A.W. Nanostructure and device architecture engineering for high-performance quantum-dot light-emitting diodes. *J. Mater. Chem. C* **2018**, *6*, 10958–10981. [CrossRef]
33. Yang, Y.; Li, J.Z.; Peng, X.G. An efficient and surface-benign purification scheme for colloidal nanocrystals based on quantitative assessment. *Nano Res.* **2015**, *8*, 3353–3364. [CrossRef]
34. Zhang, X.Y.; Lin, H.; Huang, H.; Reckmeier, C.; Zhang, Y.; Choy, W.C.H.; Rogach, A.L. Enhancing the brightness of cesium lead halide perovskite nanocrystal based green light-emitting devices through the interface engineering with perfluorinated ionomer. *Nano Lett.* **2016**, *16*, 1415–1420. [CrossRef]
35. Zou, Y.T.; Ban, M.Y.; Cui, W.; Huang, Q.; Wu, C.; Liu, J.W.; Wu, H.H.; Song, T.; Sun, B.Q. A general solvent selection strategy for solution processed quantum dots targeting high performance light-emitting diode. *Adv. Funct. Mater.* **2017**, *27*, 1603325. [CrossRef]
36. Yang, Y.; Zheng, Y.; Cao, W.; Titov, A.; Hyvonen, J.; Manders, J.R.; Xue, J.; Holloway, P.H.; Qian, L. High-efficiency light-emitting devices based on quantum dots with tailored nanostructures. *Nat. Photonics* **2015**, *9*, 259–266. [CrossRef]
37. Cao, W.; Xiang, C.; Yang, Y.; Chen, Q.; Chen, L.; Yan, X.; Qian, L. Highly stable QLEDs with improved hole injection via quantum dot structure tailoring. *Nat. Commun.* **2018**, *9*, 2608. [CrossRef]
38. Chiba, T.; Hayashi, Y.; Ebe, H.; Hoshi, K.; Sato, J.; Sato, S.; Pu, Y.J.; Ohisa, S.; Kido, J. Anion-exchange red perovskite quantum dots with ammonium iodine salts for highly efficient light-emitting devices. *Nat. Photonics* **2018**, *12*, 681–687. [CrossRef]
39. Wang, J.P.; Wang, N.N.; Jin, Y.Z.; Si, J.J.; Tan, Z.K.; Du, H.; Cheng, L.; Dai, X.L.; Bai, S.; He, H.P.; et al. Interfacial control toward efficient and low-voltage perovskite light-emitting diodes. *Adv. Mater.* **2015**, *27*, 2311–2316. [CrossRef]
40. Li, G.; Rivarola, F.W.; Davis, N.J.; Bai, S.; Jellicoe, T.C.; Pena, F.; de la Hou, S.; Ducati, C.; Gao, F.; Friend, R.H.; et al. Highly efficient perovskite nanocrystal light-emitting diodes enabled by a universal crosslinking method. *Adv. Mater.* **2016**, *28*, 3528–3534. [CrossRef]
41. Wang, N.; Cheng, L.; Ge, R.; Zhang, S.T.; Miao, Y.F.; Zou, W.; Yi, C.; Sun, Y.; Cao, Y.; Yang, R.; et al. Perovskite light-emitting diodes based on solution-processed self-organized multiple quantum wells. *Nat. Photonics* **2016**, *10*, 699–704. [CrossRef]

42. Yuan, M.J.; Quan, L.N.; Comin, R.; Walters, G.; Sabatini, R.; Voznyy, O.; Hoogland, S.; Zhao, Y.B.; Beauregard, E.M.; Kanjanaboos, P.; et al. Perovskite energy funnels for efficient light-emitting diodes. *Nat. Nanotechnol.* **2016**, *11*, 872–877. [CrossRef] [PubMed]
43. Cho, H.; Jeong, S.H.; Park, M.H.; Kim, Y.H.; Wolf, C.; Lee, C.L.; Heo, J.H.; Sadhanala, A.; Myoung, N.S.; Yoo, S.; et al. Overcoming the electroluminescence efficiency limitations of perovskite light-emitting diodes. *Science* **2015**, *350*, 1222–1225. [CrossRef] [PubMed]
44. Protesescu, L.; Yakunin, S.; Bodnarchuk, M.I.; Krieg, F.; Caputo, R.; Hendon, C.H.; Yang, R.X.; Walsh, A.; Kovalenko, M.V. Nanocrystals of cesium lead halide perovskites ($CsPbX_3$, X = Cl, Br, and I): Novel optoelectronic materials showing bright emission with wide color gamut. *Nano Lett.* **2015**, *15*, 3296–3692. [CrossRef]
45. Nedelcu, G.; Protesescu, L.; Yakunin, S.; Bodnarchuk, M.I.; Grotevent, M.J.; Kovalenko, M.V. Fast anion-exchange in highly luminescent nanocrystals of cesium lead halide perovskites ($CsPbX_3$, X = Cl, Br, I). *Nano Lett.* **2015**, *15*, 5635–5640. [CrossRef]
46. Song, J.; Li, J.; Li, X.; Xu, L.; Dong, Y.; Zeng, H. Quantum dot light-emitting diodes based on inorganic perovskite cesium lead halides ($CsPbX_3$). *Adv. Mater.* **2015**, *27*, 7162. [CrossRef]
47. Wang, Y.; Li, X.M.; Song, J.Z.; Xiao, L.; Sun, H.D. All-inorganic colloidal perovskite quantum dots: A new class of lasing materials with favorable characteristics. *Adv. Mater.* **2015**, *27*, 7101–7108. [CrossRef]
48. Owen, J. The coordination chemistry of nanocrystal surfaces. *Science* **2015**, *347*, 615–616. [CrossRef]
49. Bai, Z.L.; Ji, W.Y.; Han, D.B.; Chen, L.L.; Chen, B.K.; Shen, H.B.; Zou, B.S.; Zhong, H.Z. Hydroxyl-terminated $CuInS_2$ based quantum dots: Toward efficient and bright light emitting diodes. *Chem. Mater.* **2016**, *28*, 1085–1091. [CrossRef]
50. Li, J.H.; Xu, L.M.; Wang, T.; Song, J.Z.; Chen, J.W.; Xue, J.; Dong, Y.H.; Cai, B.; Shan, Q.S.; Han, B.N.; et al. 50-Fold EQE improvement up to 6.27% of solution-processed all-inorganic perovskite $CsPbBr_3$ QLEDs via surface ligand density control. *Adv. Mater.* **2017**, *29*, 1603885. [CrossRef]
51. Dai, X.L.; Deng, Y.Z.; Peng, X.G.; Jin, Y.Z. Quantum-dot light-emitting diodes for large-area displays: Towards the dawn of commercialization. *Adv. Mater.* **2017**, *29*, 1607022. [CrossRef] [PubMed]
52. Yang, Y.; Qin, H.; Jiang, M.; Lin, L.; Fu, T.; Dai, X.; Zhang, Z.; Niu, Y.; Cao, H.; Jin, Y.; et al. Entropic ligands for nanocrystals: From unexpected solution properties to outstanding processability. *Nano Lett.* **2016**, *16*, 2133–2138. [CrossRef] [PubMed]
53. Yang, Y.; Qin, H.; Peng, X. Intramolecular entropy and size-dependent solution properties of nanocrystal-ligands complexes. *Nano Lett.* **2016**, *16*, 2127–2132. [CrossRef] [PubMed]
54. Shang, Y.Q.; Ning, Z.J. Colloidal quantum-dots surface and device structure engineering for high-performance light-emitting diodes. *Natl. Sci. Rev.* **2017**, *4*, 170–183. [CrossRef]
55. Pan, J.; Shang, Y.Q.; Yin, J.; Bastiani, M.D.; Peng, W.; Dursun, I.; Sinatra, L.; El-Zohry, A.M.; Hedhili, M.N.; Emwas, A.H.; et al. Bidentate ligand-passivated $CsPbI_3$ perovskite nanocrystals for stable near-unity photoluminescence quantum yield and efficient red light-emitting diodes. *J. Am. Chem. Soc.* **2018**, *140*, 562–565. [CrossRef]
56. Fokina, A.; Lee, Y.; Chang, J.H.; Park, M.; Sun, Y.; Bae, W.K.; Char, K.; Lee, C.; Zentel, R. The role of emission layer morphology on the enhanced performance of light-emitting diodes based on quantum dot-semiconducting polymer hybrids. *Adv. Mater. Interfaces* **2016**, *3*, 1600279. [CrossRef]
57. Brown, P.R.; Kim, D.; Lunt, R.R.; Zhao, N.; Bawendi, M.G.; Grossman, J.C.; Bulovic, V. Energy level modification in lead sulfide quantum dot thin films through ligand exchange. *ACS Nano* **2014**, *8*, 5863–5872. [CrossRef]
58. Chuang, C.H.M.; Brown, P.R.; Bulovic, V.; Bawendi, M.G. Improved performance and stability in quantum dot solar cells through band alignment engineering. *Nat. Mater.* **2014**, *13*, 796–801. [CrossRef]
59. Yang, Z.; Voznyy, O.; Liu, M.; Yuan, M.; Ip, A.H.; Ahmed, O.S.; Levina, L.; Kinge, S.; Hoogland, S.; Sargent, E.H. All-quantum-dot infrared light-emitting diodes. *ACS Nano* **2015**, *9*, 12327–12333. [CrossRef]
60. Chen, O.; Zhao, J.; Chauhan, V.P.; Cui, J.; Wong, C.; Harris, D.K.; Wei, H.; Han, H.S.; Fukumura, D.; Jain, R.K.; et al. Compact high-quality CdSe-CdS core-shell nanocrystals with narrow emission linewidths and suppressed blinking. *Nat. Mater.* **2013**, *12*, 445–451. [CrossRef]
61. Pal, B.N.; Ghosh, Y.; Brovelli, S.; Laocharoensuk, R.; Klimov, V.I.; Hollingsworth, J.A.; Htoon, H. 'Giant' CdSe/CdS core/shell nanocrystal quantum dots as efficient electroluminescent materials: Strong influence of shell thickness on light-emitting diode performance. *Nano Lett.* **2012**, *12*, 331–336. [CrossRef] [PubMed]

62. Li, Z.; Chen, F.; Wang, L.; Shen, H.; Guo, L.; Kuang, Y.; Wang, H.; Li, N.; Li, L.S. Synthesis and evaluation of ideal core/shell quantum dots with precisely controlled shell growth: Nonblinking, single photoluminescence decay channel, and suppressed FRET. *Chem. Mater.* **2018**, *30*, 3668–3676. [CrossRef]
63. Bae, W.K.; Padilha, L.A.; Park, Y.S.; McDaniel, H.; Robel, I.; Pietryga, J.M.; Klimov, V.I. Controlled alloying of the core-shell interface in CdSe/CdS quantum dots for suppression of auger recombination. *ACS Nano* **2013**, *7*, 3411–3419. [CrossRef] [PubMed]
64. Park, Y.S.; Bae, W.K.; Padilha, L.A.; Pietryga, J.M.; Klimov, V.I. Effect of the core/shell interface on auger recombination evaluated by single-quantum-dot spectroscopy. *Nano Lett.* **2014**, *14*, 396–402. [CrossRef]
65. Cao, H.; Ma, J.; Huang, L.; Qin, H.; Meng, R.; Li, Y.; Peng, X. Design and synthesis of antiblinking and antibleaching quantum dots in multiple colors via wave function confinement. *J. Am. Chem. Soc.* **2016**, *138*, 15727–15735. [CrossRef]
66. Kim, D.; Fu, Y.; Kim, S.; Lee, W.; Lee, W.; Lee, K.H.; Chung, H.K.; Lee, H.J.; Yang, H.; Chae, H. Polyethylenimine ethoxylated-mediated all-solution-processed high-performance flexible inverted quantum dot-light-emitting device. *ACS Nano* **2017**, *11*, 1982–1990. [CrossRef]
67. Lin, Q.; Wang, L.; Li, Z.; Shen, H.; Guo, L.; Kuang, Y.; Wang, H.; Li, L.S. Nonblinking quantum-dot-based blue light-emitting diodes with high efficiency and a balanced charge-injection process. *ACS Photonics* **2018**, *5*, 939–946. [CrossRef]
68. Cai, J.G.; Qi, L.M. Recent advances in antireflective surfaces based on nanostructure arrays. *Mater. Horiz.* **2015**, *2*, 37–53. [CrossRef]
69. Liu, S.Y.; Ho, S.H.; So, F. Novel patterning method for silver nanowire electrodes for thermal- evaporated organic light emitting diodes. *ACS Appl. Mater. Interfaces* **2016**, *8*, 9268–9274. [CrossRef]
70. Rishinaramangalam, A.K.; Nami, M.; Fairchild, M.N.; Darryl, M.; Shima, D.M.; Balakrishnan, G.; Brueck, S.R.J.; Daniel, F.; Feezell, D.F. Semipolar InGaN/GaN nanostructure light-emitting diodes on c-plane sapphire. *Appl. Phys. Express* **2016**, *9*, 032101. [CrossRef]
71. Djavid, M.; Mi, Z.T. Enhancing the light extraction efficiency of AlGaN deep ultraviolet light emitting diodes by using nanowire structures. *Appl. Phys. Lett.* **2016**, *108*, 051102. [CrossRef]
72. Zhang, C.; Marvinney, C.M.; Xu, H.Y.; Liu, W.Z.; Wang, C.L.; Zhang, L.X.; Wang, J.N.; Ma, J.G.; Liu, Y.C. Enhanced waveguide-type ultraviolet electroluminescence from ZnO/MgZnO core/shell nanorod array light-emitting diodes via coupling with Ag nanoparticles localized surface plasmons. *Nanoscale* **2015**, *7*, 1073–1080. [CrossRef] [PubMed]
73. Yao, Y.C.; Yang, Z.P.; Hwang, J.M.; Chuang, Y.L.; Lin, C.C.; Haung, J.Y.; Chou, C.Y.; Sheu, J.K.; Tsai, M.T.; Lee, Y.J. Enhancing UV-emissions through optical and electronic dual-function tunings of Ag nanoparticles hybridized with n-ZnO nanorods/p-GaN heterojunction light-emitting diodes. *Nanoscale* **2016**, *8*, 4463–4473. [CrossRef] [PubMed]
74. Zheng, Q.; Zhou, W.C.; Peng, Y.H.; Yin, Y.L.; Zhong, M.Y.; Zhuang Zhao, Z.; Zhang, Q.L.; Tang, D.S.; Zeng, R.S.; Zou, B.S. Surface polarons and optical micro-cavity modulated broad range multi-mode emission of Te-doped CdS nanowires. *Nanotechnology* **2018**, *29*, 465709. [CrossRef]
75. You, D.T.; Xu, C.X.; Qin, F.F.; Zhu, Z.; Manohari, A.G.; Xu, W.; Zhao, J.; Liu, W. Interface control for pure ultraviolet electroluminescence from nano-ZnO-based heterojunction devices. *Sci. Bull.* **2018**, *63*, 38–45. [CrossRef]
76. Casu, M.B. Nanoscale studies of organic radicals: Surface, interface, and interface. *Acc. Chem. Res.* **2018**, *51*, 753–760. [CrossRef]
77. Guo, Y.N.; Yan, J.C.; Zhang, Y.; Wang, J.X.; Lia, J.M. Enhancing the light extraction of AlGaN-based ultraviolet light-emitting diodes in the nanoscale. *J. Nanophotonics* **2018**, *12*, 043510. [CrossRef]
78. Lee, D.; Lee, J.W.; Jang, J.; Shin, I.S.; Jin, L.; Park, J.H.; Kim, J.; Lee, J.; Noh, H.S.; Kim, Y.; et al. Improved performance of AlGaN-based deep ultraviolet light-emitting diodes with nano-patterned AlN/sapphire substrates. *Appl. Phys. Lett.* **2017**, *110*, 191103. [CrossRef]
79. Yun, J.; Hirayama, H. Investigation of the light-extraction efficiency in 280 nm AlGaN-based light-emitting diodes having a highly transparent p-AlGaN layer. *J. Appl. Phys.* **2017**, *121*, 013105. [CrossRef]
80. Nguyen, H.P.T.; Zhang, S.; Connie, A.T.; Kibria, M.G.; Wang, Q.; Shih, I.; Mi, Z. Breaking the carrier injection bottleneck of phosphor-free nanowire white light-emitting diodes. *Nano Lett.* **2013**, *13*, 5437–5442. [CrossRef]

81. Nguyen, H.P.T.; Djavid, M.; Mi, Z. Nonradiative recombination mechanism in phosphor-free GaN-based nanowire white light emitting diodes and the effect of ammonium sulfide surface passivation. *ECS Trans.* **2013**, *53*, 93–100. [CrossRef]
82. Philip, M.R.; Choudhary, D.D.; Djavid, M.; Bhuyian, M.N.; Piao, J.; Pham, T.T.; Misra, D.; Nguyen, H.P.T. Controlling color emission of InGaN/AlGaN nanowire light-emitting diodes grown by molecular beam epitaxy. *J. Vac. Sci. Technol. B* **2017**, *35*, 02B108. [CrossRef]
83. Ledig, J.; Wang, X.; Fündling, S.; Schuhmann, H.; Seibt, M.; Jahn, U.; Wehmann, H.H.; Waag, A. Characterization of the internal properties of InGaN/GaN core-shell LEDs. *Phys. Status Solidi A* **2016**, *213*, 11–18. [CrossRef]
84. Müller, M.; Veit, P.; Krause, F.F.; Schimpke, T.; Metzner, S.; Bertram, F.; Mehrtens, T.; Muller-Caspary, K.; Avramescu, A.; Strassburg, M.; et al. Nanoscopic insights into InGaN/GaN core-shell nanorods: Structure, composition, and luminescence. *Nano Lett.* **2016**, *16*, 5340–5346. [CrossRef]
85. Philip, M.R.; Choudhary, D.D.; Djavid, M.; Le, K.Q.; Piao, J.; Nguyen, H.P.T. High efficiency green/yellow and red InGaN/AlGaN nanowire light-emitting diodes grown by molecular beam epitaxy. *J. Sci. Adv. Mater. Devices* **2017**, *2*, 150–155. [CrossRef]
86. Sadaf, S.M.; Zhao, S.; Wu, Y.; Ra, Y.H.; Liu, X.; Vanka, S.; Mi, Z. An AlGaN core-shell tunnel junction nanowire light-emitting diode operating in the ultraviolet-C band. *Nano Lett.* **2017**, *17*, 1212–1218. [CrossRef]
87. Sim, Y.C.; Lim, S.H.; Yoo, Y.S.; Jang, M.H.; Choi, S.H.; Yeo, H.S.; Woo, K.Y.; Lee, S.W.; Song, H.G.; Cho, Y.H. Three-dimensional GaN dodecagonal ring structures for highly efficient phosphor-free warm white light-emitting diodes. *Nanoscale* **2018**, *10*, 4686–4695. [CrossRef]
88. Ra, Y.H.; Lee, C.R. Understanding the p-type GaN nanocrystals on InGaN nanowire heterostructures. *ACS Photonics* **2019**, *6*, 2397–2404. [CrossRef]
89. Bui, H.Q.T.; Velpula, R.T.; Jain, B.; Aref, O.H.; Nguyen, H.D.; Lenka, T.R.; Nguyen, H.P.T. Full-color InGaN/AlGaN nanowire micro light-emitting diodes grown by molecular beam epitaxy: A promising candidate for next generation micro displays. *Micromachines* **2019**, *10*, 492. [CrossRef]
90. Cheng, S.B.; Langelier, B.; Ra, Y.H.; Rashid, R.T.; Mi, Z.T.; Botton, G.A. Structural origin of the high-performance light-emitting InGaN/AlGaN quantum disks. *Nanoscale* **2019**, *11*, 8994–8999. [CrossRef]
91. Pimputkar, S.; Speck, J.S.; DenBaars, S.P.; Nakamura, S. Prospects for LED lighting. *Nat. Photonics* **2009**, *3*, 180–182. [CrossRef]
92. Oliver, R.A.; Massabuau, F.C.P.; Kappers, M.J.; Phillips, W.A.; Thrush, E.J.; Tartan, C.C.; Blenkhorn, W.E.; Badcock, T.J.; Dawson, P.; Hopkins, M.A.; et al. The impact of gross well width fluctuations on the efficiency of GaN-based light emitting diodes. *Appl. Phys. Lett.* **2013**, *103*, 141114. [CrossRef]
93. Barettin, D.; Maur, M.A.D.; Carlo, A.D.; Pecchia, A.; Tsatsulnikov, A.F.; Sakharov, A.V.; Lundin, W.V.; Nikolaev, A.E.; Usov, S.O.; Cherkashin, N.; et al. Influence of electromechanical coupling on optical properties of InGaN quantum-dot based light-emitting diodes. *Nanotechnology* **2016**, *28*, 015701. [CrossRef] [PubMed]
94. Barrettes, D.; Maur, M.A.D.; Carlo, A.D.; Pecchia, A.; Tsatsulnikov, A.F.; Lundin, W.V.; Sakharov, A.V.; Nikolaev, A.E.; Korytov, M.; Cherkashin, N.; et al. Carrier transport and emission efficiency in InGaN quantum-dot based light-emitting diodes. *Nanotechnology* **2017**, *28*, 275201.
95. Hinzmann, C.; Magen, O.; Hofstetter, Y.J.; Hopkinson, P.E.; Tessler, N.; Vaynzof, Y.N. Effect of injection layer sub-bandgap states on electron injection in organic light-emitting diodes. *ACS Appl. Mater. Interfaces* **2017**, *9*, 6220–6227. [CrossRef] [PubMed]
96. Borges, B.G.A.L.; Veiga, A.G.; Gioti, M.; Laskarakis, A.; Tzounis, L.; Logothetidis, S.; Rocco, M.L.M. Surface, interface and electronic properties of f8:f8bt polymeric thin films used for OLED applications. *Polym. Int.* **2018**, *67*, 691–699. [CrossRef]
97. Thomas, J.R.; David, G.L.; Alastair, R.B. Ultrasonic spray coating as an approach for large-area polymer OLEDs: The influence of thin film processing and surface roughness on electrical performance. *AIP Adv.* **2019**, *9*, 015330.
98. Zhao, L.; Guo, Z.X.; Zhang, M.; Yang, S.; Zhao, L. Surface-interface analysis of $In_xGa_{1-x}As$/InP heterostructure in positive and negative mismatch system. *Surf. Interface Anal.* **2019**, *51*, 498–505. [CrossRef]
99. Wang, J.; Fu, W.F.; Jariwala, S.; Sinha, I.; Jen, A.K.Y.; Ginger, D.S. Reducing surface recombination velocities at the electrical contacts will improve perovskite photovoltaics. *ACS Energy Lett.* **2019**, *4*, 222–227. [CrossRef]

100. Jin, Y.H.; Li, Q.Q.; Chen, M.; Li, G.H.; Zhao, Y.D.; Xiao, X.Y.; Wang, J.P.; Jiang, K.L.; Fan, S.S. Study of carbon nanotubes as etching masks and related applications in the surface modification of GaAs-based light-emitting diodes. *Small* **2015**, *11*, 4111–4116. [CrossRef]
101. Seo, T.H.; Lee, S.; Min, K.H.; Chandramohan, S.; Park, A.H.; Lee, G.H.; Park, M.; Suh, E.K.; Kim, M.J. The role of graphene formed on silver nanowire transparent conductive electrode in ultraviolet light emitting diodes. *Sci. Rep.* **2016**, *6*, 29464. [CrossRef] [PubMed]
102. Lim, J.T.; Kim, J.; Lee, H.; Moon, J.; Kwon, B.H.; Ahn, S.; Cho, N.S.; Ahn, B.W.; Lee, J.I.; Ihm, K.; et al. Unraveled face-dependent effects of multilayered graphene embedded in transparent organic light-emitting diodes. *ACS Appl. Mater. Interfaces* **2017**, *9*, 43105–43112. [CrossRef] [PubMed]
103. Lee, B.R.; Lee, S.; Park, J.H.; Jung, E.D.; Yu, J.C.; Nam, Y.S.; Heo, J.; Kim, J.Y.; Kim, B.S.; Song, M.H. Amine-based interfacial molecules for inverted polymer-based optoelectronic devices. *Adv. Mater.* **2015**, *27*, 3553–3559. [CrossRef] [PubMed]
104. Yang, X.L.; Zhang, X.W.; Deng, J.X.; Chu, Z.M.; Jiang, Q.; Meng, J.H.; Wang, P.Y.; Zhang, L.Q.; Yin, Z.G.; You, J.B. Efficient green light-emitting diodes based on quasi-two-dimensional composition and phase engineered perovskite with surface passivation. *Nat. Commun.* **2018**, *9*, 570. [CrossRef] [PubMed]
105. Zhang, X.Y.; Sun, C.; Zhang, Y.; Wu, H.; Ji, C.Y.; Chuai, Y.H.; Wang, P.; Wen, S.P.; Zhang, C.F.; Yu, W.W. Bright perovskite nanocrystal films for efficient light-emitting devices. *J. Phys. Chem. Lett.* **2016**, *7*, 4602–4610. [CrossRef]
106. Chen, Z.M.; Zhang, C.Y.; Jiang, X.F.; Liu, M.Y.; Xia, R.X.; Shi, T.T.; Chen, D.C.; Xue, Q.F.; Zhao, Y.J.; Su, S.J.; et al. High-performance color-tunable perovskite light emitting devices through structural modulation from bulk to layered film. *Adv. Mater.* **2017**, *29*, 1603157. [CrossRef]
107. Zhang, X.W.; Mao, J.; Shao, Z.B.; Diao, S.L.; Hu, D.; Tang, Z.J.; Wu, H.H.; Jie, J.S. Efficient photovoltaic devices based on p-ZnSe/n-CdS core-shell heterojunctions with high open-circuit voltage. *J. Mater. Chem. C* **2017**, *5*, 2107–2113. [CrossRef]
108. Varun, V.; Takuya, A. Molecular orientation of conjugated polymer chains in nanostructures and thin films: Review of processes and application to optoelectronics. *J. Nanomater.* **2017**, *2017*, 3624750.
109. Liu, X.Y.; Wang, X.H.; Zhang, Y.G.; Wei, K.; Zheng, Y.K.; Xuanwu Kang, X.W.; Jiang, H.J.; Li, J.F.; Wang, W.W.; Wu, X.B.; et al. Insight into the near-conduction band states at the crystallized interface between GaN and SiNx grown by low-pressure chemical vapor deposition. *ACS Appl. Mater. Interfaces* **2018**, *10*, 21721–21729. [CrossRef]
110. Chen, Y.J.; Lee, W.K.; Chen, Y.T.; Lin, C.Y.; Wen, S.W.; Jiao, M.; Su, G.D.; Lin, H.Y.; Visser, R.J.; Kwak, B.L.; et al. A vision toward ultimate optical out-coupling for organic light-emitting diode displays: 3D pixel configuration. *Adv. Sci.* **2018**, *5*, 1800467. [CrossRef]
111. Muhammad, S.I.; Shanmugan, S.; Wan, M.W.A.K.; Mutharasu, D. Synthesis of MgO thin film on aluminum and copper substrates as thermal interface materials. *IEEE Trans. Electron. Dev.* **2019**, *66*, 1450–1457.
112. Bulling, A.F.; Venter, P.J. Improved light extraction efficiency of complementary metal-oxide semiconductor hot carrier lights sources with the use of improved back-end-of-line light directing structures. *Opt. Eng.* **2019**, *58*, 065105. [CrossRef]
113. Lei, P.H.; Yang, C.D.; Po-Chun Huang, P.C.; Yeh, S.J. Enhancement of light extraction efficiency for InGaN/GaN light-emitting diodes using silver nanoparticle embedded ZnO thin films. *Micromachines* **2019**, *10*, 239. [CrossRef] [PubMed]

micromachines

MDPI

Review

Epitaxy of III-Nitrides on β-Ga_2O_3 and Its Vertical Structure LEDs

Weijiang Li [1,2,3], Xiang Zhang [1,2,3], Ruilin Meng [1,2,3], Jianchang Yan [1,2,3], Junxi Wang [1,2,3], Jinmin Li [1,2,3] and Tongbo Wei [1,2,3,*]

1 State Key Laboratory of Solid-State Lighting, Institute of Semiconductors, University of Chinese Academy of Sciences, Beijing 100083, China; wjli18@semi.ac.cn (W.L.); zhangxiang@semi.ac.cn (X.Z.); mengruilin@semi.ac.cn (R.M.); yanjc@semi.ac.cn (J.Y.); jxwang@semi.ac.cn (J.W.); jmli@semi.ac.cn (J.L.)

2 Center of Materials Science and Optoelectronics Engineering, University of Chinese Academy of Sciences, Beijing 100049, China

3 Beijing Engineering Research Center for the 3rd Generation Semiconductor Materials and Application, Beijing 100083, China

* Correspondence: tbwei@semi.ac.cn; Tel.: +86-010-8230-5430

Received: 7 April 2019; Accepted: 8 May 2019; Published: 13 May 2019

Abstract: β-Ga_2O_3, characterized with high n-type conductivity, little lattice mismatch with III-Nitrides, high transparency (>80%) in blue, and UVA (400–320 nm) as well as UVB (320–280 nm) regions, has great potential as the substrate for vertical structure blue and especially ultra violet LEDs (light emitting diodes). Large efforts have been made to improve the quality of III-Nitrides epilayers on β-Ga_2O_3. Furthermore, the fabrication of vertical blue LEDs has been preliminarily realized with the best result that output power reaches to 4.82 W (under a current of 10 A) and internal quantum efficiency (IQE) exceeds 78% by different groups, respectively, while there is nearly no demonstration of UV-LEDs on β-Ga_2O_3. In this review, with the perspective from materials to devices, we first describe the basic properties, growth method, as well as doping of β-Ga_2O_3, then introduce in detail the progress in growth of GaN on (1 0 0) and (−2 0 1) β-Ga_2O_3, followed by the epitaxy of AlGaN on gallium oxide. Finally, the advances in fabrication and performance of vertical structure LED (VLED) are presented.

Keywords: β-Ga_2O_3; III-Nitrides; monoclinic; hexagonal arrangement; high-power; current distribution; vertical structure LED

1. Introduction

Due to their unique and very desirable properties, such as the direct tunable wide bandgap that span the infrared, ultraviolet, and whole visible spectra region, high chemical and thermal stability, and high mobility, III-Nitrides materials (InN, GaN, and AlN) have attracted worldwide attention both in research and industry. Applications of III-Nitrides in photoelectronic devices and high-power electronic devices, such as laser diodes (LDs), visible and ultraviolet (UV) detectors, and light emitting diodes (LEDs), have been realized [1–6]. LEDs therein are the most mature devices with high brightness, compact structure, low energy consumption, high switching speed, long lifetime, and environmental friendliness.

GaN-based blue-LEDs were first commercially introduced in the 1990s [7]. Over the past two decades, due to the huge advances in the crystalline quality of materials and configuration of devices, a wall-plug efficiency reaching around 80% has been realized in industry [8], which has led to a massive market of GaN-based solid-state lighting. Compared with traditional mercury-based UV sources, $Al_xGa_{1-x}N$ (x from 0 to 1) based UV-LEDs are more attractive for potential applications of non-line-of-sight communications, water purification, food or medical equipment sterilization, phototherapy, UV curing, detection as well as identification of biological or chemical agents, plant growth lighting, and so on [9–16]. Rapid progress in the development of III-Nitrides based UV-LEDs has been forecasted. Yole Développement predicted that there will be an annual growth rate of more than 28% for UV-LED components in the world market, and the aggregate volume will reach $520 million US dollars by 2019 [17].

Up to now, despite the fact that plentiful efforts have been made, there is still a challenge to obtain high-power III-Nitrides based LEDs, especially in the UV region [18–20]. The efficiency of LEDs will decrease with increased injection current density, which is the so called efficiency droop effect. Therefore, high-power LEDs are still beyond realization and cannot completely replace the traditional illumination techniques due to the degraded efficiency and rising cost when it comes to high-power applications [1,2,21,22].

The efficiency droop effect can be ascribed to various reasons, including the self-heating effect, carrier delocalization, electron overflow due to the different effective masses of electrons and holes, auger recombination due to the high concentration of carriers, and poor hole injection due to the poor p-type doping efficiency [3–8,23–28]. It is worth noting that the current density is defined as dividing injection current by chip area, which is not the authentic value when the injection current localizes in a portion of the chip. Consequently, the electron overflow and auger recombination will take place earlier due to the current crowding effect. An earlier overflow and Auger recombination may lead to an earlier self-heating effect that further decrease the output power of LEDs. Thus, uniformizing the current distribution in the active region is of great significance to inhibit the droop effect and improve the performance of devices [11,29].

The widely used sapphire (Al_2O_3) substrate has large lattice mismatch (14% with GaN) and large thermal expansion coefficient mismatch (30% with GaN) with III-Nitrides materials [30–32], which will introduce high threading dislocation density (TDD) [32,33] acting as nonradiative recombination and scattering centers that deteriorate the performance of LEDs [34]. Furthermore, LEDs are fabricated in a horizontal structure since the sapphire substrates are insulated, leading to that both the n- and p-contact must be established at the top surface of the devices. In this way, the total emission area is reduced, and the fabrication, the encapsulation, as well as the integration procedure become more complex. Moreover, the efficiency droops rapidly with an increased current as the aftermath of current crowding and poor heat distribution due to lateral injection.

Vertical structure is prospective to overcome the problems mentioned above for LEDs fabricated on sapphire [35]. Due to its straightforward configuration, the series resistance of vertical structure LEDs (VLEDs) is less than that of conventional LEDs. Thus, lower forward operating voltage of VLEDs can be obtained, leading to a reduced thermal load. Doan et al. found that the forward I–V curves of GaN-based VLEDs was steeper compared with the lateral structure, indicating a reduced series resistance [36]. A forward voltage of vertical LEDs around 4.5 V at 200 mA compared to 5.8 V for lateral LED on sapphire was reported by Cao et al. [37]. Also, the reduced thermal resistance (Rth) of GaN-based VLEDs can lead to an enhanced heat dissipation ability. Doan et al. demonstrated that the Rth of GaN-based LEDs could be reduced by 55% employing a vertical configuration [38]. A vertical path of current injection can result in a more uniform current distribution, avoiding the localized overheating. Theoretical calculation performed by Li et al. manifested that the injected current of lateral LEDs mostly concentrated in the region below the p electrode, while it was more dispersive for vertical LEDs [39]. In addition, given that there is only one electrode on both the top and bottom sides of VLEDs, the light extraction efficiency can be improved due to the decreased absorption. Additionally, the simplified structure of VLEDs can lower the cost of device fabrication. All these advantages together make it possible for vertical LEDs to work at a high injection current. The GaN-based VLEDs reported by Cao et al. showed higher output power compared with conventional LEDs [37]. A delayed saturation of output power with increased current was obtained due to the reduced efficiency droop effect of VLEDs. The optical degradation of VLEDs was less than 5% after stress compared to 12% for lateral LEDs, indicating an improved reliability with vertical configuration. Thus, vertical structure provides a potential way to realize high brightness and high power LEDs.

So far there are two prevalent methods to manufacture vertical LEDs, one of which is the lift-off and wafer-bonding technique. The usual approach for separating the III-Nitrides epilayer from the sapphire substrate is the laser lift-off (LLO) technique [40], which will, however, introduce a rough surface to the epilayer and substrate due to the physical and thermal impact [41,42]. Recently, the chemical lift-off (CLO) technique for GaN has attracted much attention due to its excellent surface morphology [43]. Materials (e.g., CrN, ZnO, carbon nanotube and Ga_2O_3 [43–48]), which are successfully employed in the growth of GaN as not only a buffer layer but also a sacrificial layer, have been reported. The CLO process offers a possible approach for III-Nitrides-based vertical structure LEDs. However, it is noteworthy that both the LLO and CLO methods will increase the complexity of fabrication, as will the cost. On this ground, the direct growth of III-Nitride semiconductors on conductive substrates are preferable. The commonly used conductive substrates in vertical LEDs include Si, GaN, and SiC [49]. Silicon is currently the most widely used semiconductor material due to its low cost, large size (6–12 inches), high quality, and a high thermal conductivity compared with sapphire. However, due to the large lattice mismatch (16.9% with GaN) and thermal mismatch (57% with GaN) between Si and III-Nitrides [50], the epilayer will generate a large number of defects, and even cracks, making it quite difficult to grow high-quality III-Nitrides films on silicon substrates. Moreover, silicon is non-transparent in the whole region of UV spectra, deteriorating the light extraction efficiency of UV-LEDs on silicon. SiC has poly type structures including cubic phase (3C-SiC), hexagonal phase (2H-SiC, 4H-SiC, 6H-SiC), and rhombic phase (15R-SiC). Among the substrates mentioned above for heteroepitaxy of III-Nitrides, 6H-SiC exhibits the smallest lattice mismatch (3.4%) and thermal mismatch with GaN, leading to a better crystalline quality of epilayers on SiC than that on silicon [51]. Compared with the GaN layer, a lower refractive index (2.65) of SiC can improve the light extraction efficiency due to the suppressed total internal reflection. Also, SiC has a thermal conductivity of 3–5 W/cm·K, which is three times that of silicon, and can furthermore enhance the heat dissipation and inhibit the self-heating effect of VLEDs. However, an absorption edge of 380 nm will limit the application of SiC as a substrate for vertical LEDs in the deep UV region. The high cost also impedes the extensive use of SiC. Homoepitaxy of GaN can be realized using GaN itself as a substrate. Due to the elimination of lattice mismatch and thermal mismatch, ultrahigh crystalline quality and ultralow density of threading dislocations can be obtained for GaN. In addition, the lattice mismatch between the GaN substrate and AlGaN alloy is also quite low (between 0% and 2.4%, depending on Al-molar fraction). However, being similar to the SiC substrate, an absorption edge of 365 nm, and a high cost, will act as obstacles to the application of GaN substrate, especially in deep UV spectra. Thus, the exploration of novel material as a conductive substrate for VLEDs with a low cost to prepare and high transmittance in visible, especially UV, region is quite necessary.

Recently, due to its unique properties, β-phase gallium oxide semiconductor attracts more and more attention in various fields. β-Ga_2O_3 based power electronic devices, including Schottky diodes and metal-semiconductor field-effect transistors have been reported, and photoelectronic devices including x-ray detectors and solar-blind deep-ultraviolet Schottky photodetectors with bulk-like, film-like, as well as nanowire-like gallium oxides, also have been carried out [52–55]. In addition, β-Ga_2O_3 based gas sensors for H_2, O_2, CO, or CH_4 detection, photocatalytic devices for water splitting and gas degradation, and Er doped β-Ga_2O_3 for photoresponse or luminescence have been demonstrated [56–62]. Single crystal bulk β-Ga_2O_3 has a wide bandgap of 4.8 eV and an absorption edge of 260 nm, resulting in a high transmittance in visible (>80%) [49], UVA, and UVB region. A high n-type conductivity of β-Ga_2O_3 with resistivity of 0.02 Ω·cm can be realized via doping [45]. Also, a low lattice mismatch between β-Ga_2O_3 and GaN has been presented [49]. Thus, combined with the advantages of sapphire and SiC, β-Ga_2O_3 is a promising candidate as the substrate of vertical structure LEDs.

In this review, we first introduce the properties, the growth method, as well as the doping of β-Ga_2O_3, and describe the epitaxial relationship between III-Nitrides and β-Ga_2O_3. Then the progresses in epitaxy of GaN on (1 0 0) and (−2 0 1) β-Ga_2O_3 substrates are discussed, followed by preliminary attempts to grow AlGaN alloys on gallium oxide. Subsequently, the advances in vertical structure LEDs are introduced. Finally, we present a brief conclusion of the development of vertical blue, especially ultra violet, LEDs on β-Ga_2O_3 substrate.

2. Properties and Growth Method of Bulk β-Ga_2O_3

2.1. Structure and Properties

In 1952, various polymorphs of gallium oxide including α, β, γ, δ, ε, and a transient κ phase, were experimentally demonstrated through performing research on phase equilibration in the Al_2O_3-Ga_2O_3-H_2O system [63]. Among them, β-Ga_2O_3 is the most stable one under different conditions. Compared with other polymorphs, β-Ga_2O_3 is the only stable one at any temperature below the melting point [64]; other polymorphs of gallium oxide will transform to the beta phase due to their metastability above 750–900 °C, as shown in Figure 1a [65]. The crystal structure of β-Ga_2O_3, which belongs to the monoclinic system and the space group C2/m (C_{2h}^3), was first reported in 1960 [66]. The lattice parameters of β-Ga_2O_3 are a = 12.21 Å, b = 3.04 Å, c = 5.80 Å, and β = 103.8°. Figure 1b shows the unit cell of β-Ga_2O_3, which contains four formula units occupied by two inequivalent Ga sites and three inequivalent O sites. Two Ga cations are characterized with tetrahedral and octahedral positions, respectively. Ga(I) tetrahedral formula units connect with others only by the corners, while Ga(II) octahedral formula units do it with the edges [64]. As for O anions, three crystallographically inequivalent sites are labeled as O(I), O(II), and O(III), respectively. Two of them are in trigonal coordination and the rest are in tetrahedral coordination. The symmetry of β-Ga_2O_3 is quite low with only a 2-fold symmetry along the *b*-axis. Unlike III-Nitrides, the existence of a symcenter leads to the inexistence of spontaneous polarization and piezoelectric polarization for β-Ga_2O_3. As illustrated in Figure 1c, β-Ga_2O_3 has a nature of cleavage of two planes including the primary (1 0 0) and the subordinate (0 0 1) due to the weak bond. Thus, β-Ga_2O_3 usually cracks into needles or plates as a result of fragility.

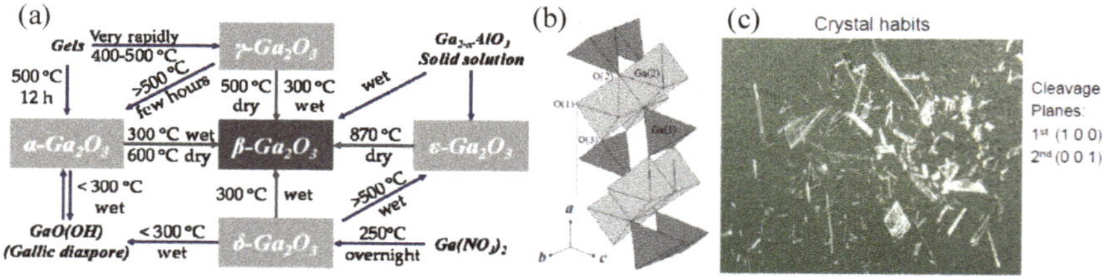

Figure 1. (**a**) Transformation relationships among Ga_2O_3 in different crystalline phases and their hydrates. (**b**) Structural schematic illustration of the β-Ga_2O_3 unit cell, manifesting the two gallium locations and three oxygen locations. (**c**) Cleavage nature of single crystal β-Ga_2O_3. Reprinted with permission from reference [64]. Copyright 2014 SPIE.

β-Ga_2O_3 is a wide bandgap (4.8 eV) [67], oxide semiconductor characterized with an absorption edge of 260 nm, leading to a remarkable transmittance in the visible (more than 80%), UVA, and UVB range [49]. It has stable chemical natures, such as strong acid as well as alkali resistance, high mechanical strength, and is stable at any temperature below the melting point of 1725 °C. In addition, it has a Vickers hardness of 12.5 GPa along (−2 0 1) plane and a density of 5.95 g/cm^{-3}. Due to its low symmetry, the thermal conductivity of β-Ga_2O_3 manifests a significant anisotropy, which is 13.6 W/mK along [1 0 0], 22.8 W/mK along [0 1 0], 13.3 ± 1.0 W/mK along [−2 0 1], 14.7 ± 1.5 W/mK along [0 0 1], and 27.0 ± 2.0 W/mK along [1 1 0] [68]. An intentional doping of β-Ga_2O_3 will decrease its thermal conductivity resulting from an increased phonon-impurity scattering.

2.2. Growth Methods

As mentioned above, β-Ga_2O_3 is stable under the whole range of temperature below the melting point, leading to single crystal bulk β-Ga_2O_3 that can be prepared by melt growth with a high growth rate, controllable size, and high quality, which also means it can be cheap to grow and process in the future [49,68]. However, there are two main challenges to produce single crystal bulk β-Ga_2O_3. One is the quite high melting point. The other is, as the following chemical equations show, the decomposition of β-Ga_2O_3 in insufficient oxygen ambient at elevated temperature, which will become more severe above 1200 °C as reported by Ueda et al. [69]:

$$Ga_2O_3\ (l,s) \rightarrow 2GaO\ (g) + 1/2O_2\ (g),$$

$$2GaO\ (g) \rightarrow Ga_2O\ (g) + 1/2O_2\ (g),$$

$$Ga_2O\ (g) \rightarrow 2Ga\ (g) + 1/2O_2\ (g).$$

Figure 2 illustrates the current techniques of melt growth for β-Ga_2O_3, including the Verneuil method, floating zone (FZ) method, Czochralski (CZ) method, vertical Bridgeman (VB) method, and edge-defined film-fed growth (EFG) method [70–80]. In 1902, Auguste Verneuil first developed the Verneuil method. As shown Figure 2a, the Verneuil method uses powders as raw materials, and melts them via an oxyhydrogen flame. Then the melting powders cool down and crystallize on the seed crystal. A β-Ga_2O_3 single crystal prepared by the Verneuil method was first reported by Chase in 1964 with a size of 3/8 inch in diameter and 1 inch in length [70]. The growth of β-Ga_2O_3 using this method was mainly along the *b*-axis. Lorenz et al. demonstrated that single crystal gallium oxide could be prepared under oxidizing conditions, whereas a reducing growth condition would lead to a blue conducting crystal [71]. When doped with Mg and Zr, the single crystal β-Ga_2O_3 showed colorless and light blue, respectively [81]. The main disadvantage of the Verneuil method is the inadequate size of as-grown crystals, which is too small for various applications.

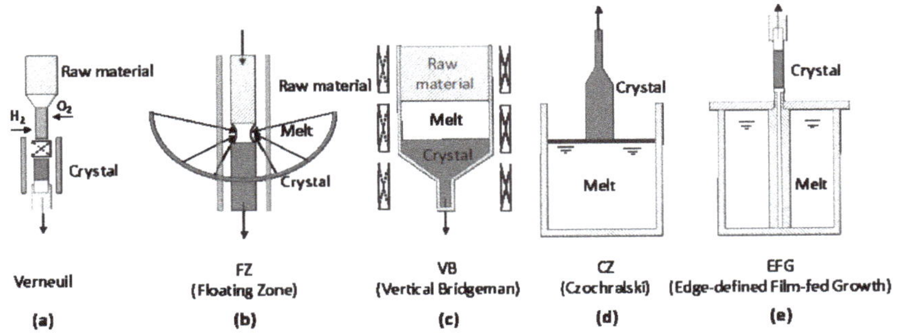

Figure 2. Schematic illustrations of melt growth of β-Ga_2O_3: (**a**) Verneuil method; (**b**) floating zone method; (**c**) vertical Bridgeman method; (**d**) Czochralski method; (**e**) edge-defined film-fed growth method.

The floating zone method employs a high-frequency coil to heat a partial region of the rod-like polycrystalline raw materials, which will suspend on the crystals via surface tension after melting, as shown in Figure 2b. Then the growth of crystals is realized by the upward movement of the coil. Being free of crucible, high purity of β-Ga_2O_3 can be obtained due to the reduction of contaminations. The possibility of introducing an oxidizing ambient during the crystal growth by the FZ method can inhibit the decomposition of β-Ga_2O_3. In view of the stability of the melt zone, it is difficult to obtain crystals with a large size, and cracks may generate during the cooling down process due to thermal stress. Ueda et al. employed an ordinary ceramic process to prepare the feed rods and realized the Sn doping of β-Ga_2O_3 by pressing the SnO_2 powder into the rods [69]. In 2001, Tomm et al. demonstrated the FZ method growth of β-Ga_2O_3 with a growth speed of 5 mm/h. The undoped and Sn-doped crystals were colorless, while the Ti-doped one was pale purple [82]. Crystals with a size of 1 inch along the (1 0 0), (0 1 0), and (0 0 1) axes prepared by the FZ method were demonstrated by Villora et al.

in 2004 [72]. In 2006, Zhang et al. reported a FZ method-grown single crystal β-Ga_2O_3 along (0 1 0) with a diameter of 6 mm and a length of 20 mm [73].

As shown in Figure 2c, the vertical Bridgeman method uses a crucible to contain the raw material, the melt, as well as the crystal. The raw material is melted through a radio-frequency induction heating furnace. By moving the crucible, while fixing the thermal field in a crystal furnace, the directional solidification process will lead to the growth of β-Ga_2O_3 single crystal on seeds. The commonly used material for crucible is an alloy of platinum and rhodium (70% and 30%), which is stable under oxidizing conditions. The main advantages of the VB method are the controllable shape and stoichiometry of crystals due to the use of crucible, which will, however, introduce impurities into the as-grown crystals. Hoshikawa et al. demonstrated the VB growth of β-Ga_2O_3 with a diameter of 25 mm along the direction perpendicular to (1 0 0) plane [74]. The seed-free preparation of crystals was realized without adhesion to the crucible wall. Also, the growth of β-Ga_2O_3 with different crucible shapes, including a full-diameter type and a conical type, was studied. Ohba et al. characterized the defects of β-Ga_2O_3 single crystals, and found no regions of high dislocation density existed near the wafer edge due to the lack of adhesion [75]. In addition, low mean density of dislocation was obtained, which could be ascribed to low thermal gradient during the directional solidification growth of the VB method.

The Czochralski method uses a seed crystal dipping into the melted source materials in a crucible. When pulling and rotating the seed, the reduction of temperature induces the phase transition from liquid state to solid state at the interface between the seed and the melt, as shown in Figure 2d. Due to the high stability of the liquid melt, a large size crystal can be obtained using the CZ method. Tomm et al. first demonstrated the CZ growth of a β-Ga_2O_3 single crystal with a growth rate of 2 mm/h [76]. A modified ambient of 10% CO_2 and 90% Ar gas mixture was found to suppress the evaporation of melted β-Ga_2O_3 as CO_2 would decompose and provide oxygen partial pressures. Galazka et al. grew a 2-inch β-Ga_2O_3 single crystal using a CO_2-containing ambient [77]. High crystalline quality was obtained with the rocking curve FWHM (full width half maximum) values less than 50 arcsec. Dislocations were found to mostly propagate parallel to (1 0 0) plane. In addition, Galazka et al. further studied the segregation of dopants in β-Ga_2O_3 single crystal prepared by the CZ method and their influence of optical properties [83]. Dopants such as Ce and Al were found to have a thermodynamically stabilizing effect during the growth of crystals by inhibiting decomposition. The doping of Ce would not degrade the transparency of β-Ga_2O_3, while the doping of Cr would lead to three more absorption bands. When doped with Al, the absorption edge of β-Ga_2O_3 would shift due to the formation of $Ga_{2(1-x)}Al_{2x}O_3$.

Edge-defined film-fed growth method is the most prevalent technique for β-Ga_2O_3 single crystal preparation. As illustrated in Figure 2e, compared with the CZ method, EFG method additionally employs a shaper or a die located in the crucible. Through the channel, the capillarity will force the transport of melt from the crucible to the top surface of the shaper. The melt will spread out until the edge of the shaper. Thus, the shape and the size of EFG crystals can be precisely controlled. The main advantages of the EFG method are the high speed of growth and the possibility of preparing complicated shapes, while the main disadvantages are the geometry and material of the shaper. In 2008, Aida et al. first reported the EFG of β-Ga_2O_3 crystal with a rocking curve FWHM value of 70–160 arcsec. Typically, the as-grown crystals had a length of 70 mm, a width of 50 mm, and a thickness of 3 mm with a growth reaching to 10 mm/h [84]. Mu et al. employed a gas mixture of 50% Ar and 50% CO_2 to grow a high crystalline quality 1-in β-Ga_2O_3 with a rocking curve FWHM value of 43.2 arcsec [85]. Kuramata et al. succeeded in obtaining a large size, including 4-in and 6-in, β-Ga_2O_3 single crystals by the EFG method [78,79]. Silicon was found to be the main residual impurity for the unintentionally doped EFG crystals, and the observation of etching pits manifested a dislocation density of 10^3 cm^{-3}. By using an ambient of 20% Ar and 80% CO_2, the crystalline quality of β-Ga_2O_3 prepared by the EFG method was further improved by Zhang et al. with FWHM of rocking curve reaching to 19.06 arcsec [80].

2.3. Conductivity Control and Doping

A controllable conductivity for semiconductors is the key to various applications. Thus, the realization of both n-type and p-type β-Ga_2O_3 is of great importance. In this section, the doping techniques for β-Ga_2O_3 are introduced.

2.3.1. N-Type β-Ga_2O_3

β-Ga_2O_3 is an intrinsic insulator due to its wide bandgap. However, through modifying the melt growth ambient, n-type conductivity can be obtained [86,87]. Single crystal β-Ga_2O_3 prepared by Lorenz et al. using the FZ method was insulated under oxidizing conditions, while it had a n-type conductivity under reducing growth conditions [71]. This n-type conductivity can be ascribed to the oxygen vacancies that act as donors after being ionized [86]. Thus, it is believed that there is a strong relation between the conductivity of unintentionally doped β-Ga_2O_3 and the existence of oxygen vacancies. Ueda et al. demonstrated a controllable conductivity of gallium oxides ranging from 10^{-9} $\Omega^{-1}\cdot cm^{-1}$ to 38 $\Omega^{-1}\cdot cm^{-1}$ by altering the oxygen concentrations in growth ambient [69]. Insulating crystals were obtained when the growth was performed under pure O_2 with a gas flow rate of 0.2 $m^{-3}\cdot h^{-1}$. By introducing the nitrogen gas and decreasing the oxygen contents in the ambient, the conductivity of as-grown crystals gradually increased. The conductivity was 0.63 $\Omega^{-1}\cdot cm^{-1}$ when the O_2 gas flow rate decreased to 0.05 $m^{-3}\cdot h^{-1}$, and reached a maximum value of 38 $\Omega^{-1}\cdot cm^{-1}$ with a N_2/O_2 ratio of 0.4/0.6. A similar tendency of conductivity controlled by growth atmosphere was reported by Galazka et al. [77]. A gas mixture of Ar and CO_2 or pure CO_2 was employed to inhibit the evaporation of CZ grown β-Ga_2O_3. CO_2 could provide an oxygen partial pressure due to its decomposition under high temperatures. By increasing the CO_2 contents from 30% to 50%, the electron concentration was decreased from 10 × 10^{17} cm^{-3} to 0.4–4.8 × 10^{17} cm^{-3}.

In addition, intentional doping with other impurities also can influence the conductivity of β-Ga_2O_3. The commonly used donors for n-type conductivity include Sn, Si, Ge, Cl, and F [86,88,89]. Due to the different radius, Si and Ge tend to occupy the tetrahedral Ga(I) site, while Sn tends to substitute the octahedral Ga(II) sites. Both Cl and F tend to occupy the O(I) sites. Ueda et al. doped Sn into FZ grown β-Ga_2O_3 by pressing SnO_2 power into the feed rods [69]. A conductivity of 0.96 $\Omega^{-1}\cdot cm^{-1}$ was measured for Sn-doped β-Ga_2O_3 even though the growth ambient was oxygen. Despite large amounts of Sn evaporated during the growth caused by the low efficiency of incorporation, the rest was still enough for n-type doping. FZ grown β-Ga_2O_3 crystals doped with Sn reported by Suziki et al. showed an electrical resistivity of 4.27 × 10^{-2} cm^{-2} and a carrier density of 2.26 × 10^{18} cm^{-3} [90]. The doping concentration of SnO_2 in crystals ranged from 2 to 10 mol%. Due to the evaporation during growth, the actual doping concentration of Sn atoms was in a range of 20–70 ppm.

Silicon is another potential dopant for n-type β-Ga_2O_3. Kuramata et al. experimentally studied the correlation between the effective donor concentration (N_d-N_a) and Si concentration of the FZ grown β-Ga_2O_3 [78]. The silicon doping was realized by adding SiO_2 powder into Ga_2O_3 powder. It was found that N_d-N_a for crystals annealing in nitrogen ambient was nearly equal to the Si concentration, while it kept approximately 1 × 10^{17} cm^{-3} when annealed in an oxygen atmosphere and showed no correlation with the Si concentration. Víllora et al. believed that Si was a more efficient donor dopant for β-Ga_2O_3 due to the lower vapor pressure of SiO_2 compared to SnO_2 and GeO_2 [91]. By increasing the concentration of Si impurities, the n-type conductivity of β-Ga_2O_3 gradually increased from 0.03 $\Omega^{-1}\cdot cm^{-1}$ to 50 $\Omega^{-1}\cdot cm^{-1}$, corresponding to a free carrier concentration from 10^{16} cm^{-3} to 10^{18} cm^{-3}. A saturation of carrier concentration was observed at a doping level of 0.2 mol% due to the segregation of the second-phase. The efficiency of incorporation for Si was close to unity at low doping levels, while it was only 5% of the Si atoms in the feed rod. Interestingly, a Si-ion implantation doping technique for β-Ga_2O_3 was presented by Sasaki et al. [92]. The implanted silicon ions were activated by annealing crystals under nitrogen ambient at 900–1000 °C. High activation efficiency above 60% was obtained when the injected Si ions were in range of 1 × 10^{19} cm^{-3}–5 × 10^{19} cm^{-3}, and it decreased dramatically with an increased Si^+ concentration. Additionally, an electrical resistance of 1.4 m$\Omega\cdot$cm was measured with a doping concentration of 5×10^{19} cm^{-3}. Recently, Zhou et al. demonstrated the controllable conductivity of FZ grown β-Ga_2O_3 employing Nb as dopants [93]. By increasing the Nb doping concentrations, the electrical resistance of β-Ga_2O_3 could be changed from 3.6×10^2 Ω·cm to 5.5×10^{-3} Ω·cm, corresponding to a carrier concentration from 9.55×10^{16} cm^{-3} to 1.8×10^{19} cm^{-3}.

Up to now, as illustrated in Figure 3a, the carrier concentrations of n-type β-Ga_2O_3 can be highly controlled in a range of 10^{16}–10^{19} cm^{-3}, and even can reach 10^{20} cm^{-3} by Si doping [89]. Consequently, high conducting β-Ga_2O_3 single crystals can be obtained. However, doping will not only change the electrical, but also influence the optical properties of β-Ga_2O_3. As shown in Figure 3b, the transmittance of β-Ga_2O_3 is decreased with an increasing electron concentration via Sn doping [77]. Insulating β-Ga_2O_3 is usually colorless or slightly yellow due to the slight absorption in the blue region, while n-type β-Ga_2O_3 shows a bluish-like appearance due to the enhanced absorption of free carriers in red and near infrared regions. Crystals grown in CO_2 ambient showed grey coloration which could be ascribed to the possible incorporation of carbon impurities. Thus, there should be an eclectic choice between the resistivity and transmittance of β-Ga_2O_3, since the crystalline quality and mobility will degrade after doping. Suzuki et al. found that the X-ray rocking curve FWHM value of β-Ga_2O_3 was increased from 43 arcsec to 162 arcsec with increased doping concentration of Sn from 32 ppm to 45 ppm [90]. The mobility of free carriers was decreased to 49.3 cm^2/V·s compared with 87.5 cm^2/V·s for an undoped sample. Also, an intentional doping of β-Ga_2O_3 will decrease its thermal conductivity resulting from an increased phonon-impurity scattering.

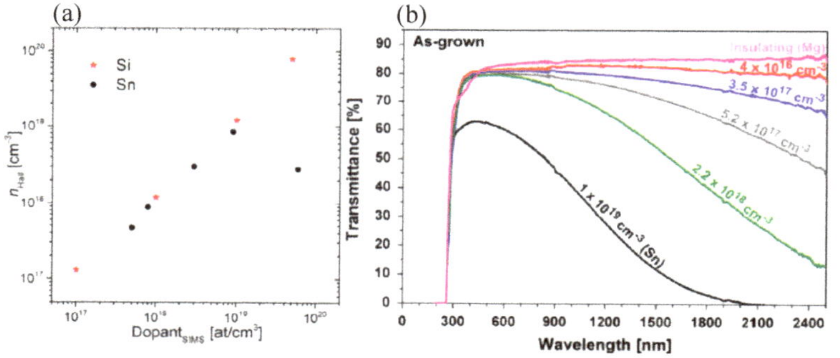

Figure 3. (**a**) Hall free carrier concentration versus the dopant (Si and Sn) concentration obtained by secondary ion mass spectrometry (SIMS). Reprinted with permission from reference [89]. Copyright 2017 The Electrochemical Society. (**b**) Transmittance spectra of β-Ga_2O_3 single crystals prepared by the Czochralski (CZ) method with different concentrations of electrons. Reprinted with permission from Reference [77]. Copyright 2014 Elsevier.

2.3.2. P-Type β-Ga_2O_3

The lack of an efficient method for p-type conductivity is now one of the major drawbacks of β-Ga_2O_3, which largely limits the fabrication of various devices. The difficulty of p-type doping for β-Ga_2O_3 can be ascribed to several reasons [94]. Due to the relatively low formation energy of oxygen vacancies, the acceptors doped into crystals can be easily compensated. The high activation energy of acceptors caused by a relatively low energy level of valence band and lack of shallow dopants leads to a poor activation efficiency of dopants. Also, the self-trapping effect and the relatively high effective masses of holes make it harder for the realization of p-type conductivity.

Tomm et al. characterized the electrical properties of FZ grown β-Ga_2O_3 doped with Ge and Ti. Hole concentrations of 2×10^5 cm^{-3} and 5×10^5 cm^{-3} were measured for Ge-doped and Ti-doped crystals, respectively, which however, was too low for any practical applications [83]. The study of Mg dopants for β-Ga_2O_3 has been carried out by several groups [77,95]. Onuma et al. reported that (0 1 0)-faceted colorless Mg-doped β-Ga_2O_3 exhibited a semi-insulating behavior [95]. The electrical resistance measured was around 6×10^{11} Ω·cm with Mg concentration in the range of 4×10^{18} cm^{-3}–2×10^{19} cm^{-3}. Galazka et al. found that Mg dopants could help to inhibit the formation of spiral for CZ grown β-Ga_2O_3 [77]. Insulating crystals were obtained with a doping concentration of Mg above 6 wt ppm. A Mg equilibrium segregation coefficient around 0.10–0.12 at the interface between liquid and solid was estimated. Unlike Sn donors, Mg dopants would lead to a stable insulating behavior of β-Ga_2O_3 that was nearly independent with an annealing ambient (oxygen or hydrogen), temperature (up to 1400 °C), time (up to 66 h), and pressure (up to 19 bar).

In addition, other impurities including Zn and N for p-type doping of gallium oxide also have been presented [96,97]. Chang et al. experimentally studied the Zn dopants for Ga_2O_3 nanowires using a diffusion doping method [96]. Due to the similar radius (0.074 nm for Zn^{2+} and 0.062 nm for Ga^{3+}), Zn^{2+} ions tended to substitute the Ga^{3+} ions, making it a possible acceptor for gallium oxide. The calculated carrier concentration and the mobility of the quasi 1D-nanowires were approximately 5.3×10^8 cm^{-1} and 3.5×10^{-2} cm^2/V·s, respectively. Liu et al. used NH_3 as sources to dope nitrogen into β-Ga_2O_3 microwires [97]. I–V curves were measured for nitrogen-doped β-Ga_2O_3 microwires and undoped β-Ga_2O_3/nitrogen-doped β-Ga_2O_3 microwires, respectively. P-type conductivity was deduced based on the I–V behavior of the homojunction.

In summary, melt growth of β-Ga_2O_3 has been established, which means that β-Ga_2O_3 can be cheap to prepare with a scalable size in the future. Despite the possibility of realizing that p-type conductivity still remains controversial, transparent and highly n-type conducting beta phase gallium oxides can be obtained already. Thus, β-Ga_2O_3 should be a promising candidate as the substrate of vertical structure LEDs.

3. Epitaxial Relationship between III-Nitrides and β-Ga_2O_3

Even though there is a significant difference in crystal structure, the epitaxy of hexagonal III-Nitrides on (1 0 0) and (−2 0 1) β-Ga_2O_3 is still achieved [98–103]. The epitaxial relationship between monoclinic β-Ga_2O_3 and the wurtzite III-Nitrides is investigated at the atomic level [45]. The (1 0 0) planes of β-Ga_2O_3 consist of weakly bonded oxygen atoms in $O_{(3)}$ sites, leading to the cleavage properties of the *a*-plane. Disregarding other atoms in consideration of clarity, Figure 4a shows the projection view normal to the *a*-plane along the <2 0 1> direction where gallium atoms in $Ga_{(1)}$ and $Ga_{(2)}$ sites are arranged in nearly a regular hexagonal structure, making it possible to grow high crystalline quality III-Nitrides on (1 0 0) β-Ga_2O_3. As shown in Figure 4b, the epitaxial relationship between (0 0 0 1) GaN and (1 0 0) β-Ga_2O_3 substrate is $<0\ 1\ 0>_{Ga2O3} \| <1\ 1\ {-2}\ 0>_{GaN}$ and $<0\ 0\ 1>_{Ga2O3} \| <-1\ 1\ 0\ 0>_{GaN}$, determined by reflection high-energy electron diffraction (RHEED) [104]. The roughly estimated lattice mismatch between (0 0 0 1) GaN and (1 0 0) β-Ga_2O_3 is −4.74 % along the *c*-axis and 5.05% along the *b*-axis.

However, such an anisotropic stress state (tensile stress parallel to the *c*-axis and compressive one parallel to the *b*-axis) was considered to be of little possibility. Víllora et al. presented that there was a deviation angle of 1° calculated by image asymmetry of an electron-diffraction pattern. The model of the epitaxial relationship was adjusted, where there was a 1° titled angle compared to the former one, as illustrated in Figure 5b. The rectificatory value of lattice mismatch was then reduced to the lowest value (2.6%) ever reported [45]. As shown in Figure 5c, the oxygen atoms on (−2 0 1) planes of single crystal β-Ga_2O_3 are also hexagonally arranged. Thus, the deposition of hexagonal III-Nitrides is allowed. The epitaxial relationship defined as $(-2\ 0\ 1)_{\beta\text{-}Ga2O3} \| (0\ 0\ 0\ 1)_{GaN}$ with a lattice mismatch of 4.7% was presented [100]. Compared with sapphire, the lattice mismatch between III-Nitrides (GaN or AlGaN) and β-Ga_2O_3 is much smaller, leading to the better crystalline quality of epilayers. Kachel et al. [105] have presented that under the same condition, GaN deposited on β-Ga_2O_3 substrate displayed a smooth surface morphology, while that on sapphire exhibited a rough and irregular morphology. The root mean square (RMS) roughness of GaN deposited on β-Ga_2O_3 and sapphire was 5 nm and 75 nm, respectively. As Figure 5d shows, the full width half maximum (FWHM) of high resolution x-ray diffraction (HR-XRD) rocking curves(RCs) around the (0 0 0 2) GaN Bragg reflection peak was 1020 arcsec on β-Ga_2O_3, while that on Al_2O_3 was 2580 arcsec, manifesting a better crystalline quality using β-Ga_2O_3 as substrate.

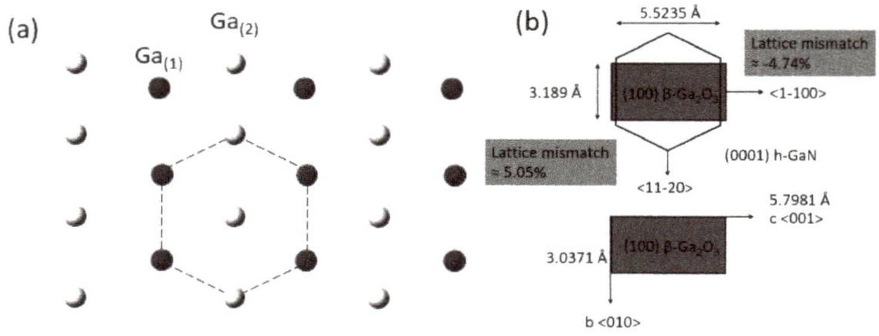

Figure 4. (**a**) Projection of β-Ga_2O_3 atomic structure perpendicular to the *a*-plane, showing the hexagonal-like arrangement of the Ga atoms bonded to $O_{(3)}$, and (**b**) schematic illustration for the epitaxial relationship between *c*-plane h-GaN and *a*-plane β-Ga_2O_3.

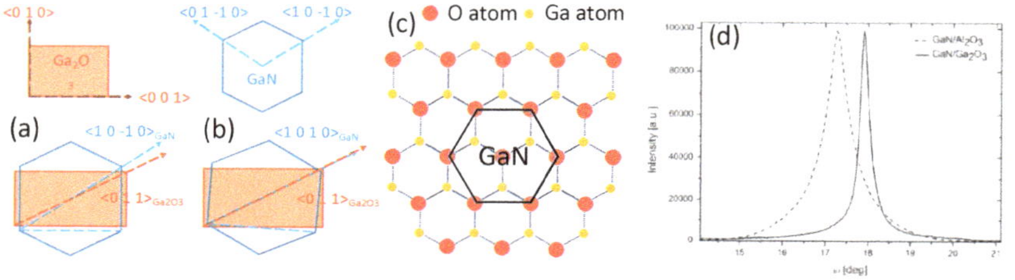

Figure 5. (**a**) Initial observation of epitaxial relationship between *c*-plane wurtzite GaN and (1 0 0) plane β-Ga_2O_3. (**b**) Refined model of epitaxial relationship between *c*-plane wurtzite GaN and (1 0 0) plane β-Ga_2O_3 with a stressed and reoriented GaN at the interface, the lattice mismatch (LM) minimum of 2.6% is given at a 1° tilted angle with respect to case (a). (**c**) The projection of β-Ga_2O_3 atomic structure perpendicular to the (−2 0 1) plane, showing the hexagonal-like arrangement of oxygen atoms. (**d**) Full width half maximum (FWHM) of HR-XRD rocking curves around the (0 0 0 2) Bragg reflection for GaN grown on Al_2O_3 (2580 arcsec), and on β-Ga_2O_3 (1020 arcsec). Reprinted with permission from Reference [105]. Copyright 2012. The Royal Society of Chemistry.

4. Epitaxy of GaN on β-Ga_2O_3

4.1. Influences of Atmosphere

Despite the high chemical stability of β-Ga_2O_3, the growth of III-Nitrides ($Al_xGa_{1-x}N$, x from 0 to 1) on β-Ga_2O_3 substrate should be carried out in a nitrogen atmosphere. As the following chemical equations show, Ga_2O_3 will decompose under a hydrogen atmosphere at the growth temperature of HT-GaN [106,107]:

$$Ga_2O_3 + 2H_2 \rightarrow Ga_2O + 2H_2O$$

$$Ga_2O + H_2 \rightarrow 2Ga + H_2O$$

Li et al. [108] heated the β-Ga_2O_3 substrates in a pure hydrogen, pure nitrogen, and mixed gas (98% N_2 + 2% NH_3) atmosphere under different temperatures including 600, 710, 830, 880, 990, 1050, 1100, and 1150 °C. Each stage was kept for 300 s. As illustrated in Figure 6a, when heated in H_2 atmosphere, the surface reflectance of β-Ga_2O_3 sample dropped quickly even below 600 °C, and didn't recover after cool down to room temperature. A decreased reflectance corresponded to a rough surface, as shown in Figure 6d,g, indicating that β-Ga_2O_3 was drastically damaged by H_2. When exposed to pure nitrogen, the reflectance was first decreased with increased temperature due to the change of refractive index and recovered after cool down, manifesting that N_2 atmosphere would not destroy the surface morphology of β-Ga_2O_3. The SEM images indicated a smooth surface of β-Ga_2O_3 after annealing, as shown in Figure 6e,h. Therefore, using N_2 as a carrier gas was necessary for the growth of III-Nitrides in order to prevent β-Ga_2O_3 from decomposition. Considering that ammonia was employed as a nitrogen source for epitaxy of III-Nitrides by the metalorganic chemical vapor deposition (MOCVD) technique, 2% NH_3 was added to the nitrogen atmosphere for heat-up process. As illustrated in Figure 6c,f,i, above 900 °C, the reflectance of β-Ga_2O_3 in mixed gas ambient dropped more quickly than in pure N_2 atmosphere. It then recovered to a higher value than in pure N_2 atmosphere when cooled down, which could be ascribed to the change of surface composition. As the following chemical equation shows, Ga_2O_3 would be nitrided into GaN when exposed to NH_3 with an elevated temperature:

$$Ga_2O_3\ (s) + 2NH_3\ (g) \rightarrow 2GaN\ (s) + 3H_2O\ (g)$$

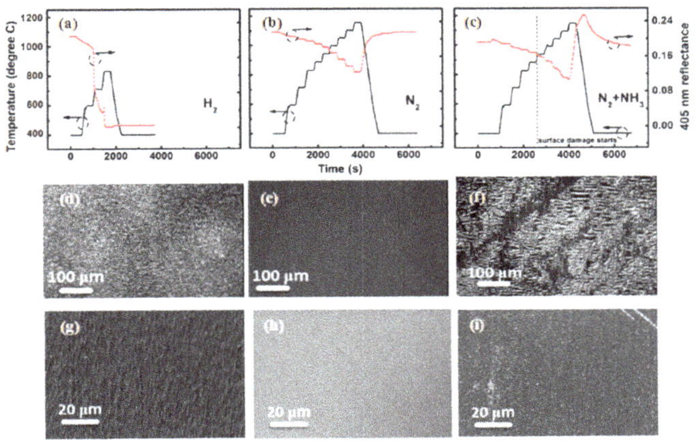

Figure 6. Temperature transients (temperature from EpiTT) and reflectance (405 nm) of (−2 0 1) β-Ga_2O_3 heated under (**a**) H_2, (**b**) N_2, and (**c**) N_2 plus NH_3 atmosphere with corresponding optical images of the resulting surfaces in (**d**), (**e**), and (**f**) and SEM images in (**g**), (**h**), and (**i**). Reprinted with permission from reference [108]. Copyright 2017 Elsevier.

Figure 7a,b illustrate the grain-like, rather than film-like, surface morphology of III-Nitrides epilayers grown in H_2 atmosphere, resulting from the etching of Ga_2O_3 during the epitaxy. As Figure 7c shows, the absence of Ga_2O_3 peaks indicate that gallium oxides will fully decompose as time goes on. However, as shown in Figure 7d, the N_2 atmosphere will deteriorate the crystalline quality albeit instrumental for preventing Ga_2O_3 from decomposition [106]. It can be attributed to the different conversion and diffusion properties of the reactants under N_2 and H_2 atmospheres. Hillocks will generate on the GaN surface under an N_2 atmosphere, resulting from the shorter mean free path length of reactant molecules compared to that under an H_2 atmosphere [107]. Smaller nuclei size and coalescence thickness will be observed when exposed to N_2 [109]. Bottcher et al. presented that threading dislocation density was inversely proportional to the average grain diameter [110]. Consequently, high density of dislocations will be generated, leading to a degraded crystalline quality of III-Nitrides epilayers. One way to obtain high crystalline quality of III-Nitrides while preventing Ga_2O_3 from etching is the regrowth method, which will be discussed in next section.

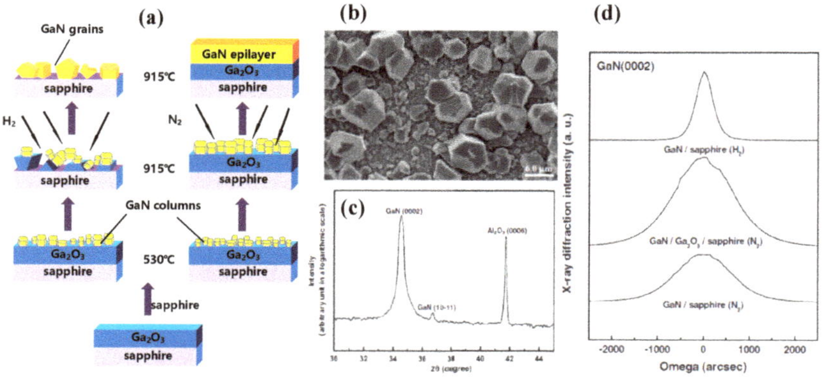

Figure 7. (**a**) Schematic illustration of GaN growth using N_2 and H_2 as carrier gases. (**b**) Top-view SEM micrograph and (**c**) double-crystal X-ray diffractometry (DC XRD) pattern of GaN deposited on Ga_2O_3 via MOCVD under an H_2 atmosphere. (**d**) DC XRD of rocking curve at the (0 0 0 2) plane of the GaN/Ga_2O_3/sapphire fabricated under an N_2 atmosphere and of GaN/sapphire fabricated under N_2 and H_2 atmospheres. Reprinted with permission from Reference [107]. Copyright 2011. The Electrochemical Society.

4.2. Orientations of β-Ga_2O_3 for Epitaxy

The epitaxy of GaN on β-Ga_2O_3 was initially studied on (1 0 0)-orientated substrates. Víllora et al. first demonstrated quasi-homoepitaxial growth of GaN on (1 0 0) β-Ga_2O_3 substrate via molecular beam epitaxy (MBE) [28]. Shimamura et al. first presented the growth of GaN on (1 0 0) β-Ga_2O_3 with a FWHM of 1200 arcsec by metal-organic vapor-phase epitaxy (MOVPE) [29]. Ohira et al. investigated the radio-frequency molecular beam epitaxy (RFMBE) growth of GaN on (1 0 0) β-Ga_2O_3 [34]. Ito et al. decreased the threading dislocation densities of GaN epilayer from 1.9×10^{10} cm^{-2} to 2.5×10^{9} cm^{-2} by utilizing a two-step growth method [45]. Kachel et al. grew GaN on (1 0 0) β-Ga_2O_3 by pseudo hydride vapor phase epitaxy (HVPE) method and demonstrated a self-separation method of bulk GaN from the β-Ga_2O_3 substrate [35]. Despite the successful deposition of GaN on β-Ga_2O_3 by various techniques including HVPE, MBE, and RFMBE, MOVPE is currently the most widely used method for the growth of GaN on gallium oxide. The epitaxy of GaN on β-Ga_2O_3 is quite different from that on sapphire due to the discrepancy of the crystal structure between wurtzite GaN and monoclinic β-Ga_2O_3.

Prior to epitaxy, a nitridation process will reconstruct the surface of (1 0 0) β-Ga_2O_3 from 2-fold symmetry to 6-fold symmetry via substituting the oxygen atoms with nitrogen atoms. As a result, the quasi-homoepitaxy of (0 0 0 1) wurtzite GaN can be realized [28]. According to the first-principles density functional theory study, GaN grown on the surface modified with nitrogen is most stable [45]. Nitridation conditions have been explored by several groups [28,35]. Víllora et al. demonstrated that the nitridation pressure was one of the critical factors during this process [28]. The nitridation procedure was performed by introducing ammonia into the chamber for a certain time at the temperature of substrate close to 800 °C. Figure 8 shows the field emission scanning electron microscope (FESEM) images of surface morphology and cross section of GaN epilayers under different nitridation pressures. GaN deposited under the nitridation pressure less than 10^2 Pa was grey and of low adhesion, indicating a non-effective nitridation process. GaN grown under pressure more than 10^3 Pa was characterized with a mirror-like surface, suggesting a sufficient nitridation process [98]. The smooth interface indicated that the precondition would not cause a reaction in depth, but only a rearrangement on the surface [45]. A non-effective nitridation step would lead to the growth of a zincblende rather than wurtzite structure of GaN. In addition, the processing time could also determine the nitridation effect. Ohira et al. demonstrated the relationship between nitridation time and the structure of the epilayer by RHEED [104]. Before nitridation, streaky patterns indicated a smooth surface of β-Ga_2O_3 substrate, and patterns remained streaky after the first five minutes of processing. However, when the nitridation time further increased, GaN turned into cubic phase and into a hexagonal structure after 60 min and 90 min, respectively. That is to say, the symmetry of GaN on β-Ga_2O_3 substrate can be adjusted by modifying the nitridation time.

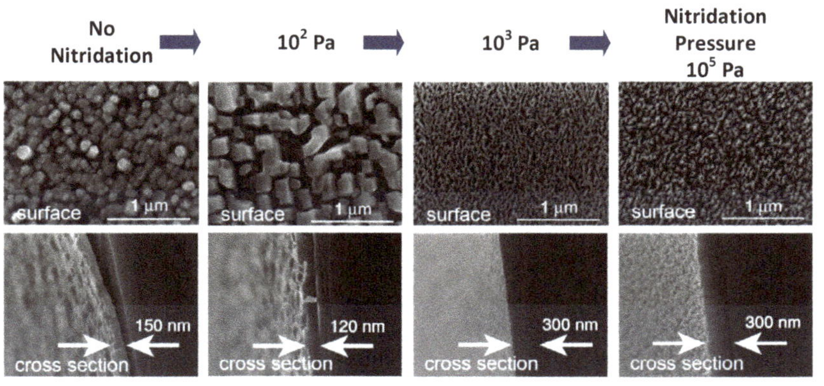

Figure 8. Field emission scanning electron microscope (FESEM) images of surface morphology and cross section of deposited GaN layers as a function of the nitridation pressure. Reprinted with permission from Reference [98]. Copyright 2006 Elsevier.

As mentioned before, the growth of GaN should be performed under a nitrogen atmosphere to protect Ga_2O_3. Tsai et al. demonstrated a regrowth method to improve the crystalline quality of epilayers that deteriorated by N_2 atmosphere [106]. After the growth of low-temperature (LT) and high-temperature (HT) GaN under N_2 at 530 °C and 915 °C, respectively, GaN was regrown under an H_2 atmosphere at a temperature of 965 °C, as illustrated in Figure 9. The x-ray rocking curves (XRC) FWHM of (0 0 0 2) GaN reflection peak was decreased from 1444 arcsec to 537 arcsec. And the FWHM of photoluminescence (PL) peak was decreased from 72.5 meV to 56.7 meV. Also, the RMS roughness of the surface decreased from 30.5 nm to 0.7 nm, indicating a better crystalline quality of GaN via a regrowth method.

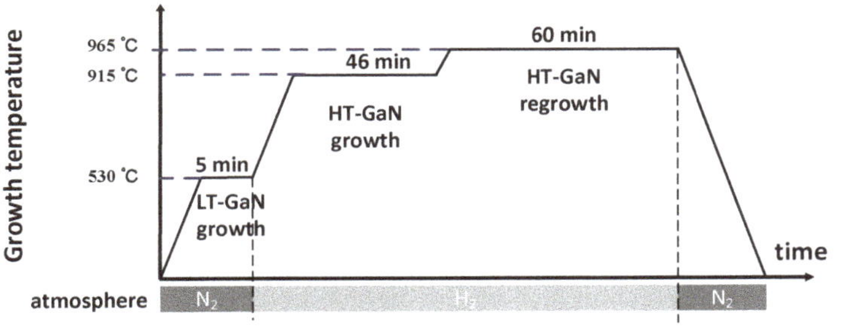

Figure 9. Procedure of HT-GaN regrowth method.

Previous efforts to grow GaN on *a*-plane β-Ga_2O_3 may be insufficient since the crystalline quality is inadequate to achieve the high-performance devices [88,89,104,105,111,112]. In addition, due to the strong cleavage properties of *a*-plane β-Ga_2O_3, GaN epilayer will be easily separated from β-Ga_2O_3 and thus complicate the process of dicing. Therefore, (−2 0 1) β-Ga_2O_3 as a substrate for GaN epitaxy was investigated [100–103]. The epitaxy of GaN on (−2 0 1) β-Ga_2O_3 with an AlN buffer layer has been reported [100–102], and the corresponding XRC FWHM of (0 0 0 2) GaN reflection peak was 0.122°

(approximately 439 arcsec), revealing an improved crystalline quality. Muhammed et al. presented an atmosphere switch and two-step growth method. It employed N_2 as the carrier gas for the growth of AlN buffer layer and switched it to H_2 during the growth of HT-GaN. Flat surface morphology and high crystalline quality of GaN epilayers were obtained [102]. In addition, Muhammed et al. also demonstrated that a GaN buffer layer instead of AlN could further improve the crystalline quality of GaN epilayers [103]. By utilizing GaN buffer layer, the surface of GaN grown on Ga_2O_3 exhibited a homogenous strain distribution, and was nearly strain-free. The E_2 (high) peak of GaN deposited on GaN buffer layer displayed a red shift (-0.07 cm^{-1}), in contrast to that of GaN on AlN buffer layer (blue shift, 1.04 cm^{-1}), suggesting a lowered strain state of GaN. Furthermore, as shown in Figure 10b,c, the FWHM of GaN (0 0 0 2) reflection peak was decreased from 430 arcsec to 330 arcsec, and the PL intensity of GaN was enhanced by a factor of 12. The calculated threading dislocation density of GaN epilayer was 1.8 ($\pm$ 0.2) $\times$ 10^8 cm^{-2} on GaN buffer layer and 4.5 ($\pm$ 0.2) $\times$ 10^8 cm^{-2} on AlN buffer layer. Consequently, the crystalline quality of a GaN epilayer was greatly improved by employing the (-2 0 1) β-Ga_2O_3 substrate and a GaN buffer layer. We succeeded to improve the crystal quality of GaN on gallium oxide by introducing a nanoscale epitaxial lateral overgrowth method through a self-assembled SiO_2 nanosphere monolayer template on (-2 0 1) β-Ga_2O_3. Compared with direct epitaxy on β-Ga_2O_3, the XRC FWHM of (0 0 2) and (1 0 2) GaN reflection peak are decreased from 550.0 arcsec to 388.4 arcsec, and from 634.4 arcsec to 356.3 arcsec, respectively [113].

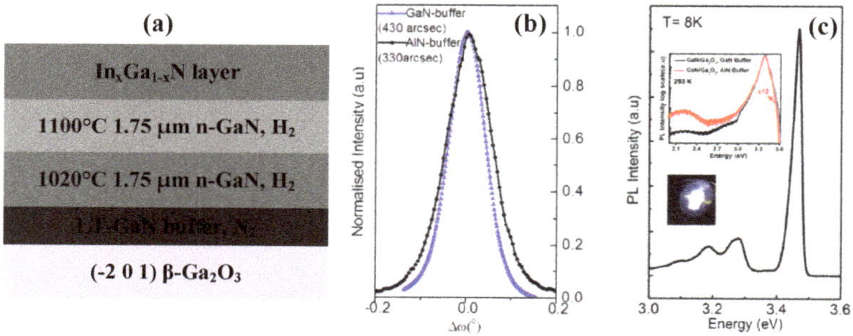

Figure 10. (**a**) Structural diagram of GaN on β-Ga_2O_3 via atmosphere switch and two-step growth. (**b**) The XRC of GaN (0 0 0 2) reflection peak for GaN grown on (-2 0 1) β-Ga_2O_3 substrate with GaN buffer layer and AlN buffer layer. (**c**) PL spectra of GaN grown on (-2 0 1) β-Ga_2O_3 substrate with a GaN buffer layer at 8 K. Reprinted with permission from Reference [103]. Copyright 2016 Springer Nature.

4.3. *Ga_2O_3 Sacrificial Layer*

As mentioned above, the laser lift-off technique is one way to fabricate vertical structure LEDs, but will lead to additional defects and a rough surface to devices resulting from the high energy laser irradiation [40–42], while chemical lift-off will not damage the interface between epilayer and substrate [43]. Recently, researches have been carried out for chemical lift-off of GaN using various interlayers [43–48,106,107,114–120]. Lin et al. demonstrated the CLO process of InGaN-based LEDs grown on triangle-shaped and truncated-triangle-striped patterned sapphire substrates employing AlN as a sacrificial layer [115,116]. However, using KOH as etchant might also damage the GaN grown on AlN. Horng et al. reported the vertical structure nitride LED fabricated on Cu substrate via CLO process employing an AlN/strip-patterned-SiO_2 interlayer as a sacrificial layer [117]. In addition, the direct growth of GaN on CrN and ZnO has been demonstrated, and the CLO process to detach the GaN epilayer has been realized using CrN and ZnO as sacrificial layers [43–48]. ZnO has a lower lattice mismatch (1.6%) with GaN compared to CrN, but will decompose above 650 °C in the atmosphere of NH_3, which is exactly the source of nitrogen in MOCVD technique.

Recently, β-Ga_2O_3 has attracted more and more attention as the buffer layer and sacrificial layer for the CLO process of GaN due to its low lattice mismatch of 2.6% and high selectivity ratio with GaN [106,107,118–120]. Tsai et al. deposited a β-Ga_2O_3 layer on sapphire substrate via pulsed laser deposition (PLD), followed by the MOCVD growth of GaN on Ga_2O_3/sapphire template. The separation of GaN epilayer from sapphire by etching β-Ga_2O_3 away using hydrofluoric solution was demonstrated [106,118,119]. Hsueh et al. [120] reported the epitaxy of high crystalline quality GaN on MOCVD-grown Ga_2O_3/Eco-GaN template using O_2 as a source of oxygen. Figure 11a shows the cross-sectional SEM image of the structure of the regrown GaN. Compared with the u-GaN grown on sapphire, the etching pits of GaN grown on a Ga_2O_3/Eco-GaN template were decreased from 2.4×10^8 cm^{-2} to 6.6×10^7 cm^{-2}, as illustrated in Figure 11b,c. Employing the regrowth method mentioned above, the crystal quality of GaN was improved, with the FWHM of GaN (0 0 0 2) reflection peak reaching to 417 arcsec, as shown Figure 11d. Thus, β-Ga_2O_3 is a promising material acting as not only a buffer layer but also as a sacrificial for the CLO process of GaN and AlGaN epilayers with high crystalline quality. The fabrication of vertical structure LEDs can be realized by the subsequent wafer bonding process.

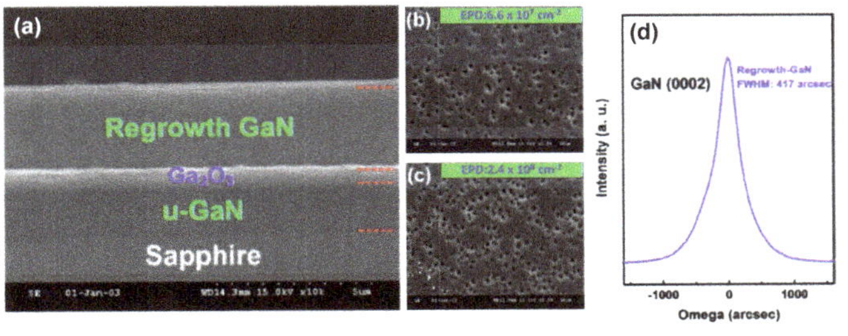

Figure 11. (**a**) Cross-sectional SEM image of the sample and (**b**) the etch-pit distribution observed by SEM image for the regrowth GaN epilayer grown on Ga_2O_3/Eco-GaN template. (**c**) The etch-pit distribution observed by SEM image for the u-GaN grown on sapphire. (**d**) XRD rocking curve of (0 0 0 2) reflection. Reprinted with permission from Reference [120]. Copyright 2015 Elsevier.

5. Epitaxy of AlGaN on β-Ga_2O_3

Although the improved crystalline quality of GaN on β-Ga_2O_3 substrate has been demonstrated, there is still a great challenge to grow AlGaN alloys on β-Ga_2O_3. Very few groups have succeeded to epitaxy AlGaN especially with high Al component alloys on β-Ga_2O_3. Shun Ito et al., utilizing a facet-controlled growth method, succeeded to improve the crystalline quality of $Al_{0.08}Ga_{0.92}N$ epilayers [111]. Figure 12a displays the timing charts of the growth of AlGaN on a facet-AlGaN layer. Thermal annealing of (1 0 0) β-Ga_2O_3 substrate was performed for three minutes at 1100 °C, followed by the growth of LT-GaN buffer layer at 550 °C and 300-nm-thick facet-AlGaN layer at 950 °C. Then the deposition of AlGaN layer was carried out at 1080 °C, utilizing the MOVPE technique with trimethylaluminum, trimethylgallium, and NH_3 as aluminum, gallium, and nitrogen sources, respectively. In addition, the deposition of LT-GaN should also be performed in nitrogen atmosphere to avoid the etching of β-Ga_2O_3 with H_2. The RMS roughness of β-Ga_2O_3 increased from 0.2 nm to 14 nm by thermal annealing with temperature up to 1100 °C. Then the AlGaN deposited on it was characterized with inclined facets due to the rough surface of the substrate, as shown in Figure 12b. The facets distributed on the AlGaN surface were controlled by thermal annealing temperature. The dislocations produced in the facet regions bended and would not penetrate or slide to the surface, leading to lower dislocation densities (4.9×10^9 cm^{-2}) compared to the counterpart (2.6×10^{10} cm^{-2}) without a facet-AlGaN layer. Moreover, as illustrated in Figure 12d, the FWHM of (0 0 0 2)- and (2 0 −2 4)-diffraction of AlGaN were decreased from approximately 2500 and 1250 arcsec to 750 arcsec and 1000 arcsec, respectively. With a facet layer, stronger PL intensity was demonstrated, indicating an improved crystalline quality of AlGaN via the facet-controlled growth method, as shown in Figure 12e.

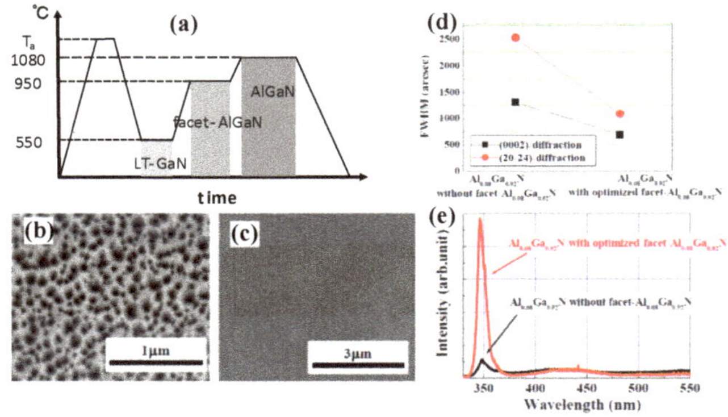

Figure 12. (**a**) Timing charts of growth temperatures of GaN using facet layers. Plan-view SEM images of (**b**) facet-$Al_{0.08}Ga_{0.92}N$ and (**c**) $Al_{0.08}Ga_{0.92}N$ grown on facet-$Al_{0.08}Ga_{0.92}N$ layer at 1080 °C. (**d**) XRC FWHM and (**e**) PL spectra of $Al_{0.08}Ga_{0.92}N$ without and with facet-$Al_{0.08}Ga_{0.92}N$ layer. Reprinted with permission from Reference [111]. Copyright 2012 Wiley.

Recently, Ajia et al. [121] demonstrated the growth of $Al_{0.3}Ga_{0.7}N$ on (−2 0 1) β-Ga_2O_3 by MOCVD. As shown in Figure 13a, the structure of the sample consisted of 2 nm AlN buffer deposited at 550 °C in the atmosphere of N_2, then a 100 nm n-$Al_{0.75}Ga_{0.25}N$ at 1020 °C in the atmosphere of H_2, followed by a 900 nm n-$Al_{0.27}Ga_{0.73}N$ at 1120 °C in H_2, and ended with 3 × 3 nm GaN/4 × 10 nm $Al_{0.2}Ga_{0.8}N$ multiple quantum wells (MQWs). Compared with the sample grown on sapphire, the FWHM of (1 0 4) rocking curves of the epilayer on β-Ga_2O_3 reduced from 0.683° to 0.469°, indicating a lower edge-type dislocation density for the sample on β-Ga_2O_3. A lower total density of V-pits and trenches was observed. The V-pits are typically generated from screw and mixed type of threading dislocations, while the trenches result from stacking faults [122,123]. The origin of V-pits and trenches can be ascribed to the release of strain in strained MQWs. Thus, the lower defect density of a sample grown on β-Ga_2O_3 manifested a relatively lower strain state compared with a sample on sapphire. Despite the demonstration of epitaxy of AlGaN on gallium oxide, the crystalline quality of epilayers are still not satisfied enough to realize high performance of UV-LEDs, which need to be improved furthermore in the future.

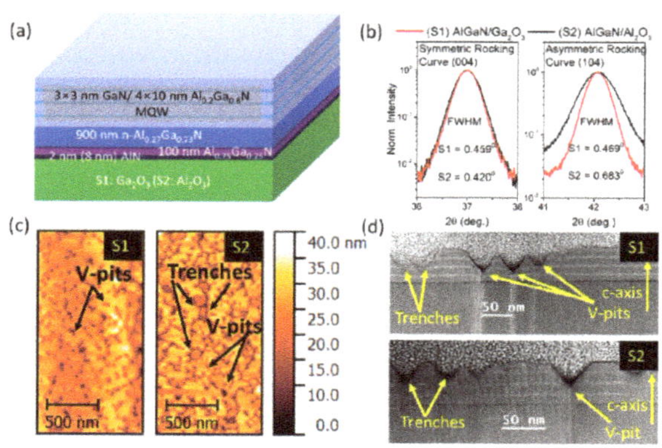

Figure 13. (**a**) Structural schematic of the samples. (**b**) Symmetric (0 0 4) and skew symmetric (1 0 4) XRD RCs. Reprinted with permission from reference [121]. Copyright 2018 AIP Publishing.

6. Vertical Structure LED

Vertical structure LEDs can be fabricated by various methods. Wang et al. employed a patterned laser lift-off process and electroplated a nickel layer for the fabrication of vertical structure GaN-based

LEDs [124]. The forward voltage of VLEDs was 3.01 V and 3.39 V at 20 mA and 80 mA, respectively, which was 10% and 21% lower than conventional LEDs. With a chip size of 300 μm, a saturation current of 520 mA for VLEDs was 4.3 times higher than conventional LEDs. Also, 2.3 times and 2.7 times higher power conversion efficiency were obtained. By using the same method, Kim et al. demonstrated a 100% enhancement of light emission of GaN-VLEDs compared to lateral LEDs [125]. As reported by Lin et al., the GaN-based VLEDs prepared by laser lift-off process showed no saturation at 500 mA and a 2.7 times increase of luminance intensity [126]. Xiong et al. presented the fabrication of vertical structure GaN LEDs by wafer bonding and chemical etching lift-off GaN from Si (1 1 1) substrate. At 20 mA, the forward voltage of VLEDs was reduced to 3.2 V compared to 4.0 V of lateral LEDs. With a chip size of 240 μm, the VLEDs showed no saturation up to 800 mA, while the output power of lateral LEDs saturated at 340 mA [127]. Kawasaki et al. obtained the vertical AlGaN deep, UV-LEDs emitting 322 nm by laser lift-off technique [128]. The differential conductance of VLEDs was improved by a factor of 5. The forward voltage was half compared to the lateral LEDs. When the injection current density increased from 0 A/cm^2 to 22 A/cm^2, the peak wavelength of lateral UV-LED was shifted from 322 nm to 328 nm, while emission peak shift of vertical UV-LEDs was only 1 nm, indicating a reduced self-heating effect for VLEDs. Adivarahan et al. used a laser lift-off method to fabricate vertical UVLEDs with a peak emission wavelength of 280 nm. An output power of 5.5 mW of single chip devices at a continuous-wave current density of 25 A/cm^2 was achieved [129]. Nishida et al. directly fabricated the 352 nm vertical UV-LEDs on bulk GaN [130]. An internal quantum efficiency of more than 80% and maximum output power of 10 mW were obtained. Also, vertical blue and ultraviolet LEDs on SiC have been demonstrated [131,132]. As mentioned above, lift-off technique will increase the complexity as well as the cost of the fabrication of VLEDs. Thus the direct deposition of III-Nitrides on conducting substrates is preferable. However, the high lattice and thermal mismatch between Si and III-Nitrides, the high cost of bulk GaN and SiC substrates impede their applications in VLEDs. The opaqueness of Si, GaN, and SiC in deep UV region also make them unsuitable for deep UV-LEDs. Thus, characterized with high transparency, high n-type conductivity and little lattice misfit with III-Nitrides, vertical LEDs especially UV-LEDs on β-Ga_2O_3 are more attractive. The relatively low cost can be expected in the future since β-Ga_2O_3 single crystal substrates can be prepared by melt growth.

Various researches have been carried out for VLEDs on β-Ga_2O_3. In 2005, Kuramata first demonstrated the InGaN/GaN based vertical LEDs (VLEDs) on β-Ga_2O_3 [133]. Also in 2005, Shimamura et al. achieved the blue emission from the vertical structure LEDs on conductive (1 0 0) β-Ga_2O_3 substrate [99]. And recently, Muhammed et al. demonstrated a high-performance vertical LED in blue region grown on (−2 0 1) β-Ga_2O_3 substrate via MOCVD technique [102]. The vertical GaN/InGaN LEDs on β-Ga_2O_3 emitting in near-UV region was first presented by Ding Li et al., with an emission wavelength around 416 nm and 410 nm on (−2 0 1) and (1 0 0) β-Ga_2O_3, respectively [134]. However, due to the poor crystalline quality of AlGaN, there are nearly no reports (as far as we know) in respect to the AlGaN based VLEDs on β-Ga_2O_3 substrate currently.

The common structure of InGaN/GaN based VLEDs on β-Ga_2O_3 is illustrated in Figure 14a. The β-Ga_2O_3 substrate, doped with Sn to implement the n-type conductivity (carrier density of 10^{18} cm^{-3}), is masked with patterned SiN_x arrays. SiN_x arrays are used to improve the crystalline quality of epilayers via hindering the penetration of threading dislocations (TDs) to the active region, and to enhance the light extraction via decreasing the total internal reflections due to the approximation of refractive index between Ga_2O_3 and SiN_x. Also, SiN_x arrays can reduce the resistivity between n-GaN and substrate since the interface displays a Schottky-like nature. Then a LT-AlN or LT-GaN buffer layer is deposited on β-Ga_2O_3. The LT-AlN (GaN) buffer layer deposited under a nitrogen atmosphere can prevent β-Ga_2O_3 from decomposition as well as offer the nuclei for GaN growth. Furthermore, the buffer layer also can reduce the lattice mismatch between GaN and substrate, and the thickness of buffer should be optimized. A thick buffer layer will introduce a resistivity in the interface that will degrade the vertical conductivity, while a relatively thin buffer layer lacks the ability to protect the substrates. Subsequent high-temperature GaN layer must be n-type conducting in order to achieve vertical current injection. An InGaN/GaN superlattices (SLS) layer is deposited prior to the $In_xGa_{1-x}N$/GaN MQWs active region in order to uniformize the current distribution [134]. Due to the high transparency of β-Ga_2O_3 substrate, the backside emission from β-Ga_2O_3 is allowed. Finally, the highly reflective p-electrode that covers the whole area and the n-electrode with small area ratio are deposited on p-GaN layer and n-Ga_2O_3, respectively, which can enhance the light extraction efficiency [135]. It is worth noting that there is an inversely proportional relationship between transmittance and conductivity of β-Ga_2O_3. A highly transparent substrate suffers an increased resistivity, while a highly conducting one suffers a degraded transmittance. Also, intentional doping can degrade the crystalline quality and thermal conductivity of the crystals, which will deteriorate the crystalline quality of epilayers and the heat dissipation ability of devices, respectively. Thus, a moderate doping level with a carrier concentration of 10^{18} cm^{-3} may be preferable.

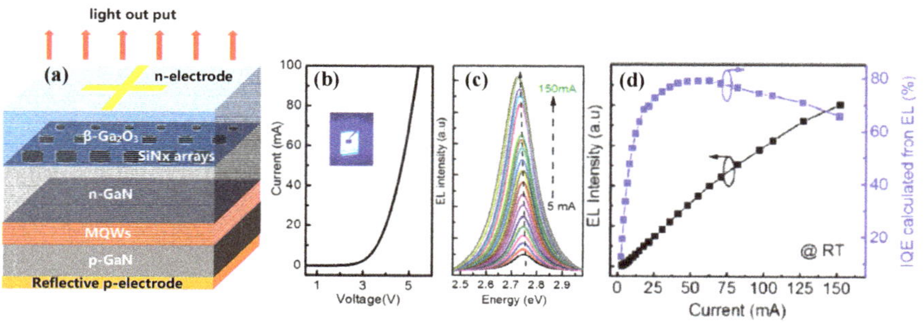

Figure 14. (**a**) Schematic structure of InGaN/GaN MQW vertical structure light emitting diodes (VLED). (**b**) I−V curve from InGaN/GaN MQWs VLED grown on β-Ga_2O_3 substrate (image of EL emission at 20 mA is shown in the inset). (**c**) EL spectra as a function of the injection current for the VLED. (**d**) EL intensity and IQE as functions of the injection current for the VLED. Reprinted with permission from reference [102]. Copyright 2017 American Chemical Society.

VLEDs presented by Muhammed et al. displayed a turn-on voltage of 2.8 V and an operating voltage of 3.7 V under an injected current of 20 mA, as shown in Figure 14b. The low turn-on voltage and operating voltage could be ascribed to the low series resistance caused by the vertical injection of the current and the high conductivity of substrates. The reverse-bias leakage current was fully suppressed below −10 V. Moreover, bright and uniform light emission was realized, indicating a uniform current distribution. An intense blue emission at approximately 452 nm with FWHM of 97 meV was obtained under a current of 20 mA. The maximum IQE determined experimentally exceeded 78%, accompanied by a relatively small efficiency droop (~17%) at 160 mA with respect to the maximum IQE, as illustrated in Figure 14d. The suppressed droop effect indicated a uniformized current distribution and a reduced self-heating effect of the devices. No down trend in the EL intensity existed in the range of measurement, indicating that radiative recombination was dominant and the auger recombination could be neglected at the high injection current. Due to the high quality of epilayers and the uniformity of injected current, VLEDs on β-Ga_2O_3 can operate at a high current without degrading the performance seriously. Thus, the saturation of VLEDs on β-Ga_2O_3 at a higher injection current can be expected, making it possible to achieve higher output power than conventional LEDs. The VLEDs presented by Kuramata [64] show a more uniform and bright emission compared to lateral LEDs, as illustrated in Figure 15a. It exhibited an operating voltage of 2.96 V under the current of 20 mA; with a chip size of 300 μm, a radiant flux of 360 mW at a current of 650 mA was achieved. With a chip size of 2 mm, a radiant flux of 4.82 W at a current of 10 A was obtained, as shown in Figure 15c. Even at a current density of 250 A/cm^2, no saturation of output power was observed, indicating the possibility of realization for high-power LEDs on β-Ga_2O_3.

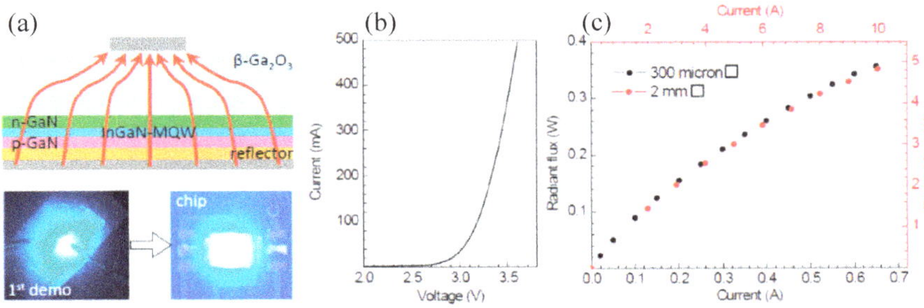

Figure 15. (**a**) Schematic of a blue-LED based on an InGaN-MQW deposited on a β-Ga_2O_3 substrate. Photograph of the initially demonstrated blue emission by vertical current injection in comparison with a current chip. (**b**) I–V characteristic of a blue-LED on β-Ga_2O_3 substrate. (**c**) Radiant flux as a function of the vertical current flow for two different chip areas, 300 μm (left-down black coordinates) and 2 mm (up-right red coordinates), respectively. The radiant fluxes were measured with an integrating sphere. Reprinted with permission from Reference [64]. Copyright 2014 SPIE.

In addition, the fabrication of VLEDs employing β-Ga_2O_3 as a sacrificial layer by CLO was also demonstrated [120]. Hsueh et al. deposited a Ga_2O_3 layer on undoped GaN/sapphire by MOCVD, followed by the epitaxy of LED structure, and transferred it to the Cu substrate via the chemical lift-off process, as shown in Figure 16a. After chemical lift-off and the fabrication of VLED, the measured forward voltage of the device under 350 mA was reduced from 3.8 V to 3.5 V due to decreased series resistance, which could be ascribed to the reduction of the lateral current path and current crowding effect, as shown in Figure 16b. As illustrated in Figure 16c, under an injection current of 350 mA, the output power was increased from 128 mW to 187 mW after the CLO process with a chip size of 45 mil × 45 mil. The 46% enhancement of output power could be ascribed to the uniform current distribution and improved extraction efficiency for vertical structure.

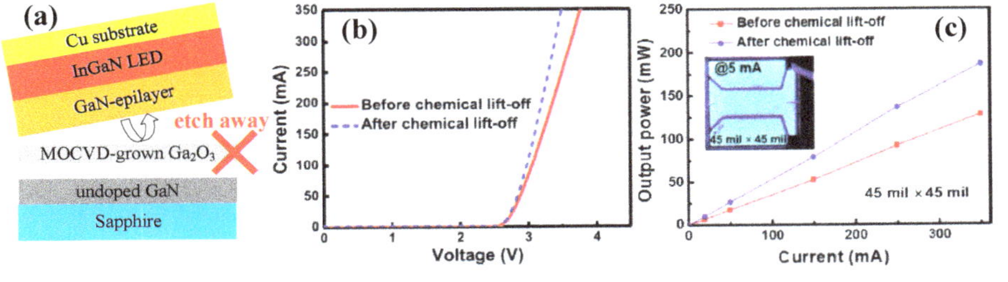

Figure 16. (**a**) Schematic diagram of the chemical lift-off (CLO) process for InGaN/GaN based vertical structure LEDs. (**b**) Current-voltage curves of the LED devices before and after performing the CLO process. (**c**) Light output powers as a function of the injection current for the LED devices before and after performing the CLO process. The inset shows the light emission image at 5 mA of the vertical-type LED (after CLO process) with an emission wavelength of 460 nm. Reprinted with permission from Reference [120]. Copyright 2015 Elsevier.

As for UV-VLEDs on gallium oxide, although the PL properties have been reported, the electrical luminescence of GaN/AlGaN multiple quantum wells is still beyond realization. It can be ascribed to the poor crystalline quality of AlGaN alloys on β-Ga_2O_3. Due to the relatively high bond energy (2.88 eV for AlN, 2.2 eV for GaN), the high adhesion coefficient will lead to a lower mobility of Al atoms compared to Ga atoms. Unlike the layer-by-layer 2D growth mechanism of GaN, AlGaN tends to grow in a 3D islandic mode. Thus, the incorporation of Al atoms tends to occur at the initial surface positions rather than the energetically favorable sites such as steps and kinks which are easier for nucleation, leading to the formation of extended defects such as threading dislocations and grain boundaries. These defects may propagate into MQWs and can act as nonradiative centers that deteriorate the internal quantum efficiency. In addition, the high chemical activity of Al atoms will result in a pre-reaction of Al sources and N sources, which further degrade the crystalline quality of AlGaN epilayers. Therefore, it is more difficult to achieve high quality AlGaN layers on β-Ga_2O_3 compared to GaN. The current crystalline quality of AlGaN is not enough for the device fabrication. Despite hardly any demonstrations, it is still a promising approach to realize high-brightness and high-power for AlGaN based UV-VLEDs on β-Ga_2O_3 substrates due to its uniform current distribution, low series resistance, and simplified fabrication process. Compared with silicon, GaN, and SiC, the high transmittance of β-Ga_2O_3 in UVA and UVB spectral regions make it more attractive as a substrate of vertical UV-LEDs with an emission wavelength longer than the absorption edge (260 nm) of β-Ga_2O_3. In view of the mature commercialization of blue LEDs on sapphire and the dilemma of high-power UV-LEDs on sapphire, the realization of vertical UV-LEDs especially in UVB region on β-Ga_2O_3 is of great significance.

7. Conclusions

Due to its unique properties, β-Ga_2O_3 is a promising conductive substrate for high-performance vertical structure blue, especially UV LEDs. The epitaxial relationships between wurtzite III-Nitrides and monoclinic β-Ga_2O_3 were explored. Efforts to improve the crystalline quality of GaN and AlGaN epilayers have been made by trying different atmospheres, different orientations of substrates, and by utilizing various growth methods. High-quality epilayer of GaN and high-performance of blue VLEDs have been demonstrated. In the future, researches will most likely concentrate on the preparation of large-size wafers with decreased cost, the optimization of growth method, the improvement in the performance of devices, and especially the epitaxy of AlGaN materials since the advantages of β-Ga_2O_3 as substrates in UV-VLED are more significant. In addition, the thermal management of VLEDs on β-Ga_2O_3 should be carefully addressed since the thermal conductivity of gallium oxide is quite poor.

Author Contributions: T.W. gave the concepts of this paper and guided the paper writing. R.M. and W.L. took part in the investigations. X.Z. provided the experimental data. W.L. wrote the paper. J.Y., J.W. and J.L. gave valuable suggestions on the manuscript.

Funding: This work was supported by the National Key R&D Program of China (No. 2018YFB0406702), the National Natural Science Foundation of China (Nos. 61474109 and 61527814), and Beijing Natural Science Foundation (No. 4182063).

Conflicts of Interest: The authors declare no conflicts of interest.

References

1. Kneissl, M.; Kolbe, T.; Chua, C.; Kueller, V.; Weyers, M. Advances in group III-nitride-based deep UV light-emitting diode technology. *J. Semicond.* **2010**, *26*, 014036. [CrossRef]
2. Neufeld, C.J.; Toledo, N.G.; Cruz, S.C.; Iza, M.; Denbaars, S.P.; Mishra, U.K. High quantum efficiency InGaN/GaN solar cells with 2.95 eV band gap. *Appl. Phys. Lett.* **2008**, *93*, 143502–143503. [CrossRef]
3. Mishra, U.K.; Likun, S.; Kazior, T.E.; Wu, Y.F. GaN-Based RF Power Devices and Amplifiers. *Proc. IEEE* **2008**, *96*, 287–305. [CrossRef]
4. Jain, S.C.; Willander, M.; Narayan, J.; Overstraeten, R.V. III−nitrides: Growth, Characterization, and Properties. *J. Appl. Phys.* **2000**, *87*, 965–1006. [CrossRef]
5. Nakamura, S.; Pearton, S.; Fasol, G. *The Blue Laser Diode: The Complete Story*; Springer: Berlin, Germany, 2013; pp. 1–28.
6. Reverchon, J.L.; Mosca, M.; Grandjean, N.; Omnes, F.; Duboz, J.Y. High performance solar blind detectors based on AlGaN grown by MBE and MOCVD. *Mrs. Proc.* **2003**, *798*, Y1.10.
7. Mukai, T. P-GaN/n-InGaN/n-GaN double-heterostructure blue-light-emitting diodes. *J. Cryst. Growth.* **1994**, *145*, 911–917.
8. Narukawa, Y.; Ichikawa, M.; Sanga, D.; Sano, M.; Mukai, T. White light emitting diodes with super-high luminous efficacy. *J. Phys. D. Appl. Phys.* **2010**, *43*, 354002. [CrossRef]

9. Hockberger, P.E. A history of ultraviolet photobiology for humans, animals and microorganisms. *Photochem. Photobiol.* **2002**, *76*, 561–579. [CrossRef]
10. Schreiner, M.; Martinez-Abaigar, J.; Glaab, J.; Jansen, M. UVB induced secondary plant metabolites. *Optik Photonik.* **2014**, *9*, 34–37. [CrossRef]
11. Vilhunen, S.; Särkkä, H.; Sillanpää, M. Ultraviolet light-emitting diodes in water disinfection. *Environ. Sci. Pollut. Res.* **2009**, *16*, 439–442. [CrossRef]
12. Banas, M.A.; Crawford, M.H.; Ruby, D.S.; Ross, M.P.; Nelson, J.S.; Allerman, A.A. Final LDRD Report: Ultraviolet Water Purification Systems for Rural Environments and Mobile Applications. 2005. Available online: https://prod-ng.sandia.gov/techlib-noauth/access-control.cgi/2005/057245.pdf (accessed on 7 May 2019).
13. Song, K.; Mohseni, M.; Taghipour, F. Application of GaN-based deep ultraviolet light emitting diodes (UV LEDs) for Water disinfection. *Water Res.* **2016**, *94*, 341–349. [CrossRef] [PubMed]
14. Kowalski, W. *Ultraviolet Germicidal Irradiation Handbook*; Springer: Berlin/Heidelberg, Germany, 2009.
15. Lui, G.Y.; Roser, D.; Corkish, R.; Ashbolt, N.; Jagals, P.; Stuetz, R. Photovoltaic powered ultraviolet and visible light-emitting diodes for sustainable point-of-use disinfection of drinking waters. *Sci. Total. Environ.* **2014**, *493*, 185. [CrossRef] [PubMed]
16. Mellqvist, J.A.; Rosen, A. DOAS for flue gas monitoring-temperature effects in the UV/visible absorption spectra of NO, NO_2, SO_2, and NH_3. *J. Quant. Spectrosc. Radiat. Transf.* **1996**, *56*, 187–208. [CrossRef]
17. UV-LED Market to Grow From $90 m to $520 m in 2019. Available online: http://www.semiconductor-today.com/features/PDF/semiconductor-today_february2015-UV-LED.pdf (accessed on 7 May 2019).
18. Taniyasu, Y.; Kasu, M.; Makimoto, T. An aluminium nitride light-emitting diode with a wavelength of 210 nanometres. *Nature* **2006**, *441*, 325–328. [CrossRef]
19. Hirayama, H.; Maeda, N.; Fujikawa, S. Recent progress and future prospects of AlGaN-based high-efficiency deep-ultraviolet light-emitting diodes. *Jpn. J. Appl. Phys.* **2014**, *53*, 100209. [CrossRef]
20. Chang, H.L.; Chen, Z.L.; Li, W.J.; Yan, J.C.; Hou, R.; Yang, S.Y.; Liu, Z.Q.; Yuan, G.D.; Wang, J.X.; Li, J.M.; Gao, P.; Wei, T.B. Graphene-assisted quasi-van der Waals epitaxy of AlN film for ultraviolet light emitting diodes on nano-patterned sapphire substrate. *Appl. Phys. Lett.* **2019**, *114*, 091107. [CrossRef]
21. Crawford, M.H. LEDs for solid-state lighting: Performance challenges and recent advances. *IEEE J. Sel. Top. Quant.* **2009**, *15*, 1028–1040. [CrossRef]
22. Laubsch, A.; Sabathil, M.; Baur, J.; Peter, M.; Hahn, B. High-power and high-efficiency InGaN-based light emitters. *IEEE T. Electron. Dev.* **2010**, *57*, 79–87. [CrossRef]
23. Keppens, A.; Ryckaert, W.R.; Deconinck, G.; Hanselaer, P. High power light-emitting diode junction temperature determination from current-voltage characteristics. *J. Appl. Phys.* **2008**, *104*, 093104–093108. [CrossRef]
24. Shen, Y.C.; Mueller, G.O.; Watanabe, S.; Gardner, N.F.; Krames, M.R. Auger recombination in InGaN measured by photoluminescence. *Appl. Phys. Lett.* **2007**, *91*, 141101–141103. [CrossRef]
25. Wei, T.B.; Ji, X.L.; Wu, K.; Zheng, H.Y.; Du, C.X.; Chen, Y.; Yan, Q.F.; Zhao, L.X.; Zhou, Z.C.; Wang, J.X.; Li, J.M. Efficiency improvement and droop behavior in nanospherical-lens lithographically patterned bottom and top photonic crystal InGaN/GaN light-emitting diodes. *Opt. Lett.* **2014**, *39*, 379–382. [CrossRef]
26. Vampola, K.J.; Iza, M.; Keller, S.; Denbaars, S.P.; Nakamura, S. Measurement of electron overflow in 450 nm InGaN light-emitting diode structures. *Appl. Phys. Lett.* **2009**, *94*, 1081. [CrossRef]
27. Monemar, B.; Sernelius, B.E. Defect related issues in the "current roll-off" in InGaN based light emitting diodes. *Appl. Phys. Lett.* **2007**, *91*, 181103. [CrossRef]
28. Ni, X.; Fan, Q.; Shimada, R.; Oezguer, U.; Morkoc, H. Reduction of efficiency droop in InGaN light emitting diodes by coupled quantum wells. *Appl. Phys. Lett.* **2008**, *93*, 171113. [CrossRef]
29. Kudryk, Y.Y.; Zinovchuk, A.V. Efficiency droop in InGaN/GaN multiple quantum well light-emitting diodes with nonuniform current spreading. *Semicond. Sci. Technol.* **2011**, *26*, 095007. [CrossRef]
30. Sugahara, T.; Sato, H.; Hao, M.; Naoi, Y.; Kurai, S.; Tottori, S. Direct Evidence that Dislocations are Non-Radiative, Recombination Centers in GaN. *Jpn. J. Appl. Phys.* **1998**, *37*, L398–L400. [CrossRef]
31. Speck, J.S.; Rosner, S.J. The role of threading dislocations in the physical properties of GaN and its alloys. *Mater. Sci. Forum.* **1999**, *353–356*, 769–778. [CrossRef]
32. Wan, Z.; He, Y.; Choi, C.; Suh, E.; Yu, Y.M.; Yi, S.N. Microstructural analysis of InGaN/GaN epitaxial layers of metal organic chemical vapor deposition on c-plane of convex patterned sapphire substrate. *Thin Solid Films* **2013**, *546*, 104–107. [CrossRef]

33. Song, J.C.; Kang, D.H.; Shim, B.Y.; Ko, E.A.; Kim, D.W.; Santhakumar, K. Characteristics comparison between GaN epilayers grown on patterned and unpatterned sapphire substrate (0 0 0 1). *Adv. Mater. Res.* **2007**, *29–30*, 355–358. [CrossRef]
34. Bennett, S.E. Dislocations and their reduction in GaN. *Metal Sci. J.* **2013**, *26*, 1017–1028. [CrossRef]
35. Mendes, M.; Fu, J.; Porneala, C. Lasers in the manufacturing of LEDs. In Proceedings of the Laser Applications in Microelectronic and Optoelectronic Manufacturing XV, San Francisco, CA, USA, 17 February 2010.
36. Doan, T.; Ferguson, I.T.; Narendran, N.; Tran, C.; Chu, C. Vertical GaN based light emitting diodes on metal alloy substrate boosts high power LED performance. *Proc. SPIE* **2007**, *6669*, 666903.
37. Cao, X.A.; Arthur, S.D. High-power and reliable operation of vertical light-emitting diodes on bulk GaN. *Appl. Phys. Lett.* **2004**, *85*, 3971–3973. [CrossRef]
38. Doan, T.; Streubel, K.P.; Yao, H.W.; Chu, C.; Chen, C.; Schubert, E.F. Vertical GaN based light emitting diodes on metal alloy substrate for solid state lighting application. *Proc. SPIE* **2006**, *6134*, 61340G.
39. Li, C.K.; Wu, Y.R. Study on the current spreading effect and light extraction enhancement of vertical GaN/InGaN LEDs. *IEEE T. Electron Dev.* **2012**, *59*, 400–407. [CrossRef]
40. Wong, W.S.; Sands, T.; Cheung, N.W.; Kneissl, M.; Bour, D.P.; Mei, P. Fabrication of thin-film InGaN light-emitting diode membranes by laser lift-off. *Appl. Phys. Lett.* **1999**, *75*, 1360–1362. [CrossRef]
41. Stach, E.A.; Kelsch, M.; Nelson, E.C.; Wong, W.S.; Sands, T.; Cheung, N.W. Structural and chemical characterization of free-standing GaN films separated from sapphire substrates by laser lift-off. *Appl. Phys. Lett.* **2000**, *77*, 1819–1821. [CrossRef]
42. Wu, Y.S.; Cheng, J.H.; Peng, W.C.; Ouyang, H. Effects of laser sources on the reverse-bias leakages of laser lift-off GaN-based light-emitting diodes. *Appl. Phys. Lett.* **2007**, *90*, 251110. [CrossRef]
43. Fujii, K.; Lee, S.; Ha, J.S.; Lee, H.J.; Lee, H.J.; Lee, S.H. Leakage current improvement of nitride-based light emitting diodes using CrN buffer layer and its vertical type application by chemical lift-off process. *Appl. Phys. Lett.* **2009**, *94*, 722–747.
44. Horng, R.H.; Pan, C.T.; Tsai, T.Y.; Wuu, D.S. Transferring Thin Film GaN LED Epi-Structure to the Cu Substrate by Chemical Lift-Off Technology. *Electrochem. Solid-State Lett.* **2011**, *14*, H281. [CrossRef]
45. Víllora, E.G.; Shimamura, K.; Kitamura, K.; Aoki, K.; Ujiie, T. Epitaxial relationship between wurtzite GaN and β-Ga_2O_3. *Appl. Phys. Lett.* **2007**, *90*, 234102. [CrossRef]
46. Wei, T.B.; Yang, J.K.; Wei, Y.; Huo, Z.Q.; Ji, X.L.; Zhang, Y. Cross-stacked carbon nanotubes assisted self-separation of free-standing GaN substrates by hydride vapor phase epitaxy. *Sci. Rep.* **2016**, *6*, 28620. [CrossRef] [PubMed]
47. Rogers, D.J.; Hosseini Teherani, F.; Ougazzaden, A.; Gautier, S.; Divay, L.; Lusson, A. Use of ZnO thin films as sacrificial templates for metal organic vapor phase epitaxy and chemical lift-off of GaN. *Appl. Phys. Lett.* **2007**, *91*, 71120. [CrossRef]
48. Liu, H.F.; Liu, W.; Chua, S.J. Epitaxial growth and chemical lift-off of GaInN/GaN heterostructures on c- and r-sapphire substrates employing ZnO sacrificial templates. *J. Vac. Sci. Technol. A* **2010**, *28*, 590–594. [CrossRef]
49. Chen, W.C.; Tang, H.-L.; Luo, P.; Ma, W.-W.; Xu, X.-D.; Qian, X.-B.; Jiang, D.-P.; Wu, F.; Wang, J.-Y.; Xu, J. Research progress of substrate materials used for GaN-Based light emitting diodes. *Acta Phys. Sin.* **2014**, *63*, 68103.
50. Wang, J.; Yan, J.; Guo, Y.; Zhang, Y.; Li, J. Recent progress of research on III-nitride deep ultraviolet light-emitting diode. *Scientia Sinica.* **2015**, *45*, 067303. [CrossRef]
51. Ponce, F.A.; Krusor, B.S.; Jr, J.S.M.; Plano, W.E.; Welch, D.F. Microstructure of GaN epitaxy on SiC using AlN buffer layers. *Appl. Phys. Lett.* **1995**, *67*, 410–412. [CrossRef]
52. Higashiwaki, M.; Sasaki, K.; Kuramata, A. Gallium oxide (Ga_2O_3) metal-semiconductor field-effect transistors on single-crystal β-Ga_2O_3 (010) substrates. *Appl. Phys. Lett.* **2012**, *100*, 013504. [CrossRef]
53. Lu, X.; Zhou, L.; Chen, L.; Ouyang, X.; Liu, B.; Xu, J. Schottky x-ray detectors based on a bulk β-Ga_2O_3 substrate. *Appl. Phys. Lett.* **2018**, *112*, 103502. [CrossRef]
54. Feng, P.; Zhang, J.Y.; Li, Q.H.; Wang, T.H. Individual β-Ga_2O_3 nanowires as solar-blind photodetectors. *Appl. Phys. Lett.* **2006**, *88*, 153107. [CrossRef]
55. Verma, A.; Jena, D. Vertical Schottky barrier diodes fabricated on un-intentionally doped and Sn-doped (−201) bulk β-Ga_2O_3 substrates. In Proceedings of the Compound Semiconductor Week, Toyama, Japan, 26–30 June 2016.

56. Wu, Z.; Bai, G.; Qu, Y.; Guo, D.; Tang, W. Deep ultraviolet photoconductive and near-infrared luminescence properties of Er^{3+}-doped β-Ga_2O_3 thin films. *Appl. Phys. Lett.* **2016**, *108*, 211903. [CrossRef]
57. Chen, Z.W.; Wang, X.; Noda, S.; Saito, K. Effects of dopant contents on structural, morphological and optical properties of Er doped Ga_2O_3 films. *Superlattice. Microst.* **2016**, *90*, 207–214. [CrossRef]
58. Chen, Z.; Saito, K.; Tanaka, T.; Guo, Q. Efficient pure green emission from Er-doped Ga_2O_3 films. *CrystEngComm.* **2017**, *19*, 4448. [CrossRef]
59. Iguchi, S.; Hasegawa, Y.; Teramura, K.; Kidera, S.; Kikkawa, S.; Hosokawa, S. Drastic improvement in the photocatalytic activity of Ga_2O_3 modified with Mg–Al layered double hydroxide for the conversion of CO_2 in water. *Sustain. Energ. Fuels.* **2017**, *1*, 1740. [CrossRef]
60. Chen, Z.; Saito, K.; Tanaka, T.; Guo, Q. Effects of Zn^{2+} and Pb^{2+} dopants on the activity of Ga_2O_3-based photocatalysts for water splitting. *Phys. Chem. Chem. Phys.* **2013**, *15*, 19380.
61. Zhang, W.; Naidu, B.S.; Ou, J.Z.; O'Mullane, A.P.; Chrimes, A.F.; Carey, B.J.; Wang, Y.; Tang, S.Y. Liquid metal/metal oxide frameworks with incorporated Ga_2O_3 for photocatalysis. *ACS Appl. Mater. Inter.* **2015**, *7*, 1943–1948. [CrossRef]
62. Pearton, S.J.; Yang, J.; Cary, P.H.; Ren, F.; Kim, J.; Tadjer, M.J.; Mastro, M.A. A review of Ga_2O_3 materials, processing, and devices. *Appl. Phys. Rev.* **2018**, *5*, 011301. [CrossRef]
63. Roy, R.; Hill, V.G.; Osborn, E.F. Polymorphism of Ga_2O_3, and the system Ga_2O_3-H_2O. *J. Am. Chem. Soc.* **1952**, *74*, 719–722. [CrossRef]
64. Encarnación, G.V.; Stelian, A.; Kiyoshi, S.; Daisuke, I.; Kazuo, A. β-Ga_2O_3 and single-crystal phosphors for high-brightness white LEDs and LDs, and β-Ga_2O_3 potential for next generation of power devices. In Proceedings of the SPIE 8987, Oxide-based Materials and Devices V, San Francisco, CA, USA, 8 March 2014.
65. Playford, H.Y.; Hannon, A.C.; Barney, E.R.; Walton, R.I. Structures of uncharacterised polymorphs of gallium oxide from total neutron diffraction. *Chem. Eur. J.* **2013**, *19*, 2803–2813. [CrossRef]
66. Geller, S. Crystal Structure of β-Ga_2O_3. *J. Chem. Phys.* **1960**, *33*, 676–684. [CrossRef]
67. Tippins, H.H. Optical Absorption and Photoconductivity in the Band Edge of β-Ga_2O_3. *Phys. Rev. D* **1965**, *140*, 316–319. [CrossRef]
68. Stepanov, S.I.; Nikolaev, V.I.; Bougrov, V.E.; Romanov, A.E. Gallium oxide: Properties and applications—A review. *Rev. Adv. Mater. Sci.* **2016**, *44*, 63–86.
69. Ueda, N.; Hosono, H.; Waseda, R.; Kawazoe, H. Synthesis and control of conductivity of ultraviolet transmitting β-Ga_2O_3 single crystals. *Appl. Phys. Lett.* **1997**, *70*, 3561–3563. [CrossRef]
70. Chase, A.O. Growth of β-Ga_2O_3 by the verneuil technique. *J. Am. Ceram. Soc.* **2006**, *47*, 470. [CrossRef]
71. Lorenz, M.R.; Woods, J.F.; Gambino, R.J. Some electrical properties of the semiconductor β-Ga_2O_3. *J. Phys. Chem. Solids* **1967**, *28*, 403. [CrossRef]
72. Víllora, E.G.; Shimamura, K.; Yoshikawa, Y. Large-size β-Ga_2O_3 single crystals and wafers. *J. Cryst. Growth* **2004**, *270*, 420–426. [CrossRef]
73. Zhang, J.; Li, B.; Xia, C.; Pei, G.; Deng, Q.; Yang, Z. Growth and spectral characterization of β-Ga_2O_3 single crystals. *J. Phys. Chem. Solids* **2006**, *67*, 2448–2451. [CrossRef]
74. Hoshikawa, K.; Ohba, E.; Kobayashi, T.; Yanagisawa, J.; Miyagawa, C.; Nakamura, Y. Growth of β-Ga_2O_3 single crystals using vertical Bridgman method in ambient air. *J. Cryst. Growth* **2016**, *447*, 36–41. [CrossRef]
75. Ohba, E.; Kobayashi, T.; Kado, M. Defect characterization of β-Ga_2O_3 single crystals grown by vertical Bridgman method. *Jpn. J. Appl. Phys.* **2016**, *55*, 1202BF. [CrossRef]
76. Tomm, Y.; Reiche, P.; Klimm, D.; Fukuda, T. Czochralski grown Ga_2O_3 crystals. *J. Cryst. Growth* **2000**, *220*, 510–514. [CrossRef]
77. Galazka, Z.; Irmscher, K.; Uecker, R.; Bertram, R.; Pietsch, M.; Kwasniewski, A. On the bulk β-Ga_2O_3 single crystals grown by the Czochralski method. *J. Cryst. Growth* **2014**, *404*, 184–191. [CrossRef]
78. Kuramata, A.; Koshi, K.; Watanabe, S.; Yamaoka, Y.; Masui, T.; Yamakoshi, S. High-quality β-Ga_2O_3 single crystals grown by edge-defined film-fed growth. *Jpn. J. Appl. Phys.* **2016**, *55*, 1202A. [CrossRef]
79. Kuramata, A.; Koshi, K.; Watanabe, S.; Yamaoka, Y.; Yamakoshi, S. Bulk crystal growth of Ga_2O_3. *Proc. SPIE Oxid. Mater. Devices IX* **2018**, *10533*, 105330E.
80. Zhang, S.N.; Lian, X.Z.; Ma, Y.C. Growth and characterization of 2-inch high quality β-Ga_2O_3 single crystals grown by EFG method. *J. Semicond.* **2018**, *39*, 083003. [CrossRef]
81. Harwig, T.; Wubs, G.J.; Dirksen, G.J. Electrical properties of β-Ga_2O_3 single crystals. *Solid State Commun.* **1976**, *18*, 1223–1225. [CrossRef]

82. Tomm, Y.; Ko, J.M.; Yoshikawa, A.; Fukuda, T. Floating zone growth of beta-Ga_2O_3: a new window material for optoelectronic device applications. *Sol. Energy Mater. Sol. Cells* **2001**, *66*, 369–374. [CrossRef]
83. Galazka, Z.; Ganschow, S.; Fiedler, A.; Bertram, R.; Klimm, D.; Irmscher, K.; Schewski, R.; Pietsch, M.; Albrecht, M.; Bickermann, M. Doping of Czochralski-grown bulk β-Ga_2O_3, single crystals with Cr, Ce and Al. *J. Cryst. Growth* **2018**, *486*, 82–90. [CrossRef]
84. Aida, H.; Nishiguchi, K.; Takeda, H. Growth of β-Ga_2O_3 single crystals by the Edge-Defined Film Fed growth method. *Jpn. J. Appl. Phys.* **2008**, *47*, 8506. [CrossRef]
85. Mu, W.; Jia, Z.; Yin, Y.; Hu, Q.; Li, Y.; Wu, B.; Tao, X. High quality crystal growth and anisotropic physical characterization of β-Ga_2O_3 single crystals grown by EFG method. *J. Alloy. Compd.* **2017**, *714*, 453–458. [CrossRef]
86. Varley, J.B.; Weber, J.R.; Janotti, A.; Van de Walle, C.G. Oxygen vacancies and donor impurities in β-Ga_2O_3. *Appl. Phys. Lett.* **2010**, *97*, 3561. [CrossRef]
87. Oshima, T.; Kaminaga, K.; Mukai, A.; Sasaki, K.; Masui, T.; Kuramata, A. Formation of semi-insulating layers on semiconducting β-Ga_2O_3 single crystals by thermal oxidation. *Jpn. J. Appl. Phys.* **2013**, *52*, 051101. [CrossRef]
88. Ahmadi, E.; Koksaldi, O.S.; Kaun, S.W.; Oshima, Y.; Short, D.B.; Mishra, U.K.; Speck, J.S. Ge doping of β-Ga_2O_3 films grown by plasma-assisted molecular beam epitaxy. *Appl. Phys. Express.* **2017**, *10*, 041102. [CrossRef]
89. Baldini, M.; Albrecht, M.; Fiedler, A.; Irmscher, K.; Schewski, R.; Wagner, G. Editors' choice—Si- and Sn-doped homoepitaxial β-Ga_2O_3 layers grown by movpe on (010)-oriented substrates. *ECS J. Solid. State. Sci. Technol.* **2017**, *6*, Q3040–Q3044. [CrossRef]
90. Suzuki, N.; Ohira, S.; Tanaka, M.; Sugawara, T.; Shishido, T. Fabrication and characterization of transparent conductive Sn-doped β-Ga_2O_3 single crystal. *Phys. Stat. Sol. C* **2007**, *4*, 2310–2313. [CrossRef]
91. Víllora, E.G.; Shimamura, K.; Yoshikawa, Y.; Ujiie, T.; Aoki, K. Electrical conductivity and carrier concentration control in β-Ga_2O_3 by Si doping. *Appl. Phys. Lett.* **2008**, *92*, A316. [CrossRef]
92. Sasaki, K.; Higashiwaki, M.; Kuramata, A.; Masui, T.; Yamakoshi, S. Si-ion implantation doping in β-Ga_2O_3 and its application to fabrication of low-resistance Ohmic contacts. *Appl. Phys. Express* **2013**, *6*, 086502. [CrossRef]
93. Zhou, W.; Xia, C.; Sai, Q. Controlling n-type conductivity of β-Ga_2O_3 by Nb doping. *Appl. Phys. Lett.* **2017**, *111*, 242103. [CrossRef]
94. Mastro, M.A.; Kuramata, A.; Calkins, J.; Kim, J.; Ren, F.; Pearton, S.J. Perspective—Opportunities and Future Directions for Ga_2O_3. *ECS J. Solid State Sci. Technol.* **2017**, *6*, 356–359. [CrossRef]
95. Onuma, T.; Fujioka, S.; Yamaguchi, T.; Higashiwaki, M.; Sasaki, K.; Masui, T. Correlation between blue luminescence intensity and resistivity in β-Ga_2O_3 single crystals. *Appl. Phys. Lett.* **2013**, *103*, 041910. [CrossRef]
96. Chang, P.C.; Fan, Z.; Tseng, W.Y.; Rajagopal, A.; Lu, J.G. β-Ga_2O_3 nanowires: synthesis, characterization, and p-channel field-effect transistor. *Appl. Phys. Lett.* **2005**, *87*, 222102. [CrossRef]
97. Liu, L.L.; Li, M.K.; Yu, D.Q.; Zhang, J.; Zhang, H.; Qian, C. Fabrication and characteristics of N-doped β-Ga_2O_3 nanowires. *Appl. Phys. A* **2010**, *98*, 831–835. [CrossRef]
98. Víllora, E.G.; Shimamura, K.; Aoki, K. Molecular beam epitaxy of c -plane wurtzite GaN on nitridized a –plane β-Ga_2O_3. *Thin Solid Films* **2006**, *500*, 209–213. [CrossRef]
99. Shimamura, K.; Víllora, E.G.; Domen, K.; Yui, K.; Aoki, K.; Ichinose, N. Epitaxial Growth of GaN on (1 0 0) β-Ga_2O_3 Substrates by Metalorganic Vapor Phase Epitaxy. *Jpn. J Appl. Phys.* **2005**, *44*, L7–L8. [CrossRef]
100. Muhammed, M.M.; Peres, M.; Yamashita, Y.; Morishima, Y.; Sato, S.; Franco, N. High optical and structural quality of GaN epilayers grown on (−2 0 1) β-Ga_2O_3. *Appl. Phys. Lett.* **2014**, *105*, 457–459. [CrossRef]
101. Roqan, I.S. (−2 0 1) β-Gallium oxide substrate for high quality GaN materials. *Proc SPIE* **2015**, *9364*, 93641K.
102. Muhammed, M.M.; Alwadai, N.; Lopatin, S. High-Efficiency InGaN/GaN Quantum Well-Based Vertical Light-Emitting Diodes Fabricated on β-Ga_2O_3 Substrate. *ACS Appl. Mater. Inter.* **2017**, *9*, 34057–34063. [CrossRef]
103. Muhammed, M.M.; Peres, M.; Yamashita, Y.; Morishima, Y.; Sato, S.; Franco, N. High-quality III-nitride films on conductive, transparent (−2 0 1)-oriented β-Ga_2O_3 using a GaN buffer layer. *Sci. Rep.* **2016**, *6*, 29747. [CrossRef]
104. Ohira, S.; Suzuki, N.; Minami, H.; Takahashi, K.; Araki, T.; Nanishi, Y. Growth of hexagonal GaN films on the nitridated β-Ga_2O_3 substrates using RF-MBE. *Phys. Stat. Sol.* **2011**, *4*, 2306–2309.
105. Kachel, K.; Korytov, M.; Gogova, D. A new approach to free-standing GaN using β-Ga_2O_3 as a substrate. *Cryst. Eng. Comm.* **2012**, *14*, 8536–8540. [CrossRef]

106. Tsai, T.Y.; Horng, R.H.; Wuu, D.S.; Ou, S.L.; Hung, M.T.; Hsueh, H.H. GaN Epilayer Grown on Ga_2O_3 Sacrificial Layer for Chemical Lift-Off Application. *Electrochem. Solid. ST.* **2011**, *14*, H434–H437. [CrossRef]
107. Tsai, T.Y.; Ou, S.L.; Hung, M.T.; Wuu, D.S.; Horng, R.H. MOCVD growth of GaN on sapphire using a Ga_2O_3 interlayer. *J. Electrochem. Soc.* **2011**, *158*, H1172–H1178. [CrossRef]
108. Li, D.; Hoffmann, V.; Richter, E.; Tessaro, T.; Galazka, Z.; Weyers, M.; Tränkle, G. MOVPE growth of violet GaN LEDs on β-Ga_2O_3 substrates. *J. Cryst. Growth* **2017**, *478*, 212–215. [CrossRef]
109. Cho, Y.S.; Hardtdegen, H.; Kaluz, N.; Veit, P.; Heidelberger, G.; Thillosen, N.; Steins, R.; Lüth, H.; Christen, J. Systematic study on the growth mechanism of GaN/sapphire epilayers grown with different H_2 and N_2 carrier gas. In Proceedings of the 13th International Conference on Metal Organic Vapor Phase Epitaxy, Miyazaki, Japan, 22–26 May 2006.
110. Bottcher, T.; Einfeldt, S.; Figge, S.; Chierchia, R.; Heinke, H.; Hommel, D. The role of high-temperature island coalescence in the development of stresses in GaN films. *Appl. Phys. Lett.* **2001**, *78*, 1976–1978. [CrossRef]
111. Ito, S.; Takeda, K.; Nagata, K.; Aoshima, H.; Takehara, K.; Iwaya, M. Growth of GaN and AlGaN on (100) β-Ga_2O_3 substrates. *Phys. Stat. Sol.* **2012**, *9*, 519–522. [CrossRef]
112. Liu, P.L.; Siao, Y.J.; Wu, Y.T.; Wang, C.H.; Chen, C.S. Structural, electronic and energetic properties of GaN[0 0 0 1]/Ga_2O_3[1 0 0] heterojunctions: A first-principles density functional theory study. *Scr. Mater.* **2011**, *65*, 465–468. [CrossRef]
113. Zhang, X.; Wei, T.B.; Ren, K.K.; Xiong, Z.; Li, W.J.; Yang, C.; Zhang, L.; Wang, J.X. High Quality GaN Epitaxial Growth on β-Ga_2O_3 Substrate enabled by Self-assembled SiO_2 Nanospheres. *J. Cryst. Growth.* (Submitted).
114. Park, J.; Song, K.M.; Jeon, S.R.; Baek, J.H.; Ryu, S.W. Doping selective lateral electrochemical etching of GaN for chemical lift-off. *Appl. Phys. Lett.* **2009**, *94*, 221907. [CrossRef]
115. Lin, C.F.; Dai, J.J.; Wang, G.M.; Lin, M.S. Chemical lift-off process for blue light-emitting diodes. *Appl. Phys. Express* **2010**, *3*, 2101. [CrossRef]
116. Lin, C.F.; Dai, J.J.; Lin, M.S.; Chen, K.T.; Huang, W.C.; Lin, C.M. An AlN sacrificial buffer layer inserted into the GaN/patterned sapphire substrate for a chemical lift-off process. *Appl. Phys. Express* **2014**, *3*, 031001. [CrossRef]
117. Horng, R.H.; Hsueh, H.H.; Ou, S.L.; Tsai, C.T.; Tsai, T.Y.; Wuu, D.S. Chemical lift-off process for nitride LEDs from an Eco-GaN template using an AlN/strip-patterned-SiO_2 sacrificial layer. *Phys. Stat. Sol.* **2017**, *214*, 1600657. [CrossRef]
118. Horng, R.H.; Tsai, T.Y.; Yen, C.Y.; Hung, M.T.; Pan, C.T.; Wuu, D.S. Novel approaches to realizing chemical lift-off of GaN epilayer from sapphire substrate. *Proc. SPIE* **2011**, *7954*, 795417.
119. Horng, R.H.; Wuu, D.S.; Hsueh, H.H. Study on GaN epilayer transferring to Cu substrate from sapphire substrate using Ga_2O_3 sacrificial layer. *Proc. SPIE Int. Soc. Opt. Eng.* **2012**, *8278*, 14.
120. Hsueh, H.H.; Ou, S.L.; Wuu, D.S.; Horng, R.H. InGaN LED fabricated on Eco-GaN template with a Ga_2O_3 sacrificial layer for chemical lift-off application. *Vacuum* **2015**, *118*, 8–12. [CrossRef]
121. Ajia, I.A.; Yamashita, Y.; Lorenz, K.; Muhammed, M.M.; Spasevski, L.; Almalawi, D. GaN/AlGaN multiple quantum wells grown on transparent and conductive (−201)-oriented β-Ga_2O_3 substrate for UV vertical light emitting devices. *Appl. Phys. Lett.* **2018**, *113*, 082102. [CrossRef]
122. Smalc-Koziorowska, J.; Grzanka, E.; Czernecki, R.; Schiavon, D.; Mike, L. Elimination of trench defects and v-pits from InGaN/GaN structures. *Appl. Phys. Lett.* **2015**, *106*, 101905. [CrossRef]
123. Massabuau, C.P.; Tartan, C.C.; Traynier, R.; Blenkhorn, W.E.; Kappers, M.J.; Dawson, P. The impact of substrate miscut on the microstructure and photoluminescence efficiency of (0001) InGaN quantum wells grown by a two-temperature method. *J. Cryst. Growth* **2014**, *386*, 88–93. [CrossRef]
124. Wang, S.J.; Uang, K.M.; Chen, S.L.; Yang, Y.C.; Chang, S.C.; Chen, T.M. Use of patterned laser liftoff process and electroplating nickel layer for the fabrication of vertical-structured GaN-based light-emitting diodes. *Appl. Phys. Lett.* **2005**, *87*, 011111. [CrossRef]
125. Kim, D.W.; Lee, H.Y.; Yoo, M.C.; Yeom, G.Y. Highly efficient vertical laser-liftoff GaN-based light-emitting diodes formed by optimization of the cathode structure. *Appl. Phys. Lett.* **2005**, *86*, 390. [CrossRef]
126. Lin, W.Y.; Wuu, D.S.; Pan, K.F.; Huang, S.H.; Lee, C.E.; Wang, W.K. High-power GaN-mirror-Cu light-emitting diodes for vertical current injection using laser liftoff and electroplating techniques. *IEEE Photon. Tech. Lett.* **2005**, *17*, 1809–1811. [CrossRef]
127. Xiong, C.; Jiang, F.; Fang, W.; Wang, L.; Mo, C.; Liu, H. The characteristics of GaN-based blue LED on Si substrate. *J. Lumin.* **2007**, *122–123*, 185–187. [CrossRef]

128. Kawasaki, K.; Koike, C.; Aoyagi, Y.; Takeuchi, M. Vertical Al2 deep ultraviolet light emitting diode emitting at 322 nm fabricated by the laser lift-off technique. *Appl. Phys. Lett.* **2006**, *89*, 261114. [CrossRef]
129. Adivarahan, V.; Heidari, A.; Zhang, B.; Fareed, Q.; Khan, A. Vertical injection thin film deep ultraviolet light emitting diodes with AlGaN multiple-quantum wells active region. *Appl. Phys. Express* **2009**, *2*, 2102. [CrossRef]
130. Nishida, T.; Saito, H.; Kobayashi, N. Efficient and high-power AlGaN-based ultraviolet light-emitting diode grown on bulk GaN. *Appl. Phys. Lett.* **2001**, *79*, 711–712. [CrossRef]
131. Edmond, J.; Abare, A.; Bergman, M.; Bharathan, J.; Bunker, K.L.; Emerson, D. High efficiency GaN-based LEDs and Lasers on SiC. *J. Cryst. Growth.* **2004**, *272*, 242–250. [CrossRef]
132. HÄRle, V.; Hahn, B.; Lugauer, H.J.; Bader, S.; Brüderl, G.; Baur, J. GaN-based LEDs and Lasers on SiC. *Phys. Stat. Sol. A* **2000**, *180*, 5–13. [CrossRef]
133. Kuramata, A. Ga_2O_3 Crystal and its LED application. *Bull. Solid State Phys. Appl. (Jpn. Soc. Appl. Phys.)* **2013**, *19*, 42. (In Japanese)
134. Ryu, H.Y.; Jun, C.W. Optimization of InGaN/GaN superlattice structures for high-efficiency vertical blue light-emitting diodes. *J. Appl. Phys.* **2013**, *114*, 160. [CrossRef]
135. Ju, I.; Kwon, Y.; Shin, C.S.; Kim, K.H.; Ko, C.G. High-Power GaN-Based Light-Emitting Diodes Using Thermally Stable and Highly Reflective Nano-Scaled Ni–Ag–Ni–Au Mirror. *IEEE Photon. Tech. Lett.* **2011**, *23*, 1685–1687. [CrossRef]

micromachines

Article

On the Luminescence Properties and Surface Passivation Mechanism of III- and N-Polar Nanopillar Ultraviolet Multiple-Quantum-Well Light Emitting Diodes

Moheb Sheikhi [1,2,†], Yijun Dai [1,2,†], Mei Cui [1,2], Liang Li [1,2], Jianzhe Liu [3], Wenan Lan [3], Rongrong Jiang [1], Wei Guo [1,2,*], Kuan W.A. Chee [4,5,*] and Jichun Ye [1,2]

1 Ningbo Institute of Materials Technology and Engineering, Chinese Academy of Sciences, Ningbo 315201, China; moheb@nimte.ac.cn (M.S.); daiyijun@nimte.ac.cn (Y.D.); cuimei@nimte.ac.cn (M.C.); lliang@nimte.ac.cn (L.L.); jiangrr@nimte.ac.cn (R.J.); jichun.ye@nimte.ac.cn (J.Y.)
2 University of Chinese Academy of Sciences, Beijing 100049, China
3 Zhe Jiang Bright Semiconductor Technology Co., Ltd., Jinhua 321016, China; jzliu@bst-group.cn (J.L.); ben@bst-group.cn (W.L.)
4 Hefei National Laboratory for Physical Sciences at Microscale, and Department of Physics, University of Science and Technology of China, Hefei 230026, China
5 Laser Research Institute, Shandong Academy of Sciences, Qingdao 226100, China
* Correspondence: guowei@nimte.ac.cn (W.G.); kuan.chee@cantab.net (K.W.A.C.)
† These authors contribute equally to this work.

Received: 6 November 2019; Accepted: 17 December 2019; Published: 5 June 2020

Abstract: The non-centrosymmetricity of III-nitride wurtzite crystals enables metal or nitrogen polarity with dramatically different surface energies and optical properties. In this work, III-polar and N-polar nanostructured ultraviolet multiple quantum wells (UV-MQWs) were fabricated by nanosphere lithography and reactive ion etching. The influence of KOH etching and rapid thermal annealing treatments on the luminescence behaviors were carefully investigated, showing a maximum enhancement factor of 2.4 in emission intensity for III-polar nanopillars, but no significant improvement for N-polar nanopillars. The discrepancy in optical behaviors between III- and N-polar nanopillar MQWs stems from carrier localization in III-polar surface, as indium compositional inhomogeneity is discovered by cathodoluminescence mapping, and a defect-insensitive emission property is observed. Therefore, non-radiative recombination centers such as threading dislocations or point defects are unlikely to influence the optical property even after post-fabrication surface treatment. This work lays solid foundation for future study on the effects of surface treatment on III- and N-polar nanostructured light-emitting-diodes and provides a promising route for the design of nanostructure photonic devices.

Keywords: III-nitride thin film; nanostructures; ultraviolet emitters; surface passivation; luminescence intensity

1. Introduction

III-nitride based ultraviolet light-emitting-diodes (UV-LEDs) are useful for many applications including UV curing, photo therapy and UV disinfection, due to the direct bandgap property of wurtzite phase III-nitride crystals [1–4]. However, the extraction efficiency of UV LEDs is relatively low due to the strong internal light reflection and strong transverse magnetic (TM) polarization of the emitted light [5–7]. Challenges in light extraction have encouraged research and development of small length scale devices for new applications, like nanostructured LEDs, for example [8–10]. The three-dimensional geometry of the nanopillars or nanoholes allows light to be extracted from the sidewalls of the nanostructures. This becomes increasingly important as the emission wavelength of the LEDs moves from visible to ultraviolet wavelength region due to increasing proportion of TM-polarized light, as well as the growing impact of light absorption in the p-AlGaN contact layer [11,12]. Additionally, the benefits from strain relaxation in nanostructured LEDs means reduced influence from quantum confined stark effect (QCSE), which has an adverse impact on the electron and hole wavefunction overlap [13].

Currently, most of the nanostructured UV-LEDs are fabricated by reactive ion etching (RIE) due to better control over morphologies and sizes of the nanostructures [14]. However, RIE introduces surface defect states that will lead to reduced external quantum efficiency (EQE) and degraded output power of LEDs through Shockley–Read–Hall (SRH) non-radiative recombination [15]. The large surface areas of GaN nanopillars can lead to dangling bonds during nanopillar formation. However, the resulting surface defects can be reduced by passivation techniques such as chemical etching or dielectric deposition [16]. It was reported that, plasma-related damages at the sidewalls of multiple quantum wells (MQWs) induced by RIE can be readily healed by thermal annealing, reducing the density of surface traps [17]. Therefore, eliminating the surface defect states is strongly required [18]. Chiu et al. reported that after a photo-enhanced chemical (PEC) wet oxidation process, emission from InGaN/GaN-based random nanorod LEDs was significantly improved, but also with a 10.5-nm blueshift with respect to the as-grown LED [19]. Sun et al. also reported a 50% enhancement in the UV light emission intensity from the KOH treated InGaN nanowires (NWs) due to the removal of surface dangling bonds [18]. Despite the above promising results, the knowledge regards to surface passivation of nanostructured LED is still not complete, since the majority of the samples are III-polar for better crystalline quality and smoother surfaces. However, N-polar LEDs have their unique advantage such as lower contact resistance, increased current injection efficiency, and an internal polarization field, which is opposite to the external bias, leading to reduced QCSE and higher radiative recombination rate [20,21]. Unfortunately, there seldom are comparative investigations on the effects of surface treatments on the optical properties of both III- and N-polar nanostructured MQWs LEDs, which serve as building blocks of next-generation high-efficiency UV emitters.

In this work, large-scale, highly periodic nanopillar UV MQWs with III- and N-polarities were fabricated by nanosphere lithography and RIE patterning. Chemical treatment and thermal annealing are acknowledged as promising methods to enhance the luminescence property of III-polar nanostructured MQWs by 2.4 times at most, but was demonstrated to show little effect on N-polar nanostructures. The underlying reason for this discrepancy was thoroughly discuss, which can be mainly attributed to intrinsic indium localization and herein defect-insensitive emissions from the N-polar nanostructures. A comprehensive investigation on the surface passivation mechanism on both III- and N-polar nanostructure emitters were provided, benefiting future development of novel nanostructured light emitters.

2. Materials and Methods

InGaN/GaN based UV-MQW was grown on 2-inch c-plane sapphire substrate via a low-pressure, high-temperature metalorganic chemical vapor deposition (MOCVD) system. Trimethylindium (TMIn), triethylgallium (TEGa) and ammonia (NH_3) were used as precursors of In, Ga and N, respectively. Hydrogen (H_2) was used as the carrier gas. To investigate the polarity influence on the luminescence property of nanopillars, both III- and N-polar planar MQWs were firstly grown. The III-nitride polarity of the MQWs was controlled by the modulation of low-temperature (LT) AlN nucleation layer (NL) prior to high-temperature (HT) epitaxial growth. Generally, III-nitride thin film is III-polarity if it is grown on LT AlN-NL. In contrast, when grown on bare sapphire substrate with proper H_2 annealing and NH_3 nitridation condition, the thin film is N-polarity. UV-MQW consists of a 3.5 μm GaN epitaxial layer, eight pairs of $In_{0.03}Ga_{0.97}N$/GaN MQWs followed by a 10 nm GaN cap layer. To fabricate nanostructured UV-MQWs, III-polar and N-polar epitaxial thin films are uniformly coated with polystyrene (PS) spheres of 2 μm diameter via a large-area micro-propulsive injection method.

The colloidal solution was injected into the water to form a Langmuir–Blodgett film in a hexagonal configuration at the air/water interface. By draining away the water, the monolayer of closely packed PS spheres is transferred onto the MQW thin film. Nanopillar arrays were then achieved by RIE using Cl_2/BCl_3 gases to pattern transfer the PS spheres to the underlying MQWs. The remaining PS spheres were removed by sonication in acetone solution. After fabrication of the nanopillar arrays, chemical and thermal treatments were applied to the MQW samples in order to heal the plasma-related damages. For chemical treatment, samples were dipped into KOH aqueous solution with a concentration of 10 wt%. The treatment was performed under room temperature (RT) or 45 °C for 40 s. For thermal treatment, rapid thermal annealing (RTA) was performed at 800 °C or 900 °C for 15 min under N_2 atmosphere. Surface morphologies of the nanostructures were characterized in the Hitachi S-4800 field-emission (FE) SEM (Hitachi, Tokyo, Japan). Dislocation densities of both samples were characterized using a point-focused high-resolution X-ray (Cu Kα1) diffractometer (HRXRD, Bruker D8 Discover, Germany) equipped with a four-bounce symmetric Ge (220) monochromator. The surface morphology of the MQW structure was characterized by atomic force microscopy (AFM) (Veeco Dimension 3100 V, Plainview, NY, USA). High-angle annular dark field scanning transmission electron microscopy (HAADF-STEM) was performed and weak beam dark field images were acquired under two beam conditions at an acceleration voltage of 300 kV (FEI Titan ST microscope, Hillsboro, OR, USA). Specimens were prepared by focus ion beam (FIB) using an FEI Helios SEM system with a Ga ion source. RT photoluminescence (PL) was performed by using an Ar-F (193 nm) excimer laser (Coherent Inc., Santa Clara, CA, USA) as an excitation pumping source, and the spectra were collected by a Horiba iHR550 spectrometer (Horiba, Kyoto, Japan). Cathodoluminescence (CL) investigations were undertaken using a Horiba MP 325 CL characterization system (Horiba, Kyoto, Japan) with voltage of 5 kV and current level of 188 μA. Strain conditions of the nanopillar MQWs were investigated by a Renishaw inVia Reflex spectrometer system (Renishaw, New Mills, UK) with a 532 nm Nd-YAG laser as the excitation source. The surface stoichiometry of the samples was further studied by Kratos Axis Ultra DLD X-ray photoelectron spectrometer (XPS; Kratos, Manchester, UK).

3. Results and Discussion

Before detailed investigations on optical properties of nanopillar samples, the structural and surface morphology information of planar III-polar and N-polar MQW samples are investigated. Figure 1 shows the HRXRD rocking curve (RC) scans of both III- and N-polar samples. The full-width-half-maximum (FWHM) values of RC scans were used to evaluate the crystalline quality of III-nitride thin films. From the fitting, the FWHM of (002) and (102) peaks of III-polar sample are 265 and 275 arcsec, respectively, which are lower than those of N-polar samples (586 and 578 arcsec for (002) and (102) peaks). The dislocation density can herein be estimated. Screw and edge type dislocation densities are 1.53×10^8 cm^{-2} and 4.59×10^8 cm^{-2} for III-polar samples, which are approximately 5 times lower than those of N-polar ones, whose values are 7.48×10^8 cm^{-2} and 1.8×10^9 cm^{-2}. This result suggests that III-polar samples exhibit superior crystalline quality than N-polar samples.

Surface morphologies of both samples were analyzed by AFM and illustrated in Figure 2. Clear bi-layer steps are illustrated in the III-polar sample, demonstrating a typical step-flow growth mode of III-polar III-nitride thin film [22]. On the other hand, N-polar sample is covered with hillocks of a few micrometers lateral size, suggesting a three-dimensional growth mode because of its much smaller surface energy and thus larger nucleation density [23]. Root-mean-square roughness for III- and N-polar samples are 0.37 and 11.0 nm, respectively. The drastic difference of surface roughness of these two polarities are also correlated with their crystalline quality. Indium content of planar MQW samples can usually be identified from symmetric and asymmetric ω-2θ scans from HRXRD. Symmetric ω-2θ scan of planar III- and N-polar samples was performed and shown in supplementary Figure S1. The most prominent peak can be identified as GaN template. Different peak positions originate from various strain states inside epitaxial thin films as a consequence of growth mode difference between III- and N-polar domains [21]. Usually, 0st order InGaN MQW peak can be found on the low-angle side of the spectrum, but can hardly be found in our study. This could possibly due to the extremely low in content of only 3% in the MQW region and thus MQW peak merges with the GaN template peak.

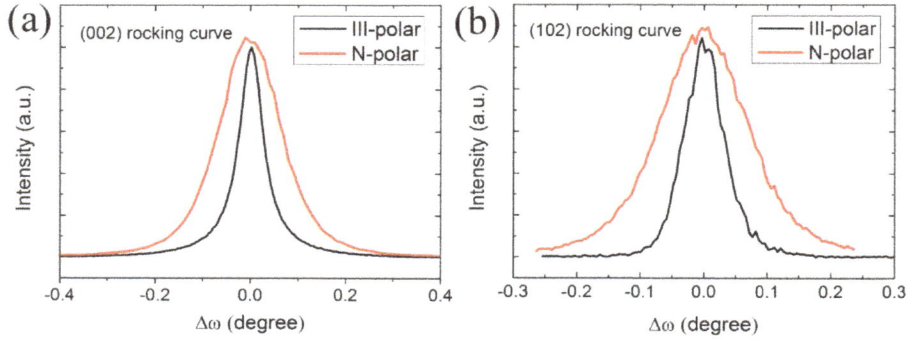

Figure 1. HRXRD (002) (**a**) and (102) (**b**) RC scans of planar III- and N-polar samples for dislocation estimation.

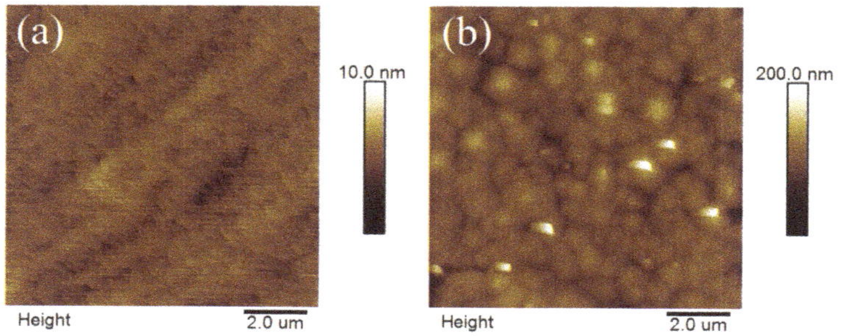

Figure 2. AFM surface morphology of planar III-polar (**a**) and N-polar (**b**) MQW samples.

RIE followed by nanosphere lithography was utilized to fabricate the nanopillar UV-MQW samples. The lattice constant of the nanopillar array is expected to match the 2 μm diameter of the PS spheres used in the nanosphere lithography. Due to ion bombardment, amorphous layers are generated on the sidewall surfaces of the nanopillars during RIE patterning, but which are expected to be removed by the KOH treatment. The SEM images in Figure 3 show the surface morphology of the uniformly distributed nanopillar hexagonal array on the as-fabricated III- and N-polar UV-MQWs after subjecting to the different treatments. It can be seen that the diameter of the nanopillars remained the same after RTA but reduced slightly to 1.98 μm after KOH treatment. Thermal annealing does not alter the surface morphologies of the nanopillars due to a lacking of energy for thin film re-crystallization or surface reconstruction. It is known that the dangling bond density for N-polar planes is higher than that of (0001) planes, leading to a much higher etch rate under KOH treatment. Therefore, elongated time of KOH etching could damage the nanopillars as shown in Figure 3e where the flat top mesa of the nanopillars is covered with small hillocks. The 60° tilted view SEM images of the as-fabricated nanopillars are shown in the inset of Figure 3a,d. After RIE process, the depths of the III and N-polar nanopillars are 0.744 and 0.748 μm, respectively, suggesting almost same RIE etching rate between these two polarities.

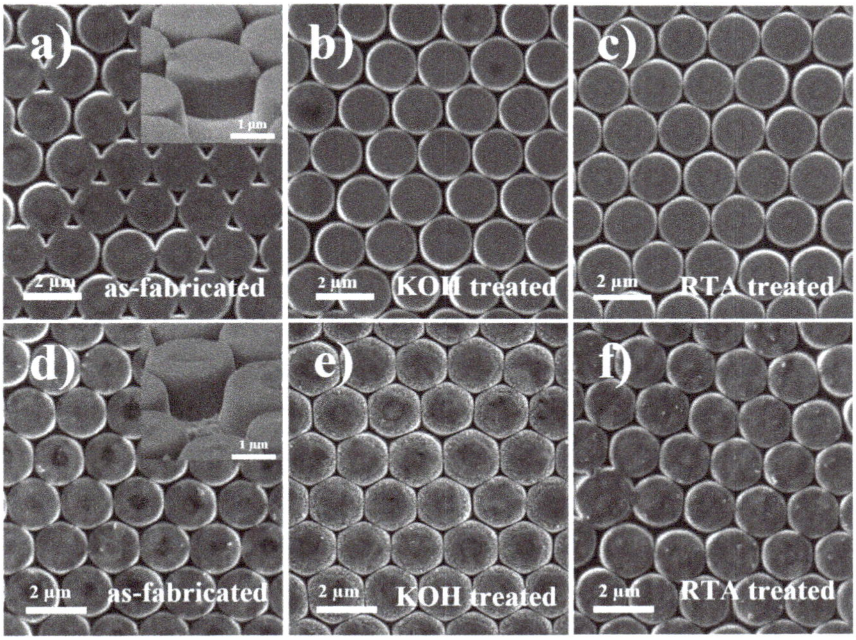

Figure 3. SEM images of III-polar (**a–c**) and N-polar (**d–f**) nanopillar multiple quantum wells (MQWs) before and after the KOH or rapid thermal annealing (RTA) treatment. Tilted-view SEM images of as-fabricated nanopillars are shown in the inset of (**a**,**d**).

RT photoluminescence (PL) measurements were carried out to determine the luminescence efficiency of the nanopillar MQWs before and after KOH or RTA treatment. First of all, PL intensities of III- and N-polar nanostructured samples are compared. III-polar nanopillars show stronger intensity than that of N-polar samples. This could because of two reasons: firstly, a strong carrier localization effect is observed in N-polar nanopillar sample as will be demonstrated later. This could originate from rough surface morphology of N-polar planar sample to begin with, and MQW thickness fluctuations as demonstrated in our previous work [20]. Even though carrier localization can greatly enhance internal quantum efficiency in LEDs, but excessive phase separation can also lead to a greatly reduced active area of the MQWs at the same time, which is detrimental to the optical property as observed in this work. Secondly, N-polar sample exhibits higher dislocations than that of III-polar samples. This greatly deteriorate the luminescence intensity of N-polar nanostructured samples. PL spectra of planar III- and N-polar MQW samples are shown in supplementary Figure S2 for comparison purpose. No obvious variation in peak position is observed between planar and nanostructured sample. However, an enhanced luminescence intensity is identified for the nanostructured sample due to enhanced light scattering effect [14].

PL spectra subject to different surface treatments were further investigated. Figure 4a,b shows the PL spectra of III- and N-polar nanopillar MQWs after the respective treatments. PL intensities dramatically increased for III-polar compared to N-polar nanopillar MQWs. The highest PL enhancement reached 2.4 times for RTA of 900 °C. This can be well explained by the passivation of surface traps and the healing of surface-related defects [14]. However, the luminescence spectra were dramatically different for N-polar nanopillar MQWs. For the N-polar nanopillar MQW, a peak and a broad shoulder were identified in the PL spectra where the left peak was located at approximately 386 nm, and the right shoulder was located at 420 nm. The left peak intensity remained roughly the same after KOH or RTA treatment, while the longer wavelength shoulder reduced, suggesting that polarity played a critical role in the influence of surface treatment on luminescence. The integrated PL intensities of the III- and N-polar nanopillars are shown in Figure 4c, indicating that the effect of surface treatment was marginal on the N-polar MQWs compared to that of the III-polar samples.

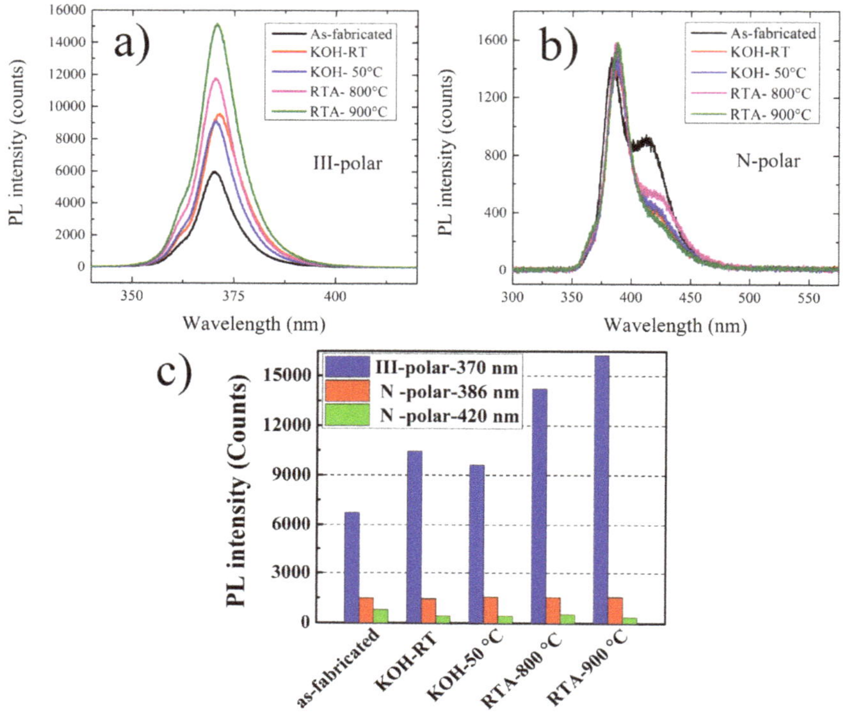

Figure 4. Room temperature (RT) photoluminescence (PL) spectra of III-polar (**a**) and N-polar (**b**) nanopillar MQWs, and integrated PL intensity of III-polar and N-polar MQW nanopillar MQWs at the specific peak wavelengths (**c**), before and after KOH treatment or rapid thermal annealing (RTA).

The strong variation of PL peak position from III- and N-polar samples is originated from either higher indium composition or thickness variation or a combination of both. An average indium content changing from 1.3% to 5.2% in the MQW is expected when composition is considered as the only factor. For the 420 nm shoulder peak in N-polar MQW sample, a further increase of indium content to 12.6% is obtained. The optical transition in the QWs was further analyzed by solving the Poisson equations and carrier transport equations. During calculation, 3/11 nm $In_{0.03}Ga_{0.97}N$ QW and GaN QB thicknesses are utilized. As shown in the calculation result illustrated in supplementary Figure S3, a broad emission peak with position located at 365 nm is identified, which is slightly shorter than the 370 nm emission peak observed in PL spectra. The internal electric field inside QW is less than 1×10^5 V/cm, which is relatively low compared to AlGaN or InGaN system reported elsewhere [24,25]. The discrepancy between the simulated electroluminescence (EL) spectrum and experimental PL spectrum can be explained by a variety of factors including different interaction volumes between injected current in EL and excited photons in PL, deviation of indium composition from target and QW thickness fluctuation. But we can safely conclude that the uniform emission peak from III-polar sample is a consequence of efficient radiative recombination inside the abrupt QWs as shown from STEM image of III-polar MQW sample observed in supplementary Figure S4, while non-uniform emission in N-polar sample is a result of carrier localization.

Finally, in order to correlate PL spectra with the influence of QW thickness variation, emission spectra are simulated based on an LED with 3 pairs of MQWs. The indium content in QW is fixed at 3%, while QW thickness was varied from 3 nm to 6 nm. The thickness of quantum barrier remains constant at 11 nm. From the simulation results shown in supplementary Figure S5, two emission peaks can be identified. This occurrence of two peaks can be explained by the shallow QW and thus weak carrier confinement. The high-energy emission peak approximately locates at 364 nm, and does not change with QW thickness. On the other hand, the low-energy peak red shifts as increasing QW thickness, in good agreement with the reduced quantum confinement and thus lower emission energy [26]. However, the red shift is only less than 7 nm, which is far less than that observed in the experiment. The coherence with a 420-nm emission cannot be simulated by only changing the QW thickness, which further demonstrates that composition inhomogeneity must play an important role in the light emission.

To further understand the PL enhancement mechanism of III-polar nanopillar MQWs, PL efficiency (PL_{eff}) as a function of excited power density of the III and N-polar MQW samples is illustrated in Figure 5. PL_{eff} is defined as the ratio of PL intensity to excited power density, and its slope is an indication of radiative recombination mechanism [10]. The negative slope of the PL_{eff} versus excited power density for planar MQWs suggests the existence of exciton-related emission together with large amount of non-radiative centers. In contrast, for III-polar nanopillar samples, the PL_{eff} increased when the pumping power density increased, which is a signature of radiative carrier recombination through free carriers. The PL_{eff} increased for KOH treatment at a higher temperature. Higher PL_{eff} was obtained after thermal annealing. The slightly larger slope of PL_{eff} vs. pumping power density after thermal annealing shows an enhanced radiative recombination rate, which might indicate annihilation of trap defects. For N-polar nanopillar MQWs, since the emission spectra had two peaks locating at 386 nm and 420 nm, thus two PL_{eff} curves are illustrated. For a 386 nm emission in N-polar MQW, PL_{eff} also increased with pumping power density similar to that of III-polar nanopillars. However, a negative slope was observed for N-polar nanopillars with 420 nm emission, again suggesting the exciton-related emission mechanism, which was a signature of carrier localization effect as will be discussed in details later in Figures 6 and 7.

Due to phase separation between In and Ga, compositional inhomogeneities in InGaN alloys lead to non-uniform optical properties of UV-LEDs, especially for N-polar surfaces where rough surface morphology promotes phase separation. Therefore, deeper understanding on the spatially resolved luminescence properties of III- and N-polar nanopillar MQWs is critically important. Panchromatic CL intensity mapping is shown in Figure 6. Figure 6a–d illustrates the CL distribution in III-polar nanopillar MQWs and Figure 6e–h illustrates the CL distribution in N-polar nanopillar MQWs. Under low magnification, bright stripes along the substrate offcut direction were clearly observed on the surface of the III-polar nanopillar MQW (Figure 6a), which could be attributed to intrinsic indium localization in the MQWs. The strong luminescence comes from the In-rich clusters along the step edges of the epitaxial thin films. Since white stripes are shown in both planar and nanostructured MQWs, nanopillar fabrication did not have an influence on the distribution of luminescence centers. A closer look at the spatially resolved CL intensity distribution of III-polar MQWs in Figure 6b–d gives us clear information that luminescence only coming from the nanopillars where MQWs were not etched away by RIE. Additionally, the emission intensity increased dramatically after thermal annealing of 900 °C, which agreed perfectly with the enhanced PL intensity as shown before in Figure 4. Different from that of III-polar samples, strong localization of luminescence centers is shown for all N-polar samples. The dot-like bright features have lateral size of 20–200 nm. White dots represent quantum-dot-like luminescence centers, which were uniformly distributed on the sample, possibly due to the rough surface morphology of N-polar MQWs and thus non-uniform light emission. Note that carrier localization is also observed in planar samples. This has been thoroughly discussed in our previous work [27,28]. Carrier localization was observed in N-polar domain of planar sample due to thickness and composition fluctuation. However, carrier localization is not observed in III-polar MQW sample.

This is due to the fact that the indium composition is too low for spinodal decomposition. In fact, carrier localization was widely reported in InGaN with high indium content above 15% [29], but seldom been reported in UV-LEDs with InGaN MQWs of indium content less than 5% as studied here.

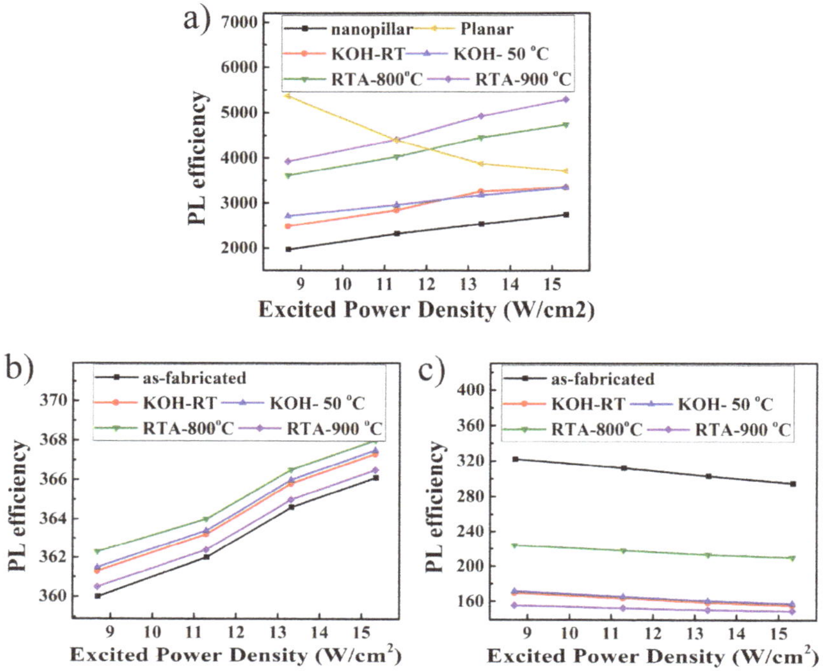

Figure 5. PL efficiency as a function of excited power density for III-polar nanopillar MQWs (**a**), and N-polar nanopillar MQWs with emission wavelength located at 386 nm (**b**) and at 420 nm (**c**) respectively.

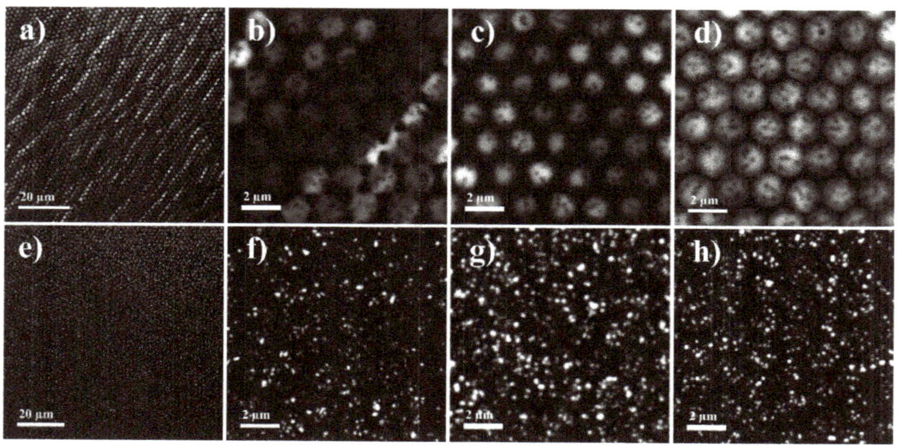

Figure 6. Low magnification (**a**) and high magnification panchromatic cathodoluminescence (CL) mapping of as-fabricated (**b**), KOH treated (**c**) and thermally treated (**d**) III-polar nanopillar MQWs. CL mapping of N-polar nanopillar MQWs are illustrated in (**e–h**), accordingly.

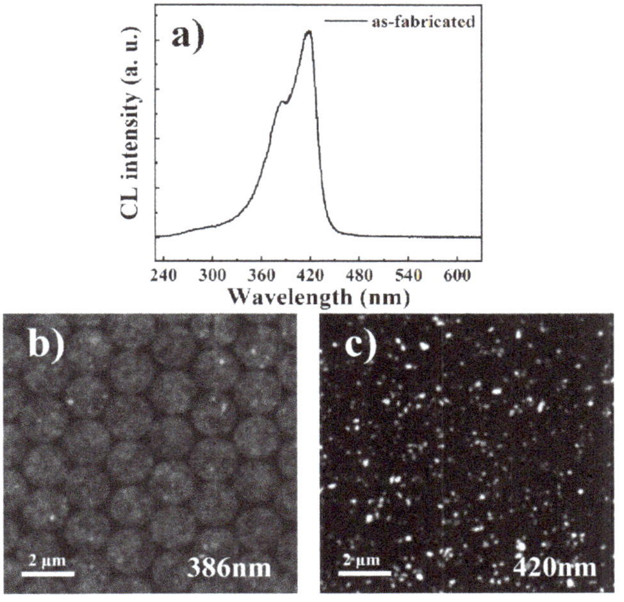

Figure 7. CL spectrum of N-polar nanopillar MQW illustrating two peaks at 386 and 420 nm, respectively (**a**). Monochromatic CL intensity distribution of N-polar nanopillar MQW at 386 nm (**b**) and 420 nm (**c**), respectively.

Additionally, the black dots were identified throughout the surface in the panchromatic CL intensity mapping shown in Figure 6b–d, representing non-radiative recombination centers like threading dislocations or point defects. The densities were 2.35×10^8 cm^2, 2.29×10^8 cm^2 and 2.38×10^8 cm^2 respectively for as-fabricated nanopillars, nanopillars subject to KOH treatment and thermal annealing, respectively. No obvious change in the defect density after thermal or KOH treatment was identified, suggesting that luminescence enhancement as shown in Figure 4 is mainly related to passivation of surface states, rather than defects in the bulk.

Figure 7a shows the CL spectrum of the as-fabricated N-polar nanopillars with two peaks located at 386 and 420 nm. The monochromatic CL mapping at 386 nm and 420 nm are shown in Figure 7b,c, respectively. Interestingly, the CL emission at 386 nm was uniformly distributed on the surface of the nanopillars, whereas 420 nm-emission originated from In-rich clusters, which are represented by the white dots. The red shift of CL spectra for N-polar nanopillars compared to that of the III-polar sample can be well explained by the in-phase separation in the MQWs leading to quantum dot-like luminescence centers at those In-rich clusters. Additionally, the carrier diffusion length decreased drastically due to indium phase separation, and the influence of point defects and threading dislocations were mostly screened by these potential minima, imposing no effect on the luminescence intensity. Therefore, even though surface trap defects were passivated by thermal and chemical treatments the luminescence intensities still remained the same for N-polar nanopillar MQWs because of carrier localization effects.

Optical properties of MQWs were also correlated to strain conditions. Therefore, Raman spectroscopy investigations were performed. Figure 8 shows the Raman spectra recorded for III- and N-polar nanopillars before and after RTA or KOH treatment. The peak position of the E_2 (high) phonon mode of III-nitride crystal is sensitive to bi-axial strains [30]. For fully relaxed GaN, the E_2 (high) peak was at 567.6 cm^{-1}, and a smaller Raman wavenumber suggests tensile strain while a larger Raman wavenumber suggests the compressive strain. In this work, since the epitaxial thin film was mainly composed of GaN or AlGaN with low Al composition, E_2 (high) peak located around 567 cm^{-1} was regarded as fully-relaxed, and was utilized for strain analysis. The peak positions for all the III-polar nanopillar MQW samples were close to each other with a value of 569 cm^{-1} indicating slightly compressive strains of 0.6 Gpa regardless of chemical or thermal treatment. Note that the E_2 (high) position of the as-fabricated planar MQW was located at a much high wavenumber of 572.1 cm^{-1} as observed in our recent study. Therefore, most of the strains were actually relaxed after nanopillar fabrication. Compared to the III-polar nanopillar samples, the compressive strain in as-fabricated N-polar nanopillar samples was found to be much smaller as the E_2 (high) position of the GaN-like Raman peak was only 1 cm^{-1} higher at maximum compared to the fully relaxed position. This could be explained by the 3D growth mode of N-polar thin film, higher dislocation density and thus more strain relaxation inside N-polar crystals. RTA treatment at 900 °C fully relaxes the compressive strain inside the as-fabricated N-polar nanopillars, even though no significant improvement in emission intensity was found.

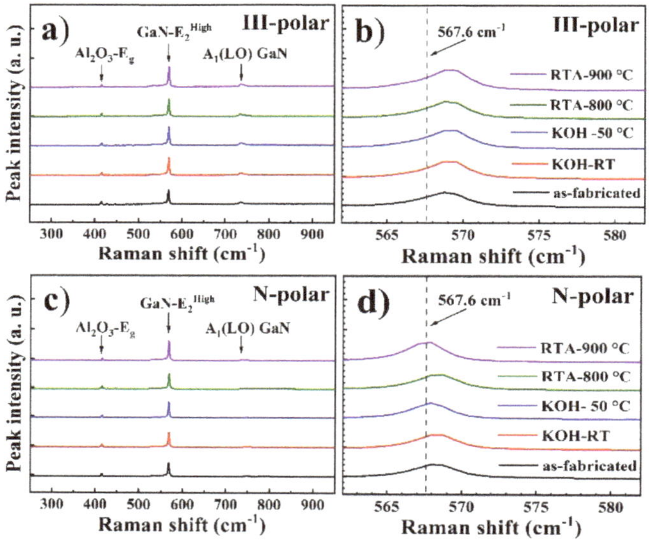

Figure 8. Full-range (**a**,**c**) and zoom-in (**b**,**d**) Raman spectra of III-polar (**a**,**b**) and N-polar (**c**,**d**) nanopillar MQWs before and after chemical and thermal treatments.

Finally, to further correlate surface stoichiometry to the enhanced UV emission of III-polar nanopillar MQWs, XPS measurements were carried out before and after RTA or KOH treatment. The de-convoluted Ga 3d peaks of the as-fabricated nanopillars, nanopillars after KOH treatment and 900 °C annealing are shown in Figure 9a. As indicated in the figure, the major Ga 3d core level peak consisted of two components at 21.0 ± 0.1 eV and 19.3 ± 0.1 eV corresponding to the binding energy of Ga–O and Ga–N, respectively. After KOH etching, the Ga 3d peak slightly shifted to higher binding energy, where the relative Ga–O component was increased, indicating that Ga cations were attacked by OH–, leading to the breakage of Ga–N bond and consequently formation of Ga–O bond. Furthermore, after RTA surface treatment, the Ga 3d peak shifts to even higher binding energy, but the total intensity was dramatically reduced. This indicates that chemical stoichiometry between Ga and N atom was deviating from 1:1 after thermal annealing. Figure 9b shows the O 1s core level peaks de-convoluted into three components with binding energies of 530.3 eV, 531.7 eV and 532.8 eV, corresponding to O^{2-}, OH^- and H_2O species, respectively. It is clearly shown that after KOH etching, the overall intensity of the O 1s peak decreased, in agreement with the removal of naturally formed surface oxide under KOH etching. After thermal annealing, there existed a strong shift of O 1s peak towards higher binding energy, indicating a transformation from the Ga–O bond to Ga–O–H bond.

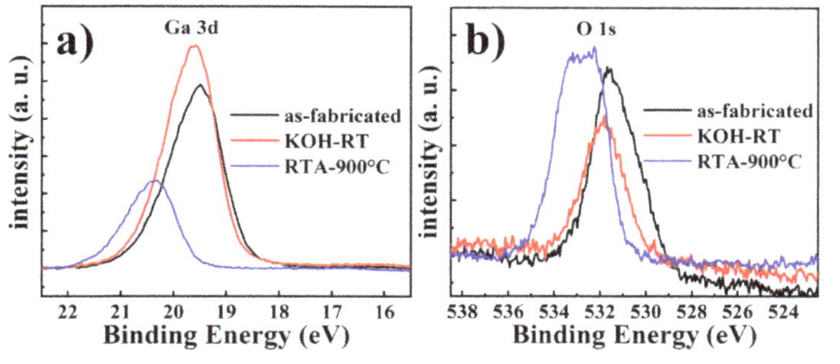

Figure 9. High resolution Ga 3d (**a**) and O 1s (**b**) XPS core level spectra of III-polar nanopillar MQWs without treatment and after KOH or RTA treatments.

4. Conclusions

In this work, the influences of KOH or RTA treatments on the optical properties of III and N-polar nanopillar UV MQWs were carefully examined. The UV emission intensity increased significantly after post-fabrication treatment and by as much as 2.4 times after RTA annealing for III-polar nanopillars, whereas almost no effect on the luminescence spectra was seen for N-polar samples. CL intensity maps indicate carrier localization effects due to In-rich clusters in the N-polar nanopillars, which dominated the light output mechanism because of the reduced carrier diffusion length scales. This led to reduced probability of carrier recombination with threading dislocations or point defects, meaning that despite the fact that surface treatment passivation methods might reduce the trap defects, they did not contribute to the PL efficiency. Slight compressive strain effects could be found in as-fabricated III-polar nanopillars, but which did not change significantly after KOH or RTA treatment. As-fabricated N-polar nanopillar samples were almost strain-free due to 3D growth mode and strain relaxation mechanism. XPS results further suggest the removal of surface oxides after KOH etching and transformation from Ga–O to Ga–O–H bond after thermal annealing for III-polar nanopillar samples. The experimental results obtained here present a deeper understanding on the different optical behaviors of nanostructured UV emitters with opposite polarities, and provide an important strategy to enhance the luminescence intensity of nanopillar LEDs, paving the way for future development of novel nanostructure-based electronic or photonic devices.

Supplementary Materials: The following are available online at http://www.mdpi.com/2072-666X/11/6/572/s1, Figure S1: HRXRD symmetric ω-2θ scans of III- and N-polar MQW samples; Figure S2: RT PL spectra of planar III-polar and N-polar MQW samples. Peak positions and shapes are almost identical to those of nanostructured samples; Figure S3: Simulated electroluminescence spectrum of UVA-LED with 3/11 nm $In_{0.03}Ga_{0.97}N$/GaN MQW; Figure S4: HAADF-STEM image of planar III-polar MQW samples with low magnification (**a**) and zoom-in view (**b**) of the individual QW and QB; Figure S5: Simulated electroluminescence spectra of UVA-LED with $In_{0.03}Ga_{0.97}N$/GaN MQWs of varying QW thicknesses. A small red shift of less than 7 nm is observed when doubling the QW thickness.

Author Contributions: Data curation, M.S., Y.D. and M.C.; Formal analysis, L.L., J.L., W.L. and R.J.; Supervision, K.W.A.C. and J.Y.; Writing—original draft, W.G. All authors have read and agreed to the published version of the manuscript.

Funding: This work is supported by National Key Research and Development Program of China (Grant 2016YFB0400802), National Natural Science Foundation of China (Grants 61974149, 61704176, 61950410617 and 61874177), Key Research and Development Program of Zhejiang Province (2020C01145).

Conflicts of Interest: The authors declare no conflict of interest.

References

1. Kneissl, M.; Seong, T.-Y.; Han, J.; Amano, H. The emergence and prospects of deep-ultraviolet light-emitting diode technologies. *Nat. Photonics* **2019**, *13*, 233. [CrossRef]
2. Hirayama, H.; Maeda, N.; Fujikawa, S.; Toyoda, S.; Kamata, N. Recent progress and future prospects of AlGaN-based high-efficiency deep-ultraviolet light-emitting diodes. *Jpn. J. Appl. Phys.* **2014**, *53*, 100209. [CrossRef]
3. Xu, H.; Sheikhi, M.; Li, L.; Yang, Z.; Hoo, J.; Guo, S.; Zeng, Y.; Guo, W.; Ye, J. Omnidirectional whispering-gallery-mode lasing in GaN microdisk obtained by selective area growth on sapphire substrate. *Opt. Express* **2019**, *27*, 16195–16205.
4. Guo, W.; Mitra, S.; Xu, H.; Sheikhi, M.; Sun, H.; Tian, K.; Zhang, Z.-H.; Jiang, H.; Roqan, I.S.; Li, X. Three-dimensional band diagram in lateral polarity junction III-nitride heterostructures. *Optica* **2019**, *6*, 1058–1062. [CrossRef]
5. Xu, H.; Long, H.; Sheikhi, M.; Li, L.; Guo, W.; Dai, J.; Chen, C.; Ye, J. Strain modulated nanostructure patterned AlGaN-based deep ultraviolet multiple-quantum-wells for polarization control and light extraction efficiency enhancement. *Nanotechnology* **2019**, *30*, 435202. [CrossRef] [PubMed]
6. Long, H.; Wang, S.; Dai, J.; Wu, F.; Zhang, J.; Chen, J.; Liang, R.; Feng, Z.C.; Chen, C. Internal strain induced significant enhancement of deep ultraviolet light extraction efficiency for AlGaN multiple quantum wells grown by MOCVD. *Opt. Express* **2018**, *26*, 680–686. [CrossRef] [PubMed]
7. Guo, W.; Yang, Z.; Li, J.; Yang, X.; Zhang, Y.; Wang, J.; Chee, K.; Gao, P.; Ye, J. Enhancing light coupling and emission efficiencies of AlGaN thin film and AlGaN/GaN multiple quantum wells with periodicity-wavelength matched nanostructure array. *Nanoscale* **2017**, *9*, 15477. [CrossRef] [PubMed]
8. Mandl, M.; Wang, X.; Schimpke, T.; Kölper, C.; Binder, M.; Ledig, J.; Waag, A.; Kong, X.; Trampert, A.; Bertram, F. Group III nitride core–shell nano-and microrods for optoelectronic applications. *Phys. Status Solidi RRL Rapid Res. Lett.* **2013**, *7*, 800–814. [CrossRef]
9. Le, B.H.; Zhao, S.; Liu, X.; Woo, S.Y.; Botton, G.A.; Mi, Z. Controlled coalescence of AlGaN nanowire arrays: An architecture for nearly dislocation-free planar ultraviolet photonic device applications. *Adv. Mater.* **2016**, *28*, 8446–8454. [CrossRef]
10. Zhang, L.; Guo, Y.; Yan, J.; Wu, Q.; Lu, Y.; Wu, Z.; Gu, W.; Wei, X.; Wang, J.; Li, J. Deep ultraviolet light-emitting diodes based on a well-ordered AlGaN nanorod array. *Photonics Res.* **2019**, *7*, B66–B72. [CrossRef]
11. Maeda, N.; Jo, M.; Hirayama, H. Improving the light-extraction efficiency of AlGaN DUV-LEDs by using a superlattice hole spreading layer and an Al reflector. *Phys. Status Solidi a Appl. Mater. Sci.* **2018**, *215*, 1700436. [CrossRef]
12. Zhang, Z.-H.; Kou, J.; Chen, S.-W.H.; Shao, H.; Che, J.; Chu, C.; Tian, K.; Zhang, Y.; Bi, W.; Kuo, H.-C. Increasing the hole energy by grading the alloy composition of the p-type electron blocking layer for very high-performance deep ultraviolet light-emitting diodes. *Photonics Res.* **2019**, *7*, B1–B6. [CrossRef]
13. Li, Q.; Westlake, K.R.; Crawford, M.H.; Lee, S.R.; Koleske, D.D.; Figiel, J.J.; Cross, K.C.; Fathololoumi, S.; Mi, Z.; Wang, G.T. Optical performance of top-down fabricated InGaN/GaN nanorod light emitting diode arrays. *Opt. Express* **2011**, *19*, 25528–25534. [CrossRef] [PubMed]
14. Guo, W.; Li, J.; Sheikhi, M.; Jiang, J.; Yang, Z.; Li, H.; Guo, S.; Sheng, J.; Sun, J.; Bo, B. Comparative study on luminescence extraction strategies of LED by large-scale fabrication of nanopillar and nanohole structures. *J. Phys. D Appl. Phys.* **2018**, *51*, 24LT01. [CrossRef]
15. Wang, Q.; Zhao, S.; Connie, A.; Shih, I.; Mi, Z.; Gonzalez, T.; Andrews, M.; Du, X.; Lin, J.; Jiang, H. Optical properties of strain-free AlN nanowires grown by molecular beam epitaxy on Si substrates. *Appl. Phys. Lett.* **2014**, *104*, 223107. [CrossRef]
16. Hartensveld, M.; Ouin, G.; Liu, C.; Zhang, J. Effect of KOH passivation for top-down fabricated InGaN nanowire light emitting diodes. *J. Appl. Phys.* **2019**, *126*, 183102. [CrossRef]
17. Sheikhi, M.; Guo, W.; Dai, Y.; Cui, M.; Hoo, J.; Guo, S.; Xu, L.; Liu, J.; Ye, J. Mechanism of improved luminescence intensity of Ultraviolet Light Emitting Diodes (UV-LEDs) under thermal and chemical treatments. *IEEE Photonics J.* **2019**. [CrossRef]
18. Chee, K.W.; Guo, W.; Wang, J.R.; Wang, Y.; Chen, Y.-E.; Ye, J. Tuning photonic crystal fabrication by nanosphere lithography and surface treatment of AlGaN-based ultraviolet light-emitting diodes. *Mater. Des.* **2018**, *160*, 661–670. [CrossRef]

19. Chiu, C.; Lu, T.-C.; Huang, H.; Lai, C.; Kao, C.; Chu, J.; Yu, C.; Kuo, H.-C.; Wang, S.; Lin, C. Fabrication of InGaN/GaN nanorod light-emitting diodes with self-assembled Ni metal islands. *Nanotechnology* **2007**, *18*, 445201. [CrossRef]
20. Guo, W.; Sun, H.; Torre, B.; Li, J.; Sheikhi, M.; Jiang, J.; Li, H.; Guo, S.; Li, K.H.; Lin, R. Lateral-polarity structure of AlGaN quantum wells: A promising approach to enhancing the ultraviolet luminescence. *Adv. Funct. Mater.* **2018**, *28*, 1802395. [CrossRef]
21. Keller, S.; Li, H.; Laurent, M.; Hu, Y.; Pfaff, N.; Lu, J.; Brown, D.; Fichtenbaum, N.; Speck, J.; DenBaars, S.; et al. Recent progress in metal-organic chemical vapor deposition of N-polar group-III nitrides. *Semicond. Sci. Technol.* **2014**, *29*, 113001. [CrossRef]
22. Bryan, I.; Bryan, Z.; Mita, S.; Rice, A.; Hussey, L.; Shelton, C.; Tweedie, J.; Maria, J.; Collazo, R.; Sitar, Z. The role of surface kinetics on composition and quality of AlGaN. *J. Cryst. Growth* **2016**, *451*, 65–71. [CrossRef]
23. Sun, Q.; Cho, Y.; Lee, I.; Han, J.; Kong, B.; Cho, H. Nitrogen-polar GaN growth evolution on c-plane sapphire. *Appl. Phys. Lett.* **2008**, *93*, 131912. [CrossRef]
24. Chichibu, S.; Sota, T.; Wada, K.; Brandt, O.; Ploog, K.; DenBaars, S.; Nakamura, S. Impact of internal electric field and localization effect on quantum well excitons in AlGaN/GaN/InGaN light emitting diodes. *Phys. Status Solidi (A)* **2001**, *183*, 91–98. [CrossRef]
25. Sánchez-Rojas, J.; Garrido, J.; Munoz, E. Tailoring of internal fields in AlGaN/GaN and InGaN/GaN heterostructure devices. *Phys. Rev. B* **2000**, *61*, 2773. [CrossRef]
26. Tan, S.; Zhang, J.; Egawa, T.; Chen, G.; Luo, X.; Sun, L.; Zhu, Y. Influence of quantum-well width on the electroluminescence properties of AlGaN deep ultraviolet light-emitting diodes at different temperatures. *Nanoscale Res. Lett.* **2018**, *13*, 1–5. [CrossRef] [PubMed]
27. Cho, Y.; Gainer, G.; Fischer, A.; Song, J. "S-shaped" temperature-dependent emission shift and carrier dynamics in InGaN/GaN multiple quantum wells. *Appl. Phys. Lett.* **1998**, *73*, 1370–1372. [CrossRef]
28. Liu, W.; Zhao, D.; Jiang, D.; Shi, D.; Zhu, J.; Liu, Z.; Chen, P.; Yang, J.; Liang, F.; Liu, S.; et al. Effect of carrier transfer process between two kinds of localized potential traps on the spectral properties of InGaN/GaN multiple quantum wells. *Opt. Express* **2018**, *26*, 3427–3434. [CrossRef]
29. Lin, Y.; Ma, K.; Hsu, C. Dependence of composition fluctuation on indium content in InGaN/GaN multiple quantum wells. *Appl. Phys. Lett.* **2000**, *77*, 2988–2990. [CrossRef]
30. Perlin, P.; Jauberthie-Carillon, C.; Itie, J.P.; Miguel, A.S.; Grzegory, I.I.; Polian, A. Raman scattering and x-ray-absorption spectroscopy in gallium nitride under high pressure. *Phys. Rev. B* **1992**, *45*, 83–89. [CrossRef]

micromachines

Article

Large-Scale Fabrication of Photonic Nanojet Array via Template-Assisted Self-Assembly

Pengcheng Zhang [1], Xi Chen [1] and Hui Yang [1,2,*]

[1] Laboratory of Biomedical Microsystems and Nano Devices, Bionic Sensing and Intelligence Center, Institute of Biomedical and Health Engineering, Shenzhen Institutes of Advanced Technology, Chinese Academy of Science, Shenzhen 518055, China; pc.zhang@siat.ac.cn (P.Z.); xi.chen@siat.ac.cn (X.C.)

[2] CAS Key Laboratory of Health Informatics, Shenzhen Institutes of Advanced Technology, Chinese Academy of Sciences, Shenzhen 518055, China

* Correspondence: hui.yang@siat.ac.cn; Tel.: +86-755-8639-2675

Received: 16 March 2020; Accepted: 28 April 2020; Published: 30 April 2020

Abstract: A large-scale homogenized photonic nanojet array with defined pattern and spacing facilitates practical applications in super-resolution imaging, subwavelength-resolution nanopatterning, nano objects trapping and detection technology. In this paper, we present the fabrication of a large-scale photonic nanojet array via the template-assisted self-assembly (TASA) approach. Templates of two-dimensional (2D) large-scale microwell array with defined pattern and spacing are fabricated. Melamine microspheres with excellent size uniformity are utilized to pattern on the template. It is found that microwells can be filled at a yield up to 95%. These arrayed microspheres on the template serve as microlenses and can be excited to generate large-scale photonic nanojets. The uniformly-sized melamine spheres are beneficial for the generation of a homogenized photonic nanojet array. The intensity of the photonic nanojets in water is as high as ~2 fold the background light signal. Our work shows a simple, robust, and fast means for the fabrication of a large-scale homogenized photonic nanojet array.

Keywords: photonic nanojet; photonic nanojet array; self-assembly; template-assisted self-assembly; patterning efficiency

1. Introduction

The photonic nanojet (PNJ) is an electromagnetic beam of high intensity that is generated by normal plane wave incidence on a dielectric microcylinder or microsphere of diameter greater than the illuminating wavelength [1,2]. These nanojets have a narrow beam waist smaller than the diffraction limit and propagate with little divergence for several wavelengths. Owning to its high intensity, narrow waist and the enhanced backscattering effect with the nanoparticles, the PNJ has been widely used in super-resolution imaging [3–7], fluorescence enhancement [8–11], enhanced Raman scattering [12], subwavelength resolution nanopatterning [13,14], nano-objects or molecules trapping and detection [15–22], and so on. In the work cited above, PNJs were mainly generated either by an individual dielectric microsphere, randomly distributed microspheres or self-assembled microsphere films. When it comes to the practical applications, there is a demand for generating large-scale homogenized PNJs simultaneously with a defined pattern and inter-PNJ spacing, especially in the fields of super-resolution imaging, nanopatterning, nano objects trapping and detection technology. The self-assembled microsphere film can generate PNJs on a large scale. However, the pattern and spacing between adjacent PNJs cannot be tuned or increased, since the microspheres in the film are mainly closely packed into a hexagonal pattern, with no room to adjust. This could be improved by utilizing microspheres with different sizes, aiming to adjust their lattice constant of the film.

Nevertheless, the range of achievable PNJ spacing is still quite limited, due to the limited number of spices or sizes of microspheres that are suitable for PNJ generation.

Numerous approaches have been developed to organize nano- or microscopic components into patterns through self-assembly [23]. Some of the approaches focus on programming of the nano- or microscopic components to organize in a predetermined way [24,25], while others tailor the environment to direct and control the assembly [26–29]. To obtain self-assembly structures with well-defined patterns, utilizing a pre-designed surface on which these nano- or microscopic components will self-assemble and eventually reside would result in a desirable assembly. Template-assisted self-assembly (TASA) is such an approach that is used to fabricate nano- or microsphere arrays with well-controlled spatial patterns on surfaces patterned with two-dimensional arrays of templates [30]. In this approach, a template with pre-designed pattern is utilized as the substrate for the self-assembly process. Nano- or microscopic components in the patterning liquid are transited over the substrate to dewet through the fine structures on the template. The components in the liquid are pushed into the physical template to form a densely packed structure. As a result, self-assembly structures with a well-defined pattern can be obtained. Here in this work, we report on the fabrication of a large-scale photonic nanojet array using the TASA approach. A microwell template of a 2D large-scale array was fabricated using conventional microfabrication technology. Melamine microspheres were then assembled on these templates. By controlling the parameters during the TASA process, up to 95% of the loading efficiency can be achieved. These well-ordered arrays of microsphere show excellent optical focusing capability. This approach provides a simple, robust, and fast means for the fabrication of a large-scale homogenized photonic nanojet array and, in future, it may develop into versatile tools to image or detect nanometric objects of biological or environmental importance.

2. Materials and Methods

2.1. Materials

Carboxyl-functionalized melamine microspheres (3 μm in diameter, standard deviation <0.1 μm) were purchased from Sigma-Aldrich (St. Louis, MO, USA). Hydrophobic polymer of carbon–fluorine (Cytop-809M) was purchased from AGC (Tokyo, Japan). pH buffer solution was purchased from Aladdin (Shanghai, China). Photoresist (AZ4620) was purchased from Merck (Darmstadt, Germany). AZ 300MIF developer was purchased from Integrated Micro Materials (Argyle, Texas, USA). Float glass wafer (Ø4 in., 550 μm in thickness) was used for microfabrication.

2.2. Methods

2.2.1. Fabrication of the Microwell Array Template

The microwell array template was fabricated using standard microfabrication technology. The main fabrication process of the templates is schematically illustrated in Figure 1a–e. The microwell array template was designed with an area of ~15 mm × 8.6 mm containing 0.17 billion microwells with three different sizes (2.4 μm, 3.3 μm and 4.3 μm in diameter). The microwells were arranged in a quadrilateral pattern with separations of 12 μm in both x- and y- directions, Figure 1f. For the fabrication process, a glass wafer was cleaned by ethanol and isopropanol twice and dried under nitrogen. Then it was exposed to an oxygen plasma to increase its affinity to the Cytop layer. A layer of Cytop–809M (2.5 μm thickness) was spin-coated on the glass wafer. Then the glass wafer with the Cytop on surface was heated on a hot plate to cure the Cytop layer, by elevating the temperature from room temperature to 180 °C in a gradient. To completely volatilize the solvent in the Cytop and to improve adhesion with the substrate, it is recommended to perform final baking at as high a temperature as possible (180 °C in our case). Afterwards, the photoresist AZ 4620 (8 μm thickness) was coated on the Cytop layer. Then the glass wafer was patterned by standard photolithography and followed by O_2 reactive ion etching (RIE). After etching, the photoresist layer was then removed and the glass wafer was diced into small chips that were ready to be used for microsphere patterning.

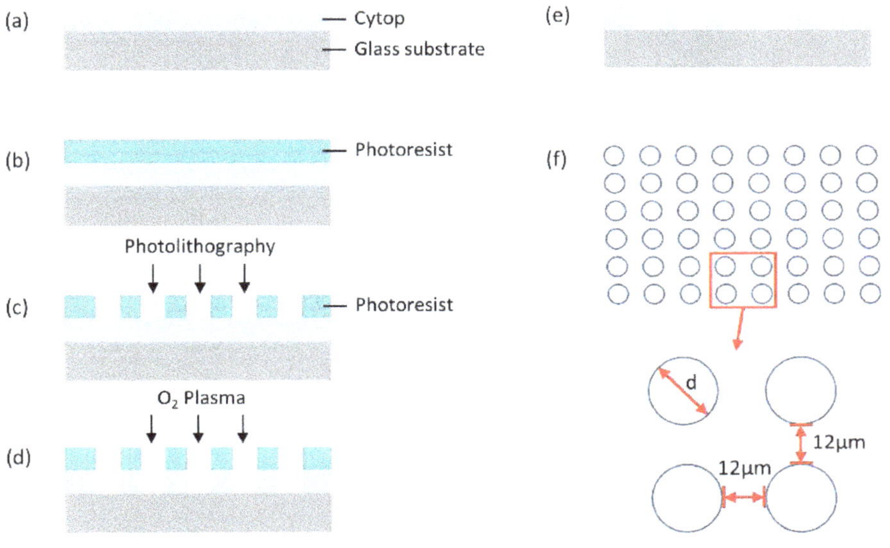

Figure 1. Schematic illustration of the microfabrication process and the design of the microwell array template. (**a**) Cytop is coated on the top of the glass wafer. (**b**) Photoresist is spin-coated on the Cytop layer. (**c**) The glass wafer is patterned by standard photolithography. (**d**) The photoresist layer and Cytop layer are etched by O_2 reactive ion etching. (**e**) Photoresist layer is removed and the microwell array template is obtained. (**f**) The microwell template is arranged in a quadrilateral pattern with separations of 12 μm in both x- and y- directions. Three different sizes of the microwell are designed (d = 2.4 μm, 3.3 μm and 4.3 μm, respectively).

2.2.2. Patterning of the Melamine Microspheres

The carboxyl-functionalized melamine microspheres with 3 μm in diameter were used to be patterned on the microwell array template (Figure 2a). To study the pH dependent patterning efficiency, the microspheres with concentration of 10% solids in aqueous solution was 10× diluted in a buffer with different pH values (pH = 1.0, 3.0, 5.0 and 7.0), respectively, as the patterning solution. A homemade self-assembly setup was used to pattern the melamine spheres on the template via template-assisted self-assembly (TASA) (Figure 2b). The self-assembly setup consists of a rectangular glass slide inclined at an angle of 40°. The glass slide is over a temperature-regulated horizontal copper plate which serves as a support for the microwell array template. The copper plate was fixed on a three-dimensional precision displacement platform. All the patterning process was carried out in an ambient environment with a surrounding temperature of 22 °C and a humidity of ~65% relative humidity (RH). The copper plate was controlled at 24 °C during the TASA process. During the TASA process, 10 μL of the patterning solution was introduced into the wedge that was formed between the glass slide and the template. The template was translated under the patterning solution for 15 patterning cycles (1 cycle = 1 passage of the patterning solution over the array) at a speed ~200 μm/s. After the patterning process, excess microspheres on the array surface were gently washed away by Milli-Q water and the surface was dried by nitrogen.

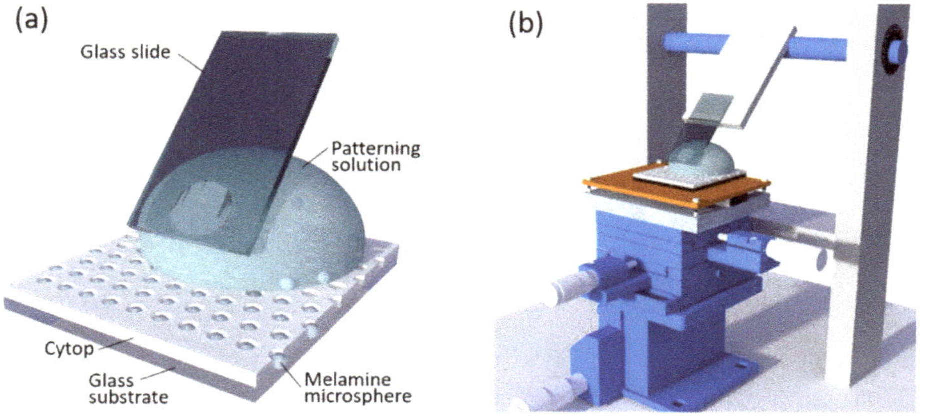

Figure 2. Schematic illustration of (**a**) patterning of the melamine microspheres on the microwell array template by (**b**) adjustable self-assembly setup.

2.2.3. Observation of the Photonic Nanojet (PNJ)

The as prepared microsphere array was configured with a glass ring on the top. Milli-Q water was introduced into the glass ring, in which the microspheres were fully immersed. The microsphere array was illuminated by a light-emitting diode (LED) white light source from the bottom of the glass substrate. A 20× objective with NA of 0.8 (Zeiss Objective Plan-Apochromat, Zeiss, Oberkochen, Germany) was utilized to image the PNJ from the top of the microsphere array, focusing on the focal plane of the microspheres (Figure 3).

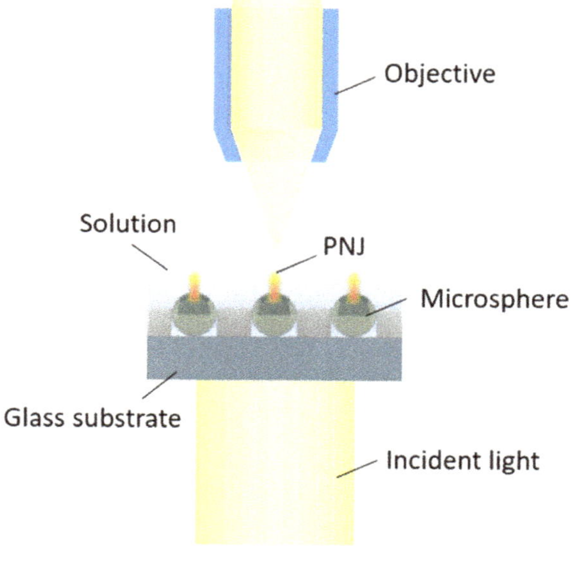

Figure 3. Schematic illustration of the optical setup used in photonic nanojet (PNJ) observation and measurements.

3. Results and Discussion

The morphology of the fabricated microwells with 2.4 μm, 3.3 μm and 4.3 μm in diameter was characterized by a light microscope and a scanning electron microscope (Zeiss Gemini scanning electron microscope (SEM), Zeiss, Oberkochen, Germany) (Figure 4). It is shown that the fabricated Cytop microwells possess uniform size and well-defined spacing in the same batch. The depth of the microwell is ~2.5 μm measured by a profile meter. Water contact angle was measured as 96.9 ± 0.57 °C on the area which contains the microwell array.

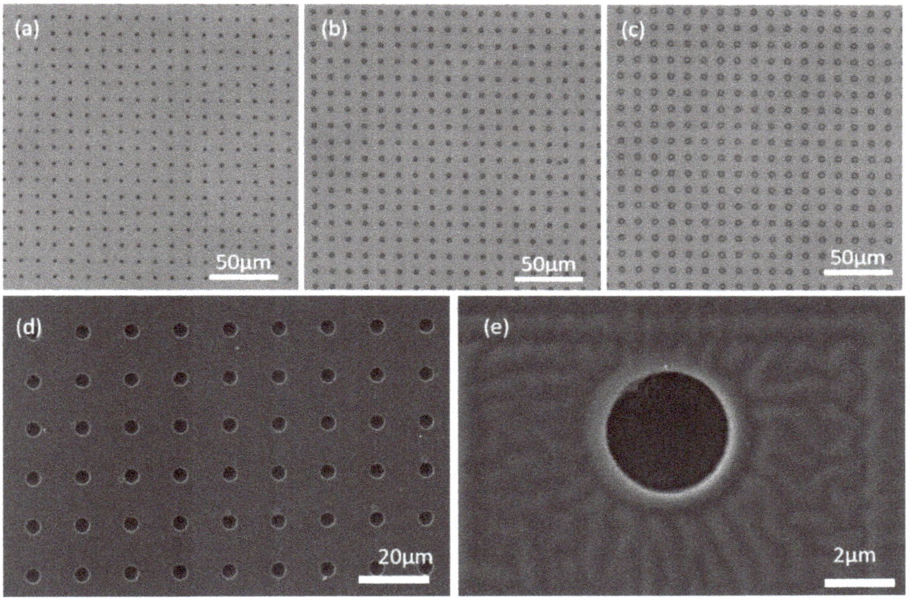

Figure 4. Morphology of the microwells. (**a**–**c**) represent the microwells with the size of 2.4 μm, 3.3 μm and 4.3 μm, respectively, characterized by a conventional optical microscope; (**d**) scanning electron microscope (SEM) image of the 3.3 μm microwells and (**e**) the enlarged SEM image of a single microwell.

After the fabrication of the microwell array, carboxyl-functionalized melamine microspheres of 3 μm were patterned via TASA approach. These commercial available melamine microspheres are practically uniform in size with standard deviation <0.1 μm. Furthermore, the previous work shows that the melamine sphere of 3 μm possesses excellent optical focusing capability than that of other sizes [11], which is beneficial to generate PNJ with best optical focusing capability in our experiment. After the patterning process, the microwell array with the patterned melamine spheres was characterized under an optical microscope and a scanning electron microscope (Figure 5). It is shown that the microspheres are precisely loaded into each individual microwell without access microspheres on the raised regions of the template. Moreover, the configuration of microspheres occupying a microwell is determined by the size of the template microwell and the size of the microspheres. For the microwells with 2.4 μm (Figure 5a), since the size of the microwell is smaller than the microsphere (3 μm), the microspheres were trapped partially on their lower surface into the microwell. This can be seen clearly on the SEM images (Figure 5d). However, for the microwells of 3.3 μm or 4.3 μm, who have larger size than the microsphere, fully entry of the microsphere into the microwell occurred. This can also be observed clearly on the template microwells of 3.3 μm and 4.3 μm (Figure 5e,f), in which the microspheres were pushed onto the sidewall, leaving excess rooms in the microwell.

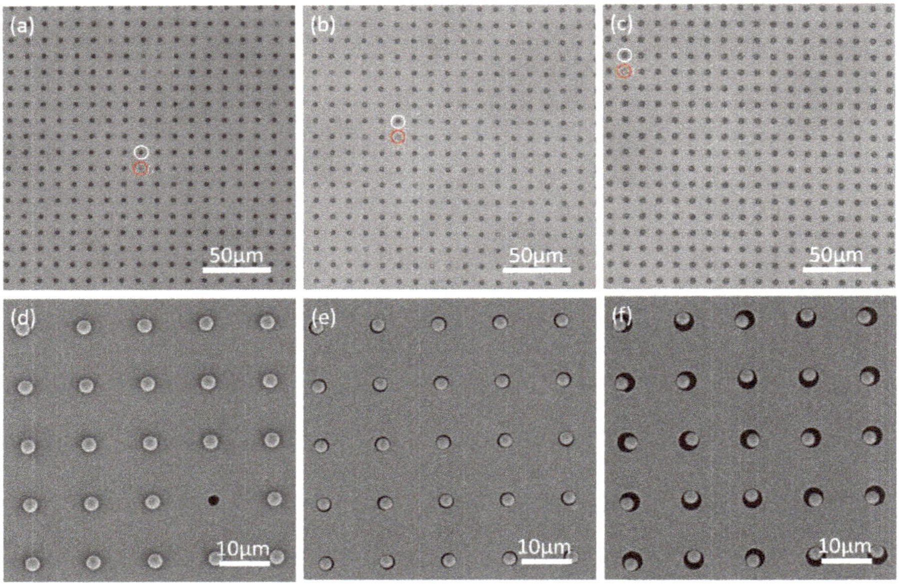

Figure 5. Microsphere array patterned on microwell array template of different sizes. (**a**–**c**) represent the microscopic image of the microspheres loaded on the microwells with the size of 2.4 μm, 3.3 μm and 4.3 μm, respectively. The white circle on each figure indicates the microwell is occupied with a microsphere and the red circle indicates the microwell without microsphere occupation. (**d**–**f**) represent the scanning electron microscope (SEM) images of the microspheres loaded on the microwells with the size of 2.4 μm, 3.3 μm and 4.3 μm, respectively.

Furthermore, the microwell with or without melamine microsphere loading can be easily distinguished, as they exhibited different shades and thicknesses on their contours. This allows us to statistically analyze the quality of the patterning, by counting the number of the microwells with or without microsphere. The patterning efficiency η, which indicates the quality of the patterning, is defined as $\eta = N_{load}/N_{tot}$, where N_{load} is the number of microwells containing the melamine microsphere and N_{tot} is the total number of microwells. It is found that the patterning efficiency η was influenced by the number of the patterning cycles, the pH value of the patterning solution, and also the size of the microwell in our experiments. For the approach of TASA we used in this work, more patterning cycles increase the probability that the microspheres encounter the microwells, thus increasing the patterning efficiency. It is found that the patterning efficiency varies from sample to sample with a relative large deviation when counted individually at the end of each patterning cycle. However, a very high efficiency (>95%) can be achieved after 15 patterning cycles for each sample in a patterning solution with proper pH condition. In our experiment, all the samples were conducted and statistically analyzed after 15 patterning cycles.

The pH value of the patterning solution exhibits a significant influence on the patterning efficiency. To study the pH effect, a patterning solution with different pH values (pH = 1.0, 3.0, 5.0 and 7.0) was used during the patterning process, respectively. We analyzed the patterning efficiency after 15 patterning cycles and the pH dependent patterning efficiency of the template with microwells of 4.3 μm is shown in Figure 6. It is shown that the patterning efficiency strongly depended on the pH value of the patterning solution. With the increase of the pH value, the patterning efficiency decreases. The patterning solution with pH value of 1.0 yielded the highest patterning efficiency (~99%). This result revealed the electrostatic nature of the patterning process. As reported from the literature [22], the pH value of the patterning solution influences the species of the charge on the surface of the microspheres and the bottom substrate, respectively, due to the isoelectric point of different materials. Thus, the electrostatic force caused by these charges has an impact on the entry of the microspheres into microwells. In our experiments, higher pattering efficiencies were obtained for pattering solution with lower pH (≤3.0). Considering the isoelectric point of the melamine microsphere (~3), the Cytop layer (~2) and glass substrate (<3), at a patterning solution of pH lower than 3.0, the microspheres actually hold the positive charges, while the substrate holds negative charges. In this case, their electrostatic interaction is positive, that is, microspheres are attracted into the microwells to a greater extent, thus promoting the patterning efficiency and stability during the patterning process.

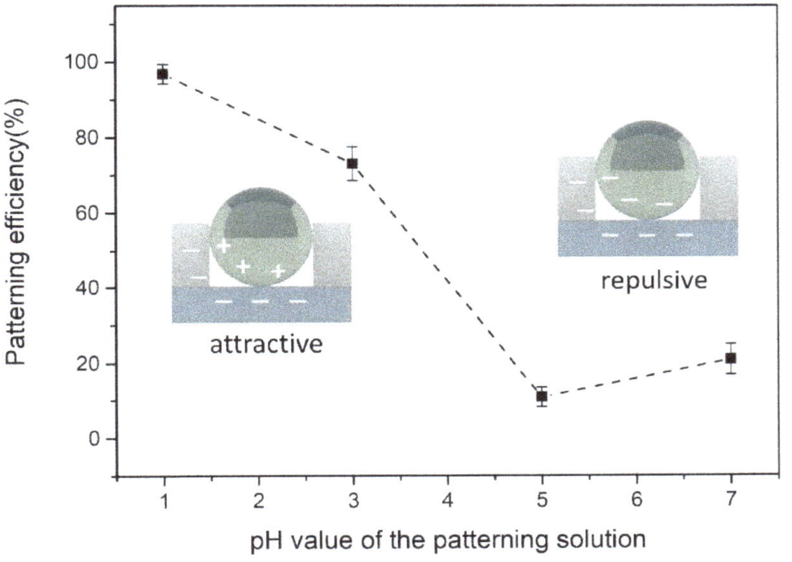

Figure 6. The patterning efficiency of the template with microwells of 4.3 μm versus the pH value of the patterning solution. Sketches illustrate the surface charges of the microspheres and bottom substrate during the patterning process at different pH conditions, indicating the electrostatic charges involved in the patterning process.

Under the optimized patterning cycle and pH value of the patterning solution, the patterning efficiency is >95% for the templates with microwells of 2.4 μm or 3.3 μm, and >99% for the templates with microwells of 4.3 μm. This slight increase of the patterning efficiency (~4%) on templates with larger size of the microwell may arise from the different configurations of microsphere occupying a microwell. In the TASA approach, the size of the microwell influences the configuration and stability of an individual microsphere in the microwell [30], as shown in Figure 5. A microwell with larger size allows the fully entry of the microsphere, thus the electrostatic attraction applies on a larger surface area, providing stronger attraction force and a more stable accommodation. Moreover, the size of the microwell determines the probability of these microspheres encountering the microwells during the patterning process, since an array with larger size of the microwell provides more capture area (interior of the microwell), thus higher chances for the microspheres to be captured. Therefore, an array with microwells of relatively large size would benefit a higher patterning efficiency, although this effect only exhibited a minor influence on the patterning efficiency in our experiment. It should be noted that the data of patterning efficiency in our experiments was collected after 15 patterning cycles. The patterning efficiency can be further increased by additional patterning cycles.

To observe the light-focusing effect of the microsphere, the microsphere array was immersed in water solution and illuminated from the bottom of the glass substrate, as illustrated in Figure 3. Under such an optical configuration, the as patterned melamine microspheres served as microlens array and could generate PNJs in the close vicinity of the top surface, which can be directly observed by the objective. Since the PNJs possess high light intensity and narrow waist, they can be observed by first focusing on the vicinity of the top surface of the microsphere, where bright white spots emerged in the view, and then followed by fine tuning the focal plane of the white spots to the highest light intensity, as in Figure 7a. It is shown that the arrayed microspheres showed excellent light focusing capability. Individual microspheres focused the light illumined from the bottom into a highly localized white spot, i.e., the PNJ. The PNJ possessed high light intensity and a peak at the center of the microsphere (Figure 7b). The peak value can be as high as ~2 fold the background light signal. It is demonstrated in Figure 7a that a large-scale PNJ array with a defined pattern and spacing was achieved. Furthermore, the arrayed melamine microspheres possess excellent size uniformity and highly consistent radial position (direction along the incident light) in the microwells. Furthermore, they possess high refractive index (1.68), low light absorbance, and optimal light-focusing capability in water [22]. Thus with these advantages, the PNJs generated by these microspheres in Figure 7a could exhibit nearly identical geometry parameters and height on the radial position. These homogenized PNJ array has great potential in practical applications, especially in super-resolution imaging, subwavelength-resolution nanopatterning, and nano objects trapping and detection.

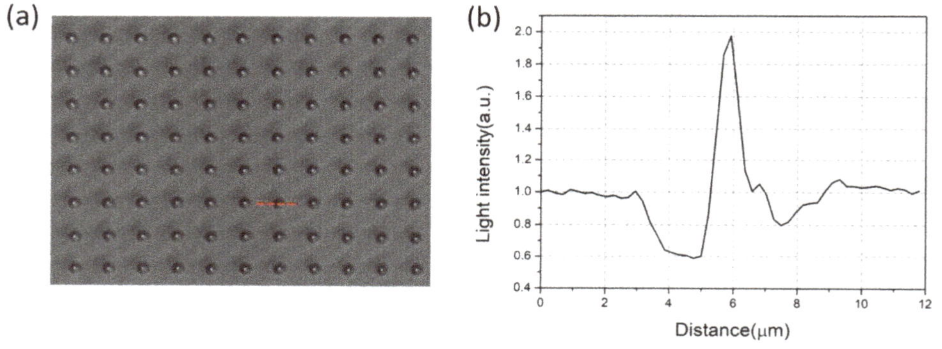

Figure 7. Microlens array and its light focusing property. (**a**) Microscopic image of the microspheres in water solution (refractive index = 1.33) illuminated by the white light from the bottom of the glass substrate. The objective is focused on the top plane of the microspheres where highest light intensity is observed. The light intensity of the dashed red line is presented in (**b**). The peak value of the light intensity at ~6 μm is ~2 times higher than the background signal from the area without microsphere.

4. Conclusions

A large-scale photonic nanojet array was fabricated via a template-assisted self-assembly approach. Templates of large-scale microwell array of three different sizes were fabricated using conventional microfabrication technology. Uniform-sized melamine microspheres of 3 μm were utilized to assemble on the templates. It is shown that the patterning efficiency is strongly depended on the pH values of the patterning solution. Under the optimized condition, the patterning efficiency can be up to 95% for the templates with different sizes. The melamine microspheres serve as the dielectric microsphere lenses and can generate homogenized PNJs over a large-scale sample surface. The intensity of the PNJs in water can be as high as ~2 fold the background light signal. This TASA approach offers a simple, robust, and fast means for the fabrication of large-scale homogenized photonic nanojet array. Therefore, we believe that our technique that exploits the properties of a large-scale PNJ array could evolve into a versatile tool for super resolution imaging and nano-object detection.

Author Contributions: Conceptualization, P.Z. and H.Y.; methodology, P.Z. and X.C.; validation, P.Z.; formal analysis, P.Z.; data curation, P.Z.; writing—original draft preparation, P.Z.; writing—review and editing, H.Y. All authors have read and agreed to the published version of the manuscript.

Funding: This research was funded by China Postdoctoral Science Foundation (2019M660218), National Natural Science Foundation of China (61805271), Key-Area Research and Development Program of Guangdong Province (2019B020226004), CAS Key Laboratory of Health Informatics (2011DP173015), Innovation Program for Excellent Young Researchers of the Shenzhen Institutes of Advanced Technology, and Basic and Applied Research Programs of Science and Technology Commission Foundation of Guangdong Province (2019A1515110974).

Acknowledgments: The authors would like to thank Rui Hao on the help of scanning electron microscope characterization.

Conflicts of Interest: The authors declare no conflict of interest.

References

1. Heifetz, A.; Kong, S.C.; Sahakian, A.V.; Taflove, A.; Backman, V. Photonic Nanojets. *J. Comput. Theor. Nanosci.* **2009**, *6*, 1979–1992. [CrossRef] [PubMed]
2. Minin, I.V.; Minin, O.V.; Geints, Y.E. Localized Em and Photonic Jets from Non-Spherical and Non-Symmetrical Dielectric Mesoscale Objects: Brief Review. *Ann. Phys.* **2015**, *527*, 491–497. [CrossRef]
3. Wang, Z.; Guo, W.; Li, L.; Luk'yanchuk, B.; Khan, A.; Liu, Z.; Chen, Z.; Hong, M. Optical Virtual Imaging at 50 Nm Lateral Resolution with a White-Light Nanoscope. *Nat. Commun.* **2011**, *2*, 218. [CrossRef] [PubMed]
4. Lecler, S.; Perrin, S.; Leong-Hoi, A.; Montgomery, P. Photonic Jet Lens. *Sci. Rep.* **2019**, *9*, 4725. [CrossRef] [PubMed]
5. Yang, H.; Moullan, N.; Auwerx, J.; Martin Gijs, A.M. Super-Resolution Biological Microscopy Using Virtual Imaging by a Microsphere Nanoscope. *Small* **2014**, *10*, 1712–1718. [CrossRef] [PubMed]
6. Yang, H.; Trouillon, R.; Huszka, G.; Gijs, M.A. Super-Resolution Imaging of a Dielectric Microsphere Is Governed by the Waist of Its Photonic Nanojet. *Nano Lett.* **2016**, *16*, 4862–4870. [CrossRef]
7. Huszka, G.; Yang, H.; Gijs, M.A.M. Microsphere-Based Super-Resolution Scanning Optical Microscope. *Opt. Express* **2017**, *25*, 15079–15092. [CrossRef]
8. Gerard, D.; Wenger, J.; Devilez, A.; Gachet, D.; Stout, B.; Bonod, N.; Popov, E.; Rigneault, H. Strong Electromagnetic Confinement near Dielectric Microspheres to Enhance Single-Molecule Fluorescence. *Opt. Express* **2008**, *16*, 15297–15303. [CrossRef]
9. Gerard, D.; Devilez, A.; Aouani, H.; Stout, B.; Bonod, N.; Wenger, J.; Popov, E.; Rigneault, H. Efficient Excitation and Collection of Single-Molecule Fluorescence Close to a Dielectric Microsphere. *J. Opt. Soc. Am. B-Opt. Phys.* **2009**, *26*, 1473–1478. [CrossRef]
10. Aouani, H.; Schon, P.; Brasselet, S.; Rigneault, H.; Wenger, J. Two-Photon Fluorescence Correlation Spectroscopy with High Count Rates and Low Background Using Dielectric Microspheres. *Biomed. Opt. Express* **2010**, *1*, 1075–1083. [CrossRef]
11. Yang, H.; Gijs, M.A. Microtextured Substrates and Microparticles Used as in Situ Lenses for on-Chip Immunofluorescence Amplification. *Anal. Chem.* **2013**, *85*, 2064–2071. [CrossRef] [PubMed]
12. Yi, K.J.; Wang, H.; Lu, Y.F.; Yang, Z.Y. Enhanced Raman Scattering by Self-Assembled Silica Spherical Microparticles. *J. Appl. Phys.* **2007**, *101*, 063528. [CrossRef]
13. Wu, W.; Katsnelson, A.; Memis, O.G.; Mohseni, H. A Deep Sub-Wavelength Process for the Formation of Highly Uniform Arrays of Nanoholes and Nanopillars. *Nanotechnology* **2007**, *18*, 485302. [CrossRef]
14. McLeod, E.; Arnold, C.B. Subwavelength Direct-Write Nanopatterning Using Optically Trapped Microspheres. *Nat. Nanotechnol.* **2008**, *3*, 413–417. [CrossRef] [PubMed]
15. Chen, Z.; Taflove, A.; Backman, V. Photonic Nanojet Enhancement of Backscattering of Light by Nanoparticles: A Potential Novel Visible-Light Ultramicroscopy Technique. *Opt. Express* **2004**, *12*, 1214–1220. [CrossRef] [PubMed]
16. Li, X.; Chen, Z.; Taflove, A.; Backman, V. Optical Analysis of Nanoparticles Via Enhanced Backscattering Facilitated by 3-D Photonic Nanojets. *Opt. Express* **2005**, *13*, 526–533. [CrossRef]
17. Chen, Z.; Taflove, A.; Li, X.; Backman, V. Superenhanced Backscattering of Light by Nanoparticles. *Opt. Lett.* **2006**, *31*, 196–198. [CrossRef]
18. Li, Y.C.; Xin, H.B.; Lei, H.X.; Liu, L.L.; Li, Y.Z.; Zhang, Y.; Li, B.J. Manipulation and Detection of Single Nanoparticles and Biomolecules by a Photonic Nanojet. *Light Sci. Appl.* **2016**, *5*, e16176. [CrossRef]

19. Li, Y.; Xin, H.; Liu, X.; Zhang, Y.; Lei, H.; Li, B. Trapping and Detection of Nanoparticles and Cells Using a Parallel Photonic Nanojet Array. *ACS Nano* **2016**, *10*, 5800–5808. [CrossRef]
20. Gu, G.; Song, J.; Chen, M.; Peng, X.; Liang, H.; Qu, J. Single Nanoparticle Detection Using a Photonic Nanojet. *Nanoscale* **2018**, *10*, 14182–14189. [CrossRef]
21. Schwartz, J.J.; Stavrakis, S.; Quake, S.R. Colloidal Lenses Allow High-Temperature Single-Molecule Imaging and Improve Fluorophore Photostability. *Nat. Nanotechnol.* **2010**, *5*, 127–132. [CrossRef] [PubMed]
22. Yang, H.; Cornaglia, M.; Gijs, M.A. Photonic Nanojet Array for Fast Detection of Single Nanoparticles in a Flow. *Nano Lett.* **2015**, *15*, 1730–1735. [CrossRef] [PubMed]
23. Whitesides, G.M.; Grzybowski, B. Self-Assembly at All Scales. *Science* **2002**, *295*, 2418–2421. [CrossRef] [PubMed]
24. Glotzer, S.C.; Solomon, M.J. Anisotropy of Building Blocks and Their Assembly into Complex Structures. *Nat. Mater.* **2007**, *6*, 557–562. [CrossRef]
25. Zhang, P.; Bousack, H.; Dai, Y.; Offenhausser, A.; Mayer, D. Shell-Binary Nanoparticle Materials with Variable Electrical and Electro-Mechanical Properties. *Nanoscale* **2018**, *10*, 992–1003. [CrossRef]
26. Rabani, E.; Reichman, D.R.; Geissler, P.L.; Brus, L.E. Drying-Mediated Self-Assembly of Nanoparticles. *Nature* **2003**, *426*, 271–274. [CrossRef]
27. Grilli, S.; Ferraro, P. Dielectrophoretic Trapping of Suspended Particles by Selective Pyroelectric Effect in Lithium Niobate Crystals. *Appl. Phys. Lett.* **2008**, *92*, 232902. [CrossRef]
28. Pagliarulo, V.; Gennari, O.; Rega, R.; Mecozzi, L.; Grilli, S.; Ferraro, P. Twice Electric Field Poling for Engineering Multiperiodic Hex-Ppln Microstructures. *Opt. Lasers Eng.* **2018**, *104*, 48–52. [CrossRef]
29. Deng, R.; Yang, L.; Bain, C.D. Combining Inkjet Printing with Emulsion Solvent Evaporation to Pattern Polymeric Particles. *ACS Appl. Mater. Interfaces* **2018**, *10*, 12317–12322. [CrossRef]
30. Matthew, R.; Pedro Camargo, H.C.; Xia, Y. Template-Assisted Self-Assembly: A Versatile Approach to Complex Micro- and Nanostructures. *Soft Matter* **2009**, *5*, 1129–1136.

micromachines

MDPI

Article

Light Extraction Analysis of AlGaInP Based Red and GaN Based Blue/Green Flip-Chip Micro-LEDs Using the Monte Carlo Ray Tracing Method

Shuyu Lan [1], Hui Wan [2], Jie Zhao [1] and Shengjun Zhou [1,2,3,*]

1. Center for Photonics and Semiconductors, School of Power and Mechanical Engineering, Wuhan University, Wuhan 430072, China; lanshuyu@whu.edu.cn (S.L.); zjie1994@whu.edu.cn (J.Z.)
2. The Institute of Technological Sciences, Wuhan University, Wuhan 430072, China; wanhui_hb@whu.edu.cn
3. State Key Laboratory of Applied Optics, Changchun Institute of Optics, Fine Mechanics and Physics, Chinese Academy of Sciences, Changchun 130033, China

* Correspondence: zhousj@whu.edu.cn; Tel.: +86-027-5085-3293

Received: 6 November 2019; Accepted: 5 December 2019; Published: 7 December 2019

Abstract: Micro-scale light emitting diodes (micro-LEDs) commonly employ a thin-film flip-chip (TFFC) structure whose substrate is lifted off by an excimer laser. However, flip-chip (FC) micro-LEDs with a substrate can provide a sharp rise on sidewall emission by increasing the sidewall area. Here, we investigate the influence of substrate thickness, encapsulation, surface texture, microstructures between the substrate and epilayer, as well as the size, cutting shape, and angle of the chip on the light extraction efficiencies (LEEs) of FC micro-LEDs by using the Monte Carlo ray tracing method. We find that the LEE of the blue FC micro-LED chip increases by 46.5% over that of the blue TFFC micro-LED chip. After the encapsulation with the epoxy lens is applied, the LEEs of the blue TFFC micro-LED and blue FC micro-LED increase by 129% and 110.5%, respectively. The underlying mechanisms for the use of surface texture, patterned sapphire substrate, air-void array, and chip shaping technologies to improve the LEEs of FC micro-LEDs are also investigated in detail. We find that the LEEs AlGaInP based red FC micro-LED and GaN based blue/green FC micro-LEDs exhibit a sharp rise when the chip size drops from 30 to 10 μm. The inverted trapezoid FC micro-LED with patterned sapphire substrate (PSS) and encapsulation shows extraordinarily strong top emission and high collimation. We believe that our study offers a promising and practical route for obtaining high efficiency micro-LEDs.

Keywords: micro-scale light emitting diode; light extraction efficiency; sapphire substrate; encapsulation

1. Introduction

Micro-scale light emitting diodes (micro-LEDs) whose size is smaller than 100 μm have been considered as the next generation display technology because of their outstanding features, such as high dynamic range, good sunlight readability, long lifetime, low power consumption, and wide color gamut [1–3]. Their deployment in wearable devices, cell phones, outdoor displays, and augmented reality is highly anticipated [4,5]. However, the external quantum efficiencies (EQEs) of micro-LEDs are still relatively low, which is far from satisfactory to support their potential applications. Therefore, the improvement in the EQE is an essential issue to enable micro-LEDs to have broader applications. The EQE equals the multiplication of the internal quantum efficiency (IQE) and the light extraction efficiency (LEE) [6]. Since micro-LEDs are obtained from an identical wafer as large scale LEDs, the fruitful methods for improving IQE of large scale LEDs are also applicable for micro-LEDs [7–11]. However, the ratio of sidewall emitting area to top emitting area is greatly different for micro-LEDs and large scale LEDs [12], so the methods applicable to large scale LEDs for improving LEE need to be reassessed in micro-LEDs.

Recently, the light extraction efficiencies (LEEs) of LEDs have experienced a continuous progress due to the use of chip shaping, patterned sapphire substrate (PSS), air-void array (AVA), and surface texture techniques [13–16]. Furthermore, the application of encapsulation also plays an important role in enhancing LEEs of LEDs, owing to the small difference in the refractive index between sapphire substrate and encapsulation. However, a detailed study about the effect of these factors on LEEs of micro-LEDs is not available. Moreover, micro-LEDs commonly employ a thin-film flip-chip (TFFC) structure whose substrate is lifted off by an excimer laser [17,18]. However, flip-chip (FC) micro-LEDs with a substrate can provide a sharp rise on sidewall emission by increasing the sidewall area [19–27]. Consequently, it is of significant importance to investigate the effect of chip shaping, patterned sapphire substrate (PSS), air-void array (AVA), and surface texture on LEE of the FC micro-LED structure. Additionally, it is essential to analyze the influence of the size effect on LEEs of FC micro-LEDs when the chip size drops down to the micron scale.

In this study, we investigated the effects of substrate thickness, encapsulation, surface texture, microstructures between the substrate and epilayer, as well as the size, cutting shape, and angle of the chip on the LEEs of AlGaInP based red FC micro-LED and GaN based blue/green FC micro-LEDs (RGB FC micro-LEDs). These factors become quite critical for the LEEs of RGB FC micro-LEDs when the length and breadth of the chip are much less than its thickness. The LEE of blue FC micro-LEDs increases by 46.5% over that of the blue TFFC micro-LED. The application of chip shaping, PSS, AVA, surface texture, and encapsulation effectively improves the LEEs of RGB FC micro-LEDs. When the chip size drops from 30 to 10 μm, the LEEs of RGB FC micro-LEDs have a sharp rise. Finally, we combine all factors that contributed to the transverse-electric (TE) mode light emission and design an inverted trapezoid FC micro-LED with PSS and encapsulation. The angle of inverted trapezoid is 25°. The diameter of hemispherically shaped PSS is 2 μm, and the refractive index of encapsulation is 1.5. The inverted trapezoid FC micro-LED with PSS and encapsulation shows extraordinarily strong top emission and high collimation. Our analysis paves the way for the realization of high efficiency RGB FC micro-LEDs for potential applications in high resolution displays and augmented reality.

2. Simulation Model

Our simulation was performed by using Light Tools (LTs) software based on the Monte Carlo ray tracing method. Monte Carlo ray tracing is one of the most accurate ways to simulate the light propagation in LEDs [28–30]. Monte Carlo ray-tracing is commonly used in the case that the spontaneous photons are arbitrarily emitted from the active layer and only the geometric optics are considered in the light–structure interaction, which is usually valid when the size of the microstructure is much larger than the wavelength of light.

Figure 1 shows four types of micro-LEDs in the simulation model: TFFC micro-LEDs with and without encapsulation and FC micro-LEDs with and without encapsulation. The dimensions of micro-LEDs were 80 × 80 μm^2. The AlGaInP based red FC micro-LED structure consisted of a metal layer, a GaInP etching stop layer, an n^+-GaAs contact layer, n-cladding AlGaInP, an n-type AlInP diffusion barrier, GaInP/AlGaInP multiple quantum wells, a p-type AlInP diffusion barrier, p-cladding AlGaInP, and a p-GaP window layer [31,32]. The sapphire substrate was bonded onto the AlGaInP based red TFFC micro-LED to form the AlGaInP based red FC micro-LED [33]. The GaN based green/blue FC micro-LEDs structure comprised a metal layer, p-type GaN, a p-AlGaN electron blocking layer, InGaN/GaN multiple quantum wells, n-type GaN, and sapphire substrate [34–38].

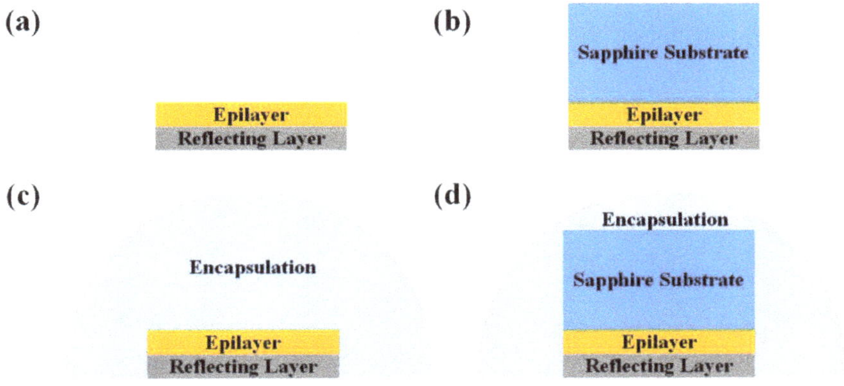

Figure 1. (**a**) Bare thin-film flip-chip (TFFC) micro-LED chip. (**b**) Bare FC micro-LED chip. (**c**) Encapsulated TFFC micro-LED. (**d**) Encapsulated FC micro-LED.

The optical parameters of the layers of RGB micro-LEDs are shown in Table 1 [1]. The central wavelengths were 622 μm, 527 μm, and 470 μm for RGB micro-LEDs. The encapsulation with the epoxy lens was designed as a hemisphere. To ensure that most of the photons could escape into the air without multiple reflections in the encapsulation, the refractive index and diameter of the epoxy lens were set to 1.5 and 1 cm [6], respectively. The bottoms of FC micro-LEDs were equipped with a reflecting layer.

Table 1. Optical parameters of the materials in the simulation model.

Materials	Thickness (μm)	Central Wavelengths (μm)	Refractive Index n	Extinction Coefficient k
Epilayer of red micro-LED	6	622	3.3315	0.02
Sapphire substrate of red micro-LED	120	622	1.7664	
Epilayer of green micro-LED	6	527	2.4263	0.003
Sapphire substrate of green micro-LED	120	527	1.7721	
Epilayer of blue micro-LED	6	470	2.4669	0.003
Sapphire substrate of blue micro-LED	120	470	1.7771	

3. Results and Discussion

Figure 2 shows the relationship between substrate thickness and the LEEs of each face of RGB micro-LEDs. The sidewall and total LEEs of RGB FC micro-LEDs remained constant when the substrate thickness was larger than 28 μm. When the substrate thickness changed from 0 to 28 μm, the sidewall and total LEEs increased with the substrate thickness. However, the top LEEs of RGB FC micro-LEDs were almost constant when the substrate thickness changed from 0 to 150 μm, indicating that the variation of total LEEs depended on the change of sidewall LEEs. Owing to the high extinction coefficient (k) of AlGaInP materials, the LEEs of AlGaInP based red FC micro-LEDs were markedly lower than those of GaN based green/blue FC micro-LEDs. In addition, the LEE of the blue FC micro-LED increased by 46.5% over that of the blue TFFC micro-LED.

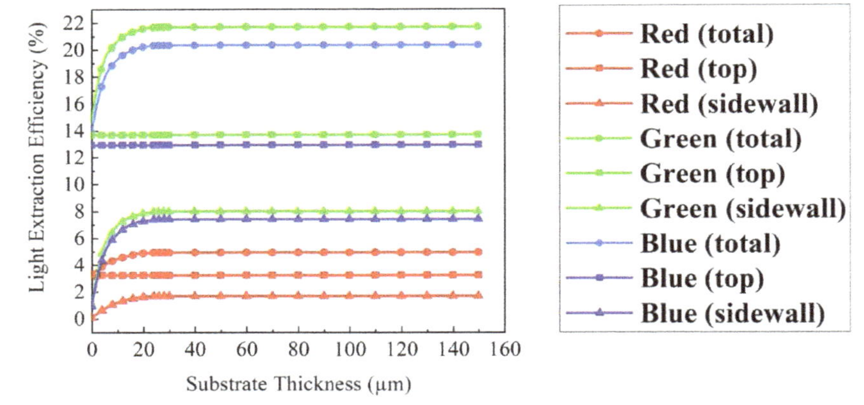

Figure 2. LEEs of each face of RGB FC micro-LEDs with variable thicknesses of sapphire substrate.

Figure 3 shows the luminous intensities of four types of RGB micro-LEDs. The luminous intensities of encapsulated RGB micro-LEDs with the epoxy lens were almost double that of bare RGB micro-LED chips. Actually, when the encapsulation was applied, the LEEs of the blue TFFC micro-LED and blue FC micro-LED increased by 129% and 110.5%, respectively. Additionally, the sidewall emissions of RGB FC micro-LEDs exhibited a significant enhancement due to the thick, transparent sapphire substrate and epoxy lens, leading to the highest peak luminous intensities of encapsulated RGB FC micro-LEDs at ± 27°.

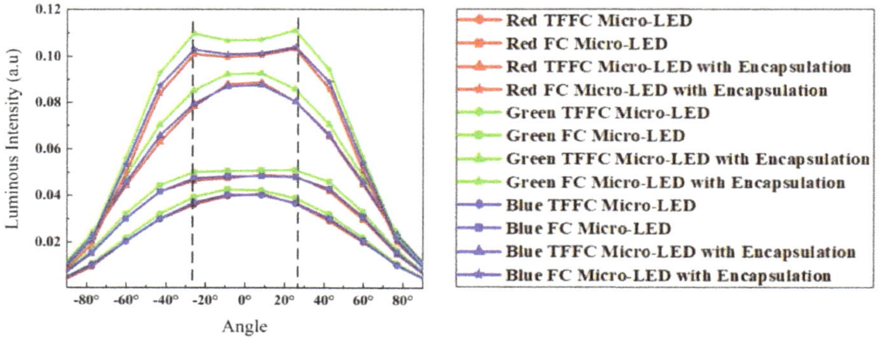

Figure 3. Luminous intensities of four types of RGB micro-LEDs: TFFC micro-LEDs, FC micro-LEDs, TFFC micro-LEDs with encapsulation, and FC micro-LEDs with encapsulation.

Figure 4a shows the relationship between the LEEs of encapsulated blue micro-LEDs and the refractive index of encapsulation. Actually, a smaller refractive index difference led to a larger escape cone, and more photons could escape into the air. Therefore, as shown in Figure 4a, the LEE of the encapsulated blue TFFC micro-LED in Figure 4b kept increasing when the refractive index of encapsulation changed from one to 2.4. However, for the encapsulated blue FC micro-LED, we could see a significant change on the rise pace at the refractive index of 1.4. The angle of total internal reflection (TIR) at the sapphire substrate–encapsulation interface was 52° when the refractive index of encapsulation was 1.4, which ensured that all photons in the sapphire substrate escaped into the encapsulation. Thereby, when the refractive index of encapsulation was smaller than 1.4, photons could escape from the top and sidewall of sapphire substrate, as well as the sidewall of the epilayer. When the refractive index of encapsulation was larger than 1.4, the remaining photons were in the epilayer of the micro-LED, and only the sidewall emission of the epilayer had a slow rise, as shown by the red lines in Figure 4c. Due to the thin epilayer and loss of top emission, the LEE of micro-LED exhibited a slow rise pace when the refractive index was larger than 1.4. Moreover, the encapsulated blue TFFC micro-LED and encapsulated blue FC micro-LED had the same LEE when the refractive index of encapsulation was 1.7771, since they were the same structure at this point. However, when the refractive index of encapsulation was larger than 1.7771, sapphire substrate would limit the rise of LEE because of the TIR at the epilayer–substrate interface.

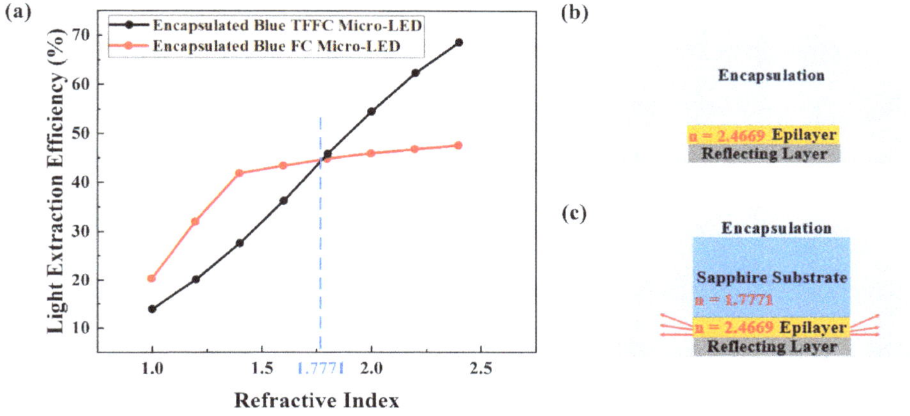

Figure 4. (**a**) LEEs of the encapsulated blue TFFC micro-LED and encapsulated blue FC micro-LED with variable refractive indices of encapsulation. (**b**) Encapsulated blue TFFC micro-LED and (**c**) encapsulated blue FC micro-LED.

In order to further enhance the LEEs of FC micro-LEDs, we introduced a special microstructure on the top surface of sapphire substrate, which was the so-called surface texture. Many photons scattered at the texture interface and emitted into the air, resulting in an enhancement of the LEEs of FC micro-LEDs. We explored the effect of the shapes of microstructures, including a circular truncated cone, circular cone, hemisphere, and pyramid, as shown in Figure 5a, on the LEEs of FC micro-LEDs. The heights and base areas of these microstructures were set to 1 μm and 3.14 μm^2, respectively. As shown in Figure 5b, the luminous intensities of blue FC micro-LEDs with textured substrate were higher than that of the blue FC micro-LED with flat substrate. Since the curved sidewall can contribute to the escape of photons, the luminous intensity of the blue FC micro-LED with a circular truncated cone, circular cone, or hemisphere as the shape of the surface texture was slightly higher than that of the blue FC micro-LED with a pyramid as the shape of the surface texture. Moreover, a circular cone was chosen to be the shape of surface texture to explore the effect of the slanted angle on the LEE of the blue FC micro-LED. In Figure 5c, we find that the slanted angle of 38° was the best angle for enhancing the LEEs of blue FC micro-LEDs.

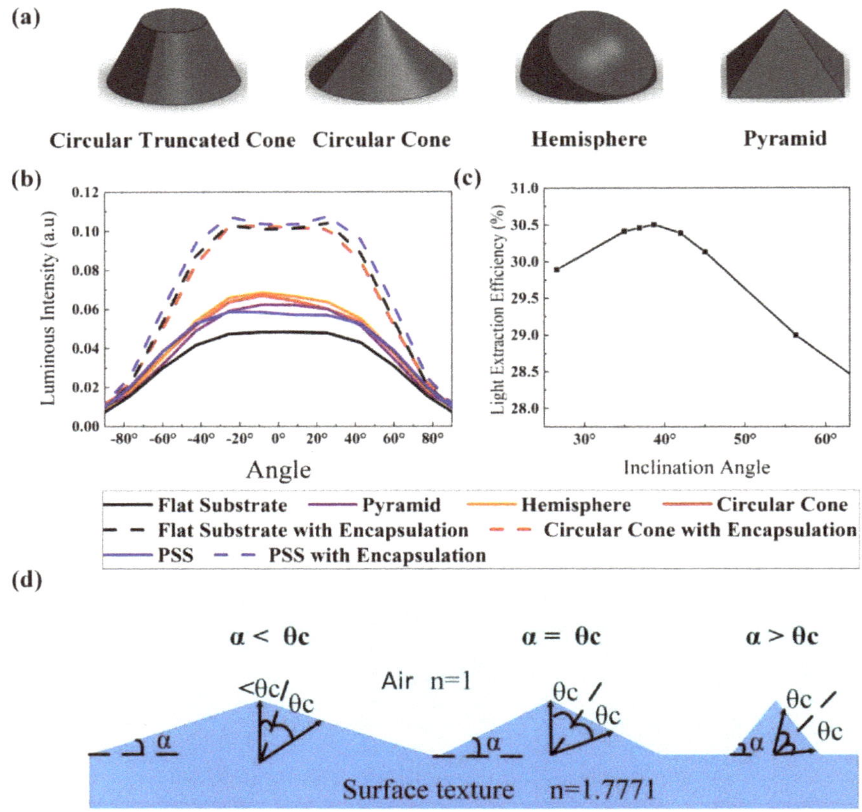

Figure 5. (**a**) Shapes of the microstructure including a circular truncated cone, circular cone, hemisphere, and pyramid. (**b**) Luminous intensities of bare and encapsulated blue FC micro-LEDs with different microstructures. (**c**) LEEs of blue FC micro-LEDs with variable slanted angles of the circular cone. (**d**) Analysis of the effect of the slanted angles of the circular cone on the LEEs of blue FC micro-LEDs. α is the slanted angles of the circular cone, and θc is the angle of total internal reflection (TIR).

As shown in Figure 5d, when the slanted angles of the circular cone (α) were smaller than the angle of TIR (θc = 34°), the range of incident angles of photons that could escape into the air was less than 2θc. When α increased from θc to 90° − θc, the LEEs would be constant since the ranges of the incident angles of photons that could escape into the air were the same (2θc) [39]. However, due to the sidewall emission of photons at a large incident angle, the photons at a small incident angle were more than those at a large incident angle at the layer of the surface texture. Thereby, the maximum LEE was only obtained at the slanted angle of θc. Taking other factors into consideration, the simulation result of 38° was reasonable. Moreover, we investigated the effect of the surface texture on the LEEs of encapsulated blue FC micro-LEDs. As shown by the dashed lines in Figure 5b, the effect of surface texture on the LEEs of encapsulated blue FC micro-LEDs was almost eliminated, since the refractive index difference between sapphire substrate and encapsulation was only 0.2. Most of the photons could escape from the substrate into the encapsulation even without the surface texture. Furthermore, we analyzed the effect of PSS on the LEEs of FC micro-LEDs. As shown in Figure 5b, the PSS could effectively increase the luminous intensity of the blue FC micro-LED with or without encapsulation, revealing that PSS was more efficient than the surface texture in an encapsulated LED.

Figure 6a shows the effect of sidewall texture on the LEEs of each face of the blue FC micro-LED. The sidewall texture significantly increased the sidewall LEE and total LEE of the bare FC micro-LED. When the encapsulation was applied, the total LEE of the FC micro-LED with the sidewall texture was the same as that of the FC micro-LED without the sidewall texture, whereas the sidewall LEE of the FC micro-LED with the sidewall texture was higher than that of the FC micro-LED without the sidewall texture. In Figure 6b,c, the top luminous intensity of the encapsulated FC micro-LED without the sidewall texture is much higher than that of the encapsulated FC micro-LED with the sidewall texture. Therefore, in the case of encapsulation, the sidewall texture was more suitable for transverse-magnetic (TM) mode light emission rather than transverse-electric (TE) mode light emission due to the great enhancement in sidewall emission rather than top emission.

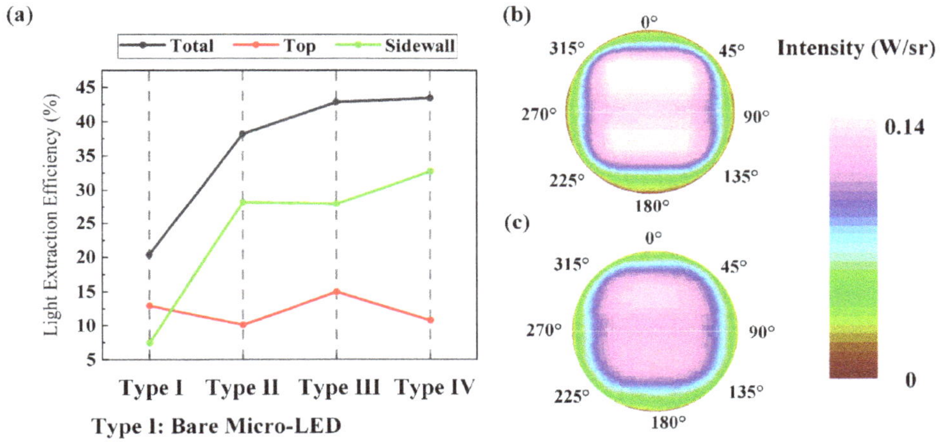

Type I: Bare Micro-LED
Type II: Bare Micro-LED with Sidewall Texture
Type III: Encapsulated Micro-LED
Type IV: Encapsulated Micro-LED with Sidewall Texture

Figure 6. (**a**) LEEs of each face (total, top, sidewall) of blue micro-LEDs. (**b**) Top luminous intensity distribution of the encapsulated blue FC micro-LED. (**c**) Top luminous intensity distribution of the encapsulated blue FC micro-LED with the sidewall texture.

Figure 7a shows the effect of PSS and AVA on the LEEs of each face of blue FC micro-LEDs. Four types of blue micro-LEDs structures were considered: a micro-LED with a flat substrate, a micro-LED with PSS, a micro-LED with AVA, and a micro-LED with PSS and AVA. The diameter of the hemispherically shaped sapphire patterns was set to 2 µm. The air-voids were set as triangular prisms with a cross-section of 3 µm in length and 1.5 µm in height. A reflecting layer was deposited onto the GaN epilayer, as shown in Figure 7b. Figure 7c shows the LEEs of each face of the blue micro-LEDs in Figure 7a,b. For the micro-LEDs without the reflecting layer, the LEEs of each face of the micro-LED with the flat substrate or with PSS were almost the same. The LEEs of each face of the micro-LED with PSS were much higher than those of micro-LED with the flat substrate, indicating that PSS improved the LEEs of each face simultaneously. The micro-LED with AVA had significantly high top and bottom LEEs, but an extremely low sidewall LEE, indicating that AVA enhanced the top and bottom LEEs, but reduced the sidewall LEE, which led to a higher collimated light emission. Moreover, the top and bottom LEEs of the micro-LED with both PSS and AVA were larger than those of the micro-LED with PSS, while the sidewall LEE of the micro-LED with both PSS and AVA was larger than that of the micro-LED with AVA. As a result, the light field of the micro-LED with both PSS and AVA exhibited a uniform distribution, as in Figure 7d, which could lead to a small angular color shift [2]. The micro-LED with AVA had a strong enhancement in top and bottom luminous intensities, while the micro-LED with PSS improved the luminous intensity of each direction simultaneously.

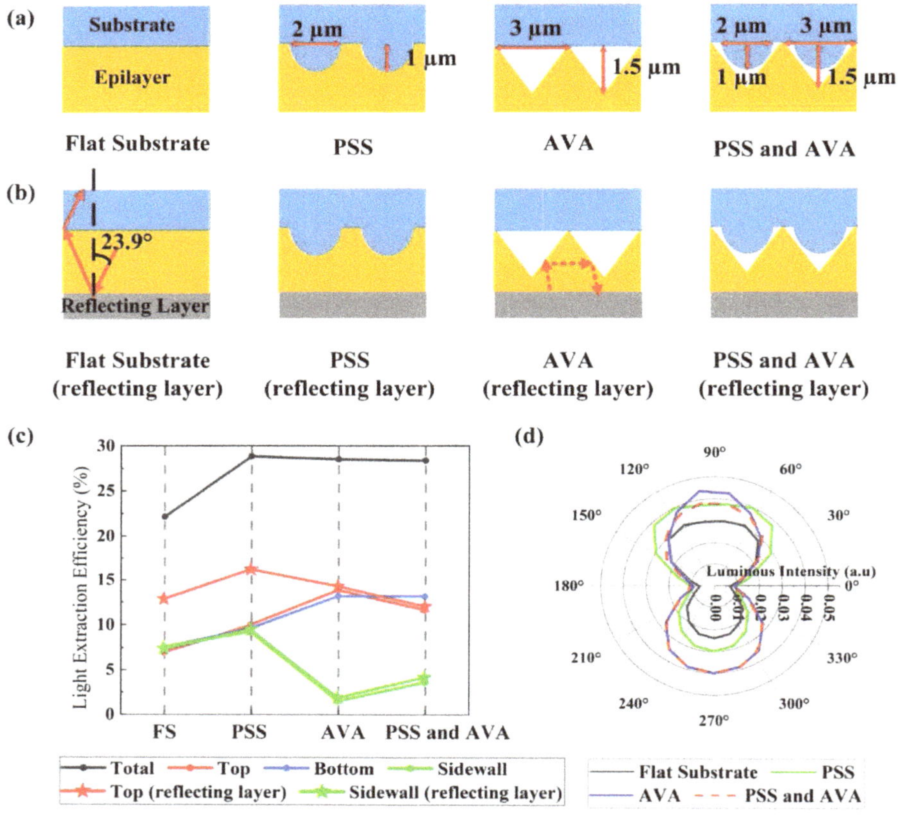

Figure 7. (**a**) Blue micro-LEDs with the flat substrate, PSS, air-void array (AVA), or both (PSS and AVA), but without the reflecting layer. (**b**) Blue micro-LEDs with the flat substrate, PSS, AVA, or both (PSS and AVA) and with the reflecting layer. (**c**) LEEs of each face (total, top, bottom, sidewall) of blue micro-LEDs. (**d**) Light fields of blue micro-LEDs without the reflecting layer.

Compared to micro-LEDs without a reflecting layer, the sidewall LEEs of micro-LEDs with the reflecting layer were almost constant. Since the TIR angle of the epilayer–air interface was 23.9° (Figure 7b), the photons with an incident angle larger than 23.9° at the bottom epilayer–air interface were reflected due to the TIR, which had no effect on the variation of LEEs when a reflecting layer was added. The photons with an incident angle from 0° to 23.9° were reflected by the reflecting layer, and their incident angles at the sidewall and top interface were in the range of 66.1°–90° and 0°–23.9°, respectively. Therefore, they could not emit from the sidewall surface, but could escape from the top surface. However, as shown by the red dashed lines in Figure 7b, the photons reflected by the reflecting layer had an incident angle between 21.1° and 45° at the slanted planes of AVA. As a result, most of the photons reflected by the reflecting layer could not escape from the top surface. Consequently, the sidewall and top LEEs of the micro-LED with AVA or the micro-LED with PSS and AVA were almost constant when a reflecting layer was added.

The cutting shape and angle of the chip also played an important role in improving the LEEs of FC micro-LEDs, since they could change the track of photon propagation. We simulated different cutting shapes of the chip to explore the effect of cutting shapes on the LEEs of FC micro-LEDs. Figure 8a shows the cross-sectional images of chip cutting shapes, including inverted trapezoid, rectangle, trapezoid, parallelogram, hexagon, triangle, and ellipse. In Figure 8b, the luminous intensity of a rectangular blue FC micro-LED was the lowest in these cases. The parallelogram-chip led to an asymmetric light field. Moreover, the light field of the blue FC micro-LED with a cutting shape of a rectangle, hexagon, or inverted trapezoid exhibited a Lambertian distribution, which was suitable for TE mode light emission, while the cutting shape of a trapezoid, triangle, or ellipse substrate was suitable for TM mode light emission due to the strong sidewall emission. We changed the cutting angle of the inverted trapezoid-chip to explore the effect of chip cutting angles on the LEEs of FC micro-LEDs. As shown in Figure 8c, the luminous intensity patterns of micro-LEDs became more collimated when the cutting angle changed from 0° to 20°, which meant more photons emitted to the air at a smaller angle. However, the luminous intensity reduced when the cutting angle was larger than 25°. Hence, it was better to choose a cutting angle in the range of 20° to 25°.

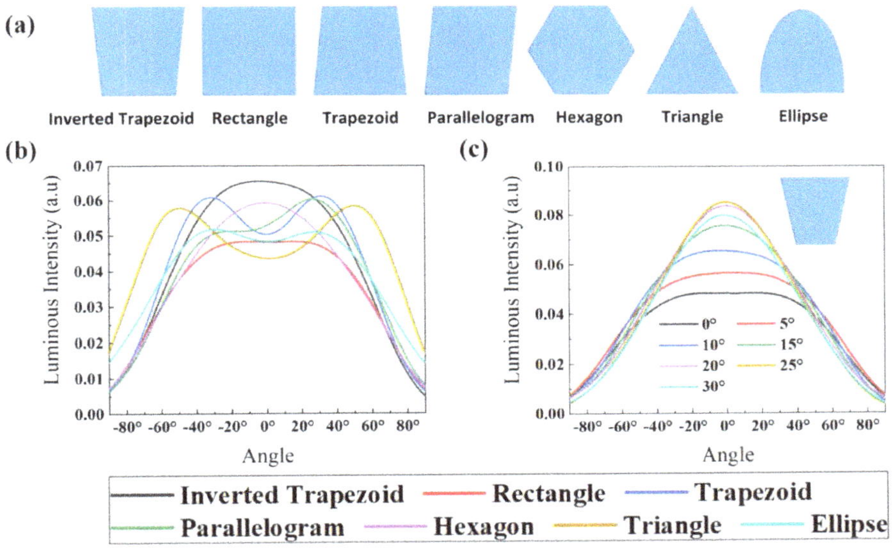

Figure 8. (**a**) Cross-sectional images of chip cutting shapes. Luminous intensities of blue FC micro-LEDs: (**b**) with different cutting shapes of the chip; (**c**) with different cutting angles of the inverted trapezoid-chip.

Figure 9a shows the effect of the chip size on the LEE of RGB FC micro-LEDs. we kept the height of RGB FC micro-LEDs at 126 μm and changed the chip size from 10 to 100 μm. When the chip size ranged from 10 to 100 μm, the sidewall and total LEEs of RGB FC micro-LEDs decreased, while the top LEEs were almost constant, indicating that the variation of total LEEs depended on the change of sidewall LEEs. Figure 9b shows the propagation pathways of photons in FC micro-LEDs with different chip sizes. As shown in Figure 9b, the light propagation pathway of sidewall emission increased with the chip size, resulting in more photons being absorbed by the epilayer of RGB FC micro-LEDs. However, the increase of the chip size had no effect on the top emission. Additionally, when the chip size dropped from 30 to 10 μm, the sidewall and total LEEs of RGB FC micro-LEDs had a sharp rise. For a deep ultraviolet LED with TM polarized light emission, this phenomenon would be more obvious.

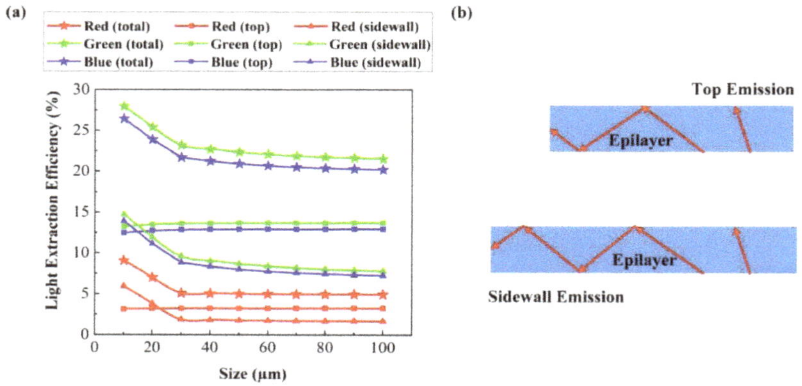

Figure 9. (**a**) LEEs of each face of RGB FC micro-LEDs with variable chip size. (**b**) Propagation pathways of photons in FC micro-LEDs with different chip sizes.

We combined all factors that contributed to the TE mode light emission and designed an inverted trapezoid FC micro-LED with PSS and encapsulation as shown in Figure 10a.The angle of the inverted trapezoid was 25°. The diameter of the hemispherically shaped PSS was 2 μm, and the refractive index of encapsulation was 1.5. Figure 10b shows the significantly high top luminous intensity of the optimized FC micro-LED. In Figure 10c, the inverted trapezoid FC micro-LED with PSS and encapsulation exhibited much stronger top emission and higher collimation than the TFFC micro-LED with or without encapsulation and the FC micro-LED with or without encapsulation. This result revealed that the inverted trapezoid FC micro-LED with PSS and encapsulation was extremely suitable for applications in high resolution displays and augmented reality. Actually, compared to these four types of micro-LEDs, the LEE of the inverted trapezoid FC micro-LED with PSS and encapsulation increased by 231.9%, 126.5%, 44.9%, and 7.6%.

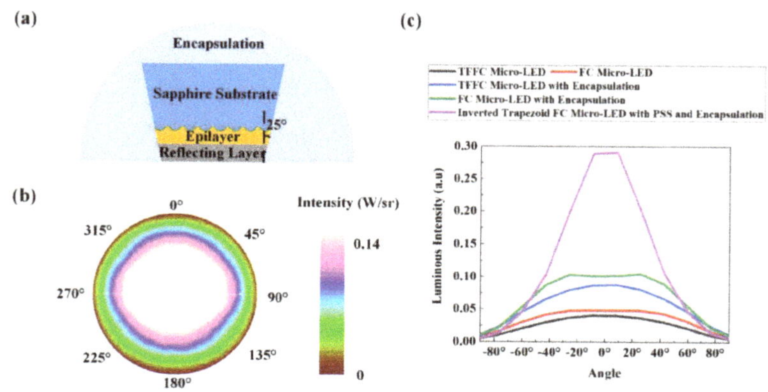

Figure 10. (**a**) Inverted trapezoid FC micro-LED with PSS and encapsulation. (**b**) Top luminous intensity distribution of the inverted trapezoid FC micro-LED with PSS and encapsulation. (**c**) Comparison between the luminous intensities of the inverted trapezoid FC micro-LED with PSS and encapsulation, TFFC micro-LED, FC micro-LED, TFFC micro-LED with encapsulation, and FC micro-LED with encapsulation.

4. Conclusions

In summary, we investigated the influence of substrate thickness, encapsulation, surface texture, microstructures between the substrate and epilayer, as well as the size, cutting shape, and angle of the chip on the LEEs of FC micro-LEDs by using the Monte Carlo ray tracing method. For GaN based blue micro-LEDs, the LEE of the bare FC micro-LED chip increased by 46.5% over that of the bare TFFC micro-LED chip. After the encapsulation with the epoxy lens was applied, the LEEs of the blue TFFC micro-LED and blue FC micro-LED increased by 129% and 110.5%, respectively. However, the effect of the surface texture on the LEEs of encapsulated micro-LEDs was almost eliminated due to the small refractive index difference between sapphire substrate and encapsulation. The PSS technology could increase the LEE when encapsulation was applied, revealing that PSS was more efficient than the surface texture in an encapsulated LED. The PSS technology improved the top, bottom, and sidewall LEEs of the micro-LED simultaneously, and the AVA had a strong enhancement of the top and bottom LEEs, but a decrease of the sidewall LEE, thereby leading to a higher collimated light emission. Moreover, the combination of PSS and AVA brought a uniform distribution on the light field whose angular color shift was small. Additionally, an inverted trapezoid chip with an angle within 20° to 25° brought a larger luminous intensity and smaller emit angle. For the AlGaInP based red FC micro-LED and GaN based blue/green FC micro-LEDs, the LEEs had a sharp rise when the chip size dropped from 30 to 10 μm. The present findings emphasized the importance of FC structure for improving the LEEs of micro-LEDs, which paves the way for high efficiency micro-LEDs.

Author Contributions: Data curation, S.L.; funding acquisition, S.Z.; investigation, S.L., H.W., and J.Z.; methodology, S.L.; project administration, S.Z.; supervision, S.Z.; writing—original draft, S.L.; writing—review, S.Z.

Funding: This work was funded by the National Natural Science Foundation of China (Grant Nos. 51775387 and 51675386) and the Natural Science Foundation of Hubei Province (Grant No. 2018CFA091).

Conflicts of Interest: The authors declare no conflict of interest.

References

1. Yang, S.M.; Wang, P.H.; Chao, C.H.; Chu, C.W.; Yeh, Y.T.; Chen, Y.S.; Chang, F.P.; Fang, Y.H.; Lin, C.C.; Wu, C.I. Angular color variation in micron-scale light-emitting diode arrays. *Opt. Express* **2019**, *27*, A1308–A1323. [CrossRef]
2. Gou, F.W.; Hsiang, E.L.; Tan, G.J.; Chou, P.T.; Li, Y.L.; Lan, Y.F.; Wu, S.T. Angular color shift of micro-LED displays. *Opt. Express* **2019**, *27*, A746–A757. [CrossRef]
3. Virey, E.H.; Baron, N. Status and Prospects of microLED Displays. *SID Symp. Dig. Tech. Pap.* **2018**, *49*, 593–596. [CrossRef]
4. Fan, Z.Y.; Lin, J.Y.; Jiang, H.X. III-nitride micro-emitter arrays: Development and applications. *J. Phys. D Appl. Phys.* **2008**, *41*, 094001. [CrossRef]
5. Kang, D.; Oh, J.T.; Song, J.O.; Seong, T.Y.; Kneissl, M.; Amano, H. Hole injection mechanism in the quantum wells of blue light emitting diode with V pits for micro-display application. *Appl. Phys. Express* **2019**, *12*, 102016. [CrossRef]
6. Lee, T.X.; Gao, K.F.; Chien, W.T.; Sun, C.C. Light extraction analysis of GaN based light-emitting diodes with surface texture and/or patterned substrate. *Opt. Express* **2007**, *15*, 6670–6676. [CrossRef] [PubMed]
7. Hu, H.; Zhou, S.; Liu, X.; Gao, Y.; Gui, C.; Liu, S. Effects of GaN/AlGaN/Sputtered AlN nucleation layers on performance of GaN based ultraviolet light-emitting diodes. *Sci. Rep.* **2017**, *7*, 44627. [CrossRef]
8. Chen, X.; Li, K.; Kong, F.; Wang, J.; Zhang, L. Improved the light extraction efficiency of GaN vertical light-emitting diodes using 3D sphere-like arrays. *Opt. Quantum Electron.* **2015**, *47*, 2957–2968. [CrossRef]
9. Zhou, S.; Yuan, S.; Liu, Y.; Guo, L.; Liu, S.; Ding, H. Highly effificient and reliable high power LEDs with patterned sapphire substrate and strip-shaped distributed current blocking layer. *Appl. Surf. Sci.* **2015**, *355*, 1013–1019. [CrossRef]
10. Zhu, P.; Tansu, N. Effect of Packing Density and Packing Geometry on Light Extraction of III-Nitride Light-Emitting Diodes with Microsphere Arrays. *Photonics Res.* **2015**, *3*, 184–191. [CrossRef]

11. Zhou, S.; Xu, H.; Hu, H.; Gui, C.; Liu, S. High quality GaN buffer layer by isoelectronic doping and its application to 365 nm InGaN/AlGaN ultraviolet light-emitting diodes. *Appl. Surf. Sci.* **2019**, *471*, 231–238. [CrossRef]
12. Tang, B.; Miao, J.; Liu, Y.; Wan, H.; Li, N.; Zhou, S. Enhanced Light Extraction of Flip-Chip Mini-LEDs with Prism-Structured Sidewall. *Nanomaterials* **2019**, *9*, 319. [CrossRef]
13. Galeotti, F.; Mróz, W.; Scavia, G.; Botta, C. Microlens arrays for light extraction enhancement in organic light-emitting diodes: A facile approach. *Org. Electron.* **2013**, *14*, 212–218. [CrossRef]
14. Wong, M.S.; Lee, C.; Myers, D.J.; Hwang, D.; Kearns, J.A.; Li, T.; DenBaars, S.P. Size-independent peak efficiency of III-nitride micro-light-emitting-diodes using chemical treatment and sidewall passivation. *Appl. Phys. Express* **2019**, *12*, 097004. [CrossRef]
15. Zhou, S.; Xu, H.; Tang, B.; Liu, Y.; Wan, H.; Miao, J. High-power and reliable GaN based vertical light-emitting diodes on 4-inch silicon substrate. *Opt. Express* **2019**, *27*, A1506–A1516. [CrossRef]
16. Zhao, P.; Zhao, H.P. Analysis of light extraction efficiency enhancement for thin-film-flip-chip InGaN quantum wells light-emitting diodes with GaN micro-domes. *Opt. Express* **2012**, *20*, A765–A776. [CrossRef]
17. Wong, W.S.; Sands, T.; Cheung, N.W.; Kneissl, M.; Bour, D.P.; Mei, P.; Romano, L.T.; Johnson, N.M. Fabrication of thin-film InGaN light-emitting diode membranes by laser lift-off. *Appl. Phys. Lett.* **1999**, *72*, 1360–1362. [CrossRef]
18. Son, K.R.; Lee, T.H.; Lee, B.R.; Im, H.S.; Kim, T.G. Nitride-Based Microlight-Emitting Diodes Using AlN Thin-Film Electrodes with Nanoscale Indium/Tin Conducting Filaments. *Small* **2018**, *14*, 1801032. [CrossRef]
19. Zhou, S.; Liu, X.; Yan, H.; Chen, Z.; Liu, Y.; Liu, S. Highly efficient GaN based high-power flip-chip light-emitting diodes. *Opt. Express* **2019**, *27*, A669–A692. [CrossRef]
20. Wierer, J.J.; Steigerwald, D.A.; Krames, M.R.; O'Shea, J.J.; Ludowise, M.J.; Christenson, G.; Shen, Y.C.; Lowery, C.; Martin, P.S.; Subramanya, S.; et al. High-power AlGaInN flip-chip light-emitting diodes. *Appl. Phys. Lett.* **2001**, *78*, 3379–3381. [CrossRef]
21. Yonkee, B.P.; Young, E.C.; DenBaars, S.P.; Nakamura, S.; Speck, J.S. Silver free III-nitride flip chip light emitting-diode with wall plug efficiency over 70% utilizing a GaN tunnel junction. *Appl. Phys. Lett.* **2016**, *109*, 191104. [CrossRef]
22. Shatalov, M.; Chitnisl, A. Thermal analysis of flip-chip packaged 280 nm nitride based deep ultraviolet light-emitting diodes. *Appl. Phys. Lett.* **2016**, *86*, 2201109. [CrossRef]
23. Zhou, S.; Xu, H.; Liu, M.; Liu, X.; Zhao, J.; Li, N.; Liu, S. Effect of Dielectric Distributed Bragg Reflector on Electrical and Optical Properties of GaN-Based Flip-Chip Light-Emitting Diodes. *Micromachines* **2018**, *9*, 650. [CrossRef]
24. Zhou, S.; Zheng, C.; Lv, J.; Gao, Y.; Wang, R.; Liu, S. GaN based flip-chip LEDs with highly reflective ITO/DBR p-type and via hole based n-type contacts for enhanced current spreading and light extraction. *Opt. Laser Technol.* **2017**, *92*, 95–100. [CrossRef]
25. Zhao, J.; Liu, X.; Xu, H.; Miao, J.; Hu, J.; Zhou, S. High-Performance Green Flip-Chip LEDs with Double-Layer Electrode and Hybrid Reflector. *ECS J. Solid State Sci. Technol.* **2019**, *8*, Q153–Q157. [CrossRef]
26. Liu, X.; Zhou, S.; Gao, Y.; Hu, H.; Liu, Y.; Gui, C.; Liu, S. Numerical simulation and experimental investigation of GaN based flip-chip LEDs and top-emitting LEDs. *Appl. Opt.* **2017**, *56*, 9502–9509. [CrossRef]
27. Zhou, S.; Liu, X.; Gao, Y.; Liu, Y.; Liu, M.; Liu, Z.; Gui, C.; Liu, S. Numerical and experimental investigation of GaN based flip-chip light-emitting diodes with highly reflective Ag/TiW and ITO/DBR Ohmic contacts. *Opt. Express* **2017**, *25*, 26615–26627. [CrossRef]
28. Ting, D.Z.; McGill, T.C. Monte Carlo simulation of light-emitting diode light extraction characteristics. *Opt. Eng.* **1995**, *34*, 3545–3553. [CrossRef]
29. Lee, S.J. Analysis of light-emitting diode by Monte Carlo photo simulation. *Appl. Opt.* **2001**, *40*, 1427–1437. [CrossRef]
30. Pan, J.W.; Tsai, P.J.; Chang, K.D.; Chang, Y.Y. Light extraction efficiency analysis of GaN based light-emitting diodes with nanopatterned sapphire substrates. *Appl. Opt.* **2013**, *52*, 1358–1367. [CrossRef]
31. Cheng, W.C.; Huang, S.Y.; Chen, Y.J.; Wang, C.S.; Lin, H.Y.; Wu, T.M.; Horng, R.H. AlGaInP Red LEDs with Hollow Hemispherical Polystyrene Arrays. *Sci. Rep.* **2018**, *8*, 911. [CrossRef]
32. Bao, X.Z.; Liang, J.Q.; Liang, Z.Z.; Wang, W.B.; Tian, C.; Qin, Y.X.; Lü, J.G. Design and fabrication of AlGaInP based micro-light-emitting-diode array devices. *Opt. Laser Technol.* **2016**, *78*, 34–41. [CrossRef]

33. Horng, R.H.; Chien, H.Y.; Chen, K.Y.; Tseng, W.Y.; Tsai, Y.T.; Tarntair, F.G. Development and fabrication of AlGaInP based flip-chip micro-LEDs. *IEEE J. Electron Dev.* **2018**, *6*, 475–479. [CrossRef]
34. Hu, H.; Zhou, S.; Wan, H.; Liu, X.; Li, N.; Xu, H. Effect of strain relaxation on performance of InGaN/GaN green LEDs grown on 4-inch sapphire substrate with sputtered AlN nucleation layer. *Sci. Rep.* **2019**, *9*, 3447. [CrossRef]
35. Zhou, S.; Liu, X.; Yan, H.; Gao, Y.; Xu, H.; Zhao, J.; Quan, Z.; Gui, C.; Liu, S. The effect of nanometre-scale V-pits on electronic and optical properties and efficiency droop of GaN based green light-emitting diodes. *Sci. Rep.* **2018**, *8*, 11053. [CrossRef]
36. Liu, M.; Zhao, J.; Zhou, S.; Gao, Y.; Hu, J.; Liu, X.; Ding, X. An InGaN/GaN superlattice to enhance the performance of green LEDs: Exploring the role of V-pits. *Nanomaterials* **2018**, *8*, 450. [CrossRef]
37. Zhou, S.; Lv, J.; Wu, Y.; Zhang, Y.; Zheng, C.; Liu, S. Reverse leakage current characteristics of InGaN/GaN multiple quantum well ultraviolet/blue/green light-emitting diodes. *Jpn. J. Appl. Phys.* **2018**, *57*, 051003. [CrossRef]
38. Ishimoto, S.; Han, D.P.; Yamamoto, K.; Mano, R.; Kamiyama, S.; Takeuchi, T.; Akasaki, I. Enhanced device performance of GaInN-based green light-emitting diode with sputtered AlN buffer layer. *Appl. Sci.* **2019**, *9*, 788. [CrossRef]
39. Zhu, P.; Zhu, H.; Thapa, S.; Adhikari, G.C. Design rules for white light emitters with high light extraction efficiency. *Opt. Express* **2019**, *27*, A1297–A1307. [CrossRef]

micromachines

MDPI

Article

Study on Modulation Bandwidth and Light Extraction Efficiency of Flip-Chip Light-Emitting Diode with Photonic Crystals

Hong Wang [1,2,3,*], Ming Zhong [1,3], Lijun Tan [2,3], Wei Shi [1,3] and Quanbin Zhou [2,3]

1 Engineering Research Center for Optoelectronics of Guangdong Province, School of Physics and Optoelectronics, South China University of Technology, Guangzhou 510640, China; 201720127668@mail.scut.edu.cn (M.Z.); phsweet121@outlook.com (W.S.)
2 Guangdong Provincial Engineering Laboratory for Wide Bandgap Semiconductor Materials and Devices, School of Electronics and Information Engineering, South China University of Technology, Guangzhou 510640, China; eeljtan@mail.scut.edu.cn (L.T.); zhouquanbin86@163.com (Q.Z.)
3 Zhongshan Institute of Modern Industrial Technology, South China University of Technology, Zhongshan 528437, China
* Correspondence: phhwang@scut.edu.cn; Tel.: +86-20-8711-1557

Received: 11 October 2019; Accepted: 9 November 2019; Published: 11 November 2019

Abstract: In this study, the photonic crystal structure is employed to increase both the light extraction efficiency and the modulation bandwidth of flip-chip GaN-based light-emitting diodes (LEDs). The finite difference time domain method is utilized to investigate the influence of structure of photonic crystals on the Purcell factor and light extraction efficiency of flip-chip GaN-based LEDs. Simulation results show that the modulation bandwidth is estimated to be 202 MHz at current densities of 1000 A/cm^2. The experimental result of modulation bandwidth is in accord with the simulation. The optical f-3dB of the device achieves 212 MHz at current densities of 1000 A/cm^2 and up to 285 MHz at current densities of 2000 A/cm^2. This design of photonic crystal flip-chip LED has the potential for applications in high-frequency visible light communication.

Keywords: visible light communication; photonic crystals; flip-chip LED; Purcell effect; light extraction efficiency

1. Introduction

Visible light communication, as a communication solution to alleviate the shortage of spectrum resources, is at the frontier of technology and the hotspot of research. GaN-based light-emitting diode (LED) chips are key devices for visible light communications. However, the narrow bandwidth of commercial LED chips, which is in the range of 20–30 MHz, limits the overall bandwidth of visible light communication systems [1,2]. The modulation bandwidth of an LED chip is limited by the resistance–capacitance (RC) time constant and carrier spontaneous emission rate [3]. Therefore, the methods to increase the modulation bandwidth mainly include decreasing the RC time constant [4–6] and the carrier lifetime [7,8]. Many studies have shown that the wavelength-sized cavity can change the local density of optical states (LDOS) and increase the spontaneous emission rate [9]. In 1946, Purcell proved that the spontaneous radiation rate of the cavity can be changed by the Purcell factor [10]. Nanostructures, such as resonant cavities, surface plasmon, and photonic crystals, can affect the Purcell factor, thereby increasing the spontaneous emission rate of carrier and the modulation bandwidth. Although there are many literatures studying photonic crystal structure to improve the modulation bandwidth of common LED [7,11,12], the literature about flip-chip LEDs with photonic crystals on the bottom is still rarely explored.

In this work, both modulation bandwidth and light extraction efficiency (LEE) of flip-chip LEDs with photonic crystals are numerically investigated by using a 3D finite-difference time-domain (FDTD) method based on Yee's algorithm with a perfectly matched layer (PML) boundary condition [13]. Three structural parameters of photonic crystals including period, height, and duty cycle were compared to study the effect of photonic crystals on the modulation bandwidth and LEE of a flip-chip LED. By optimizing the photonic crystal structure, we designed a structure that could simultaneously improve the modulation bandwidth and light extraction efficiency of the flip-chip GaN-based LEDs. The experimental result shows that photonic crystal structure can improve the modulation bandwidth due to the enhancement of the Purcell factor and the reduction of nonradiative lifetime.

2. Materials and Methods

Figure 1 shows the model of a conventional planar flip-chip LED (FCLED) and flip-chip LED with photonic crystals (PC-FCLED). The lateral dimensions of the computational domain were set to 7 μm × 7 μm × 7 μm with a mesh of 2 nm × 2 nm × 2 nm. In FCLED, the structure consisted of 3 mm sapphire substrate (extending beyond the simulation volume), a 5000 nm n-GaN layer, a 120 nm multiple quantum wells (MQWs) layer, a 20 nm p-AlGaN layer, a 110 nm p-GaN layer, and a 100 nm Ag mirror layer. The material parameters in the simulation were mostly from reference [14]. Because the radiation in the quantum well was mainly in transverse electric (TE) mode, the light source was set to the TE mode dipole source [15]. The dipole source was placed in the center of the MQWs layer of the plane of the sectional view. The wavelength of the light source was 460 nm. Six power monitors were placed near the light source on the six faces for detecting total power from the dipole. Considering the large thickness of the sapphire substrate, a monitor was placed in the sapphire substrate 1 μm from n-GaN layer, and the power exiting in the air that far away from the source was calculated by way of far-field projection. The boundary conditions were set as a perfectly matched layer (PML). The top of the FDTD simulation area was located inside the sapphire substrate, and the bottom was placed inside the mirror to ignore the cavity effect it generated [16]. As for PC-FCLED, the model was similar to FCLED except for the photonic crystals. The square lattice photonic crystals with an array of SiO_2 rods were placed in p-GaN of PC-FCLED, as shown in Figure 1b. The period, height, and radius of photonic crystals were defined as a, h, and r, respectively. The duty cycle of photonic crystals was $2r/a$. The LEE was defined as the ratio of the power exiting the structure (flux calculated by way of far-field projection) to the total power generated by the dipole inside the active region (flux through the small box in Figure 1).

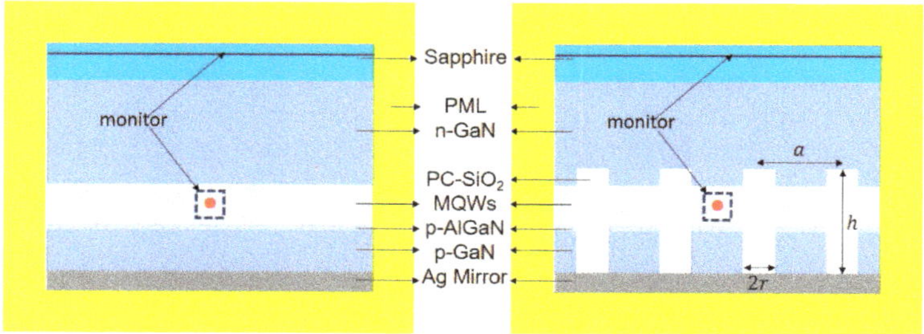

Figure 1. Model of (**a**) the conventional planar flip-chip light-emitting diodes (LED) and (**b**) the flip-chip LED with photonic crystals. Red dots inside multiple quantum wells (MQWs) represent the position of the dipole source.

In 2009, Lau et al. deduced the approximate bandwidth of a nanocavity LED [17]:

$$f_{3dB} \approx \frac{1}{2\pi} \frac{1}{\sqrt{\tau_p^2 + \tau_{eff}^2}} \tag{1}$$

where τ_{eff} is the spontaneous radiation lifetime reduced by the Purcell effect, and τ_p is the photon lifetime. For cavities with quality factors less than a few hundred, τ_p is much lower than the overall lifetime, and f_{3dB} mainly depends on τ_{eff} [18]. The relationship between f_{3dB} and τ_{eff} is as follows:

$$f_{3dB} \approx \frac{1}{2\pi} \frac{1}{\tau_{eff}} \tag{2}$$

$$\frac{1}{\tau_{eff}} = \frac{F}{\tau_r} + \frac{1}{\tau_{nr}} \tag{3}$$

where F is the Purcell factor. Then,

$$f_{3dB} \approx \frac{1}{2\pi} \left[\frac{F}{\tau_r} + \frac{1}{\tau_{nr}} \right] \tag{4}$$

For polar c-plane InGaN/GaN LEDs at current densities of 1000 A/cm^2, David et al. derived the carrier lifetime [19]: $\tau_r = 3$ ns; $\tau_{nr} = 1.5$ ns.

Thus, the modulation bandwidth of the LED can be approximated by simulating the Purcell factor.

3. Results and Discussion

Figure 2 shows the effect of p-GaN layer thickness on the Purcell factor and LEE for FCLED. Both the Purcell factor and LEE show a trend of sinusoidal function oscillating when the p-GaN layer thickness increases. Moreover, the Purcell factor tends to oscillate and attenuate, and the LEE exhibits a perfect periodic oscillation. The trends of the Purcell factor and the LEE are almost the same, which is expected to increase the LEE and the modulation bandwidth at the same time. The Purcell factor takes a peak point of 1.48 when the p-GaN thickness is 45 nm, and then takes a peak point at 140 nm and 230 nm. The periodicity of the Purcell factor with p-GaN thickness is 90 nm, which is almost the same as half the wavelength of the light source inside the p-GaN material. When the p-GaN thickness is 140 nm, the Purcell factor is 1.32 and the LEE is 51.4%.

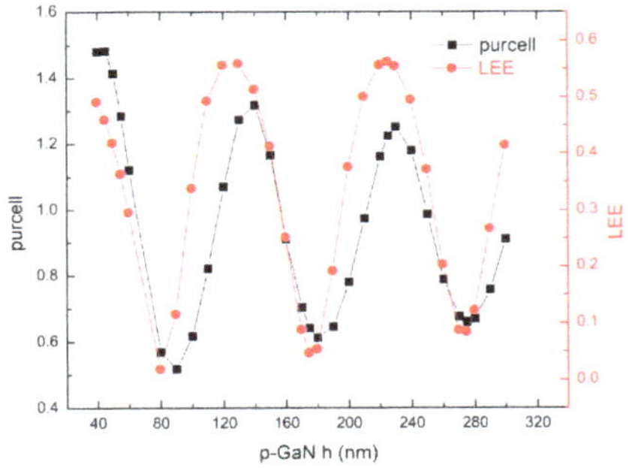

Figure 2. The effect of p-GaN thickness on the Purcell factor and light extraction efficiency (LEE) for flip-chip LED (FCLED).

Figure 3 shows the trend of the Purcell factor and LEE varying with the period and height of the photonic crystals when the duty cycle is 0.5. When the period of photonic crystals is greater than 200 nm, the Purcell factor is above 1.2. When the period and height of photonic crystals are both 500 nm, the Purcell factor is 1.8, which is better than the results of similar simulations [14,16]. Because the small period photonic crystal structure (the period 100 and 200 nm) cannot provide enough mode volume to the lower-order mode for affecting the radiative recombination, the large period photonic crystal structure in the proper duty cycle has a chance to couple with the lower-order mode with a higher spontaneous emission rate, which leads to the high Purcell factor [20]. Photonic crystal structures with different heights have a larger Purcell factor in the photonic crystal period of 400–600 nm. When the photonic crystals period is greater than 250 nm, the LEE is above 40%. The small period photonic crystal structure cannot maintain enough emission light in the extraction cone, and the emission light is almost too hard to go through the surface [21].

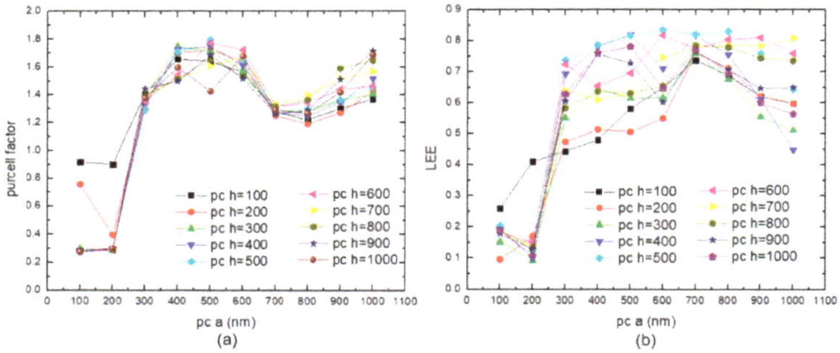

Figure 3. The trend of (**a**) Purcell factor and (**b**) LEE varying with the period and height of the photonic crystals when the duty cycle is 0.5.

Figure 4a shows the tendency of the Purcell factor and LEE to change with the duty cycle and height of the photonic crystals when the photonic crystal period is fixed at 400 nm. The Purcell factor increases first and then decreases with the duty cycle increasing. The Purcell factor is 1.81 when the height and duty cycle of photonic crystals is 400 nm and 0.3, respectively. When the duty cycle is in the range of 0.1–0.6, the Purcell factors with different photonic crystal heights are all greater than 1.2. This is because photonic crystal height cannot strongly change mode volume, compared with duty cycle and period of photonic crystals. When the duty cycle is larger than 0.8, there is not enough mode volume for the lower-order mode to achieve a high Purcell factor due to the reduction of the active region area. As shown in Figure 4b, when the photonic crystal period is fixed at 400 nm, the LEE varies with the duty cycle and height of the photonic crystals. When the photonic crystals are deep into the active region, optical modes interact with the photonic crystals and are diffracted to escape, which increases the LEE [22]. When the photonic crystal height is 400 nm and the duty cycle is 0.3, the Purcell factor and LEE of PC-FCLED are 1.81 and 68%, which is 37% and 32.3% higher than those of a conventional planar FCLED, respectively.

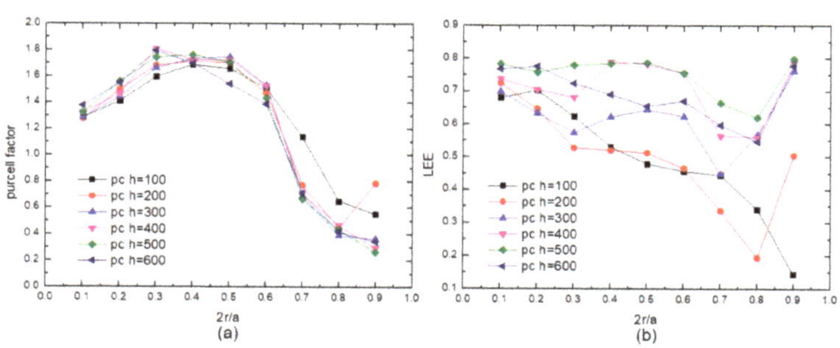

Figure 4. The tendency of (**a**) the Purcell factor and (**b**) LEE changing with the duty cycle and height of the photonic crystals when the photonic crystal period is fixed at 400 nm.

These results show that the PC-FCLED has a better LEE and Purcell factor than FCLED. When the photonic crystals are deep into the active region, the influence of the height of the photonic crystals on the Purcell factor decreases. When the photonic crystal period is more than 300 nm, the Purcell factor and LEE are better. The duty cycle, which is between 0.1–0.6, is more conducive to the improvement of the overall performance. With the decrease of the radiation carrier lifetime and the fast photon accumulation effect in the specific photonic band gap, the PC-FCLED obtains both high output optical power and large modulation bandwidth [20]. Therefore, the modulation bandwidth and LEE are increased simultaneously due to the Purcell effect in PC-FCLED. According to the above derivation and results, the trend of the modulation bandwidth of PC-FCLED changing with the Purcell factor can be obtained. When the Purcell factor is 1.81, the bandwidth is about 202 MHz at current densities of 1000 A/cm^2.

The optimized photonic crystal structure of the simulation was obtained for subsequent experimentation. The studied PC-FCLEDs were prepared by using the epitaxial wafers with a peak wavelength of 460 nm in the experiment. The LED structure consisted of a 3 μm-thick undoped GaN layer, a 1.8 μm-thick n-type GaN layer, 9 periods of InGaN/GaN MQWs, and a 140 nm-thick p-GaN layer. Then, the square lattice photonic crystal was formed in the p-GaN layer and MQWs by nanoimprint lithography and inductively coupled plasma (ICP). Figure 5a shows the scanning electron microscopy (SEM) images of the photonic crystal structure after the patterns were etched with a photonic crystal period of 400 nm, hole depths of 400 nm, and duty cycle, which is 0.3. A Spin-On-Glass (SOG) layer was deposited on the surface of p-GaN and then annealed at 400 °C for 30 min by rapid thermal annealing to fill in the hole. Subsequently, the mesa regions with a depth of 1.3 μm were transferred to the LED wafer using ICP etching. Ag and TiW metal layers were deposited on the top surface of the LED wafer, and a 1 μm-thick SiN layer was deposited to mesa sidewalls by plasma-enhanced chemical vapor deposition. Finally, Cr/Al/Ti/Au (50/800/200/200 nm) multilayer metals were sequentially deposited on the top of Ag and TiW layers and the n-GaN layer to act as the p- and n-electrodes by electron beam evaporation, respectively. The nomenclature of PC-FCLEDs with different mesa radii were PC30, PC60, and PC90. FCLEDs were also fabricated using the same procedure as the PC-FCLEDs, but without photonic crystals.

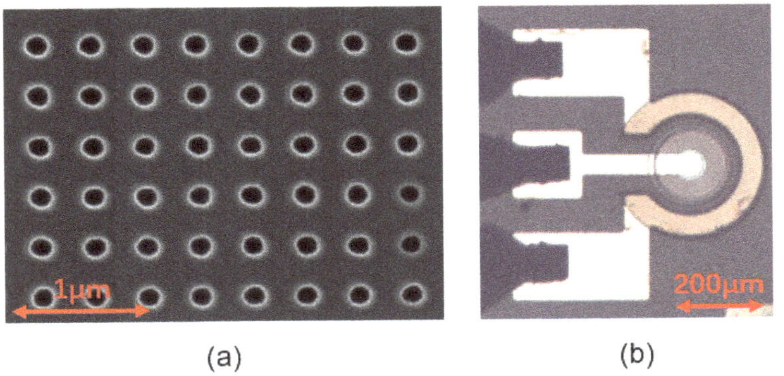

Figure 5. (**a**) The scanning electron microscopy (SEM) images of the hole-patterned photonic crystal structure. (**b**) Top view of PC90 LED chip with mesa radius of 90 μm.

Figure 6 represents the optical output power and forward current–voltage (P-I-V) characteristics of the PC-FCLEDs and the FCLEDs. The P-I curves show that the optical output power of PC-FCLED is smaller than that of the FCLED with the same mesa size. This is because, even though the photonic crystal helps light extraction, the effective light emitting area of the PC-FCLED is smaller than the FCLED, and the damage of p-GaN from the ICP procedure will strongly reduce carrier recombination efficiency.

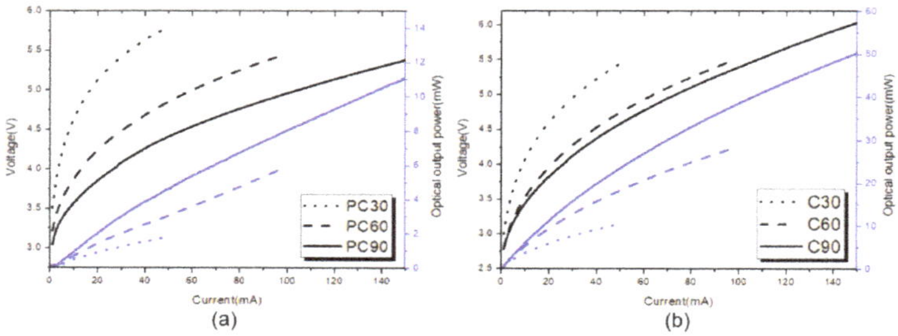

Figure 6. The optical output power and forward current–voltage (P-I-V) curves of (**a**) flip-chip LEDs with photonic crystals (PC-FCLEDs) and (**b**) FCLEDs of different mesa radii.

The optical modulation bandwidth (f-3dB) versus current density relation of the PC-FCLEDs and FCLEDs with three different radii are shown in Figure 7. The optical f-3dB in PC-FCLEDs is much higher than that of the FCLEDs due to the smaller carrier lifetime. At a 1000 A/cm^2 injected current density, the optical f-3dB of PC60 is 212 MHz, which is almost in accord with the simulation result of 202 MHz. As the mesa size of the device decreases, the current density can be improved to achieve a higher modulation bandwidth. When the injected current density is 2000 A/cm^2, the optical f−3dB of PC30 increases up to 285 MHz. However, at a 1000 A/cm^2 injected current density, the optical f−3dB of C30 is 90 MHz, which is lower than the simulation result of 176 MHz. This means that carrier lifetimes of FCLEDs are lower than the carrier lifetimes quoted, and there is another mechanism here that leads to the decrease of the carrier lifetimes of PC-FCLEDs. Therefore, the Purcell factor plays a certain role in the increase of modulation bandwidth, but cannot completely determine its implementation. In the ICP procedure for etching photonic crystal structure, many defect sites are produced in the p-GaN layer and MQWs. The increase of defect density causes a reduction of the radiative efficiency and nonradiative lifetime [23], which leads to further increasing the modulation bandwidth and reducing optical output power.

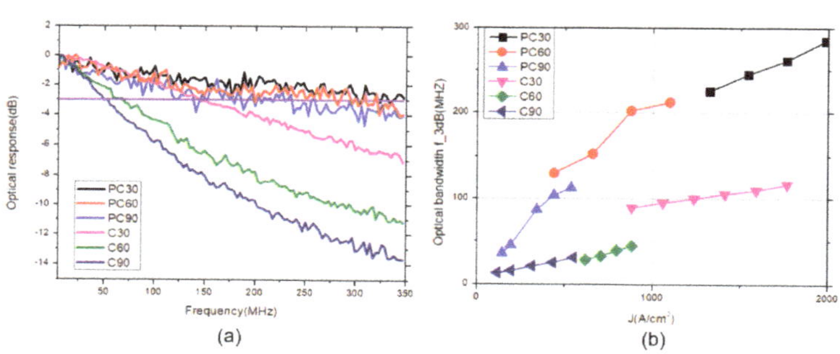

Figure 7. (**a**) Frequency responses of PC-FCLEDs and FCLEDs at an injection current of 50 mA, 100 mA, and 150 mA with mesa radii of 30 µm, 60 µm, and 90 µm, respectively. (**b**) Optical bandwidth f−3dB of PC-FCLEDs and FCLEDs at various current densities.

4. Conclusions

We investigated the Purcell factor and LEE of PC-FCLED using the FDTD method. Compared with the FCLED, the PC-FCLED shows great enhancement in the Purcell factor and LEE. When the height and duty cycle of photonic crystals is 400 nm and 0.3, the Purcell factor and LEE of PC-FCLED are 1.81 and 68%, which is 37% and 68% higher than that of FCLED, respectively. When the Purcell factor is 1.81, the modulation bandwidth is 202 MHz at current densities of 1000 A/cm^2. We constructed the devices, and the experiment results of the modulation bandwidth achieved the desired level of simulation because of the reduction of nonradiative lifetime and increase of the Purcell factor. An optical f-3dB of 285 MHz was obtained in the PC-FCLED with a mesa radius of 30 μm at a current density of 2000 A/cm^2. The PC-FCLED revealed the potential for visible light communication due to its high modulation bandwidth.

Author Contributions: Conceptualization, M.Z., W.S. and H.W.; methodology, M.Z., L.T. and H.W.; formal analysis, M.Z., L.T. and Q.Z.; investigation, M.Z. and W.S.; data curation, W.S. and M.Z.; writing—original draft preparation, M.Z., W.S. and Q.Z.; writing—review & editing, H.W.; supervision, H.W.

Funding: This work was supported by Science and Technologies Plan Projects of Guangdong Province (Nos. 2015B010127013, 2016B010123004, 2017B010112003), and by Science and Technologies Plan Projects of Guangzhou City (Nos. 201604046021, 201905010001), and by Science and Technology Development Special Fund Projects of Zhongshan City (Nos. 2017F2FC0002, 2017A1009, 2019AG014).

Conflicts of Interest: The authors declare no conflict of interest.

References

1. Grubor, J.; Lee, S.C.J.; Langer, K.; Koonen, T.; Walewski, J.W. Wireless high-speed data transmission with phosphorescent white-light leds. In Proceedings of the 33rd European Conference and Exhibition of Optical Communication—Post-Deadline Papers (published 2008), Berlin, Germany, 16–20 September 2007; pp. 1–2.
2. Tuo, J.; Shams, H.; Corbett, B. Visible light communication by using commercial phosphor based white leds. In Proceedings of the IET Irish Signals and Systems Conference (ISSC 2012), Maynooth, Ireland, 28–29 June 2012; pp. 1–4.
3. Shi, J.; Sheu, J.; Chen, C.; Lin, G.; Lai, W. High-speed GaN-based green light-emitting diodes with partially n-doped active layers and current-confined apertures. *IEEE Electron Device Lett.* **2008**, *29*, 158–160. [CrossRef]
4. Wun, J.M.; Lin, C.W.; Chen, W.; Sheu, J.K.; Lin, C.L.; Li, Y.L.; Bowers, J.E.; Shi, J.W.; Vinogradov, J.; Kruglov, R.; et al. GaN-based miniaturized cyan light-emitting diodes on a patterned sapphire substrate with improved fiber coupling for very high-speed plastic optical fiber communication. *IEEE Photon. J.* **2012**, *4*, 1520–1529. [CrossRef]
5. McKendry, J.J.D.; Massoubre, D.; Zhang, S.; Rae, B.R.; Green, R.P.; Gu, E.; Henderson, R.K.; Kelly, A.E.; Dawson, M.D. Visible-light communications using a CMOS-controlled micro-light-emitting-diode array. *J. Lightwave Technol.* **2012**, *30*, 61–67. [CrossRef]
6. Ferreira, R.X.; Xie, E.; Mckendry, J.J.D.; Rajbhandari, S.; Chun, H.; Faulkner, G.; Watson, S.; Kelly, A.E.; Gu, E.; Penty, R.V.; et al. High bandwidth GaN-based micro-LEDs for multi-Gb/s visible light communications. *IEEE Photon. Technol. Lett.* **2016**, *28*, 2023–2026. [CrossRef]
7. Yin, Y.F.; Lan, W.Y.; Lin, T.C.; Wang, C.; Feng, M.; Huang, J.J. High-speed visible light communication using GaN-based light-emitting diodes with photonic crystals. *J. Lightwave Technol.* **2017**, *35*, 258–264. [CrossRef]
8. Zhu, S.; Yu, Z.; Liu, L.; Yang, C.; Chao, H.; Xi, X.; Li, J.; Zhao, L. Enhancing the spontaneous emission rate by modulating carrier distribution in GaN-based surface plasmon light-emitting diodes. *Opt. Express* **2017**, *25*, 9617–9627. [CrossRef] [PubMed]
9. Selenyi, P. Wide-angle interferences and the nature of the elementary light sources. *Phys. Rev.* **1939**, *56*, 477–479. [CrossRef]
10. Purcell, E.M. Spontaneous emission probabilities at radio frequencies. In *Confined Electrons and Photons*; Burstein, E., Weisbuch, C., Eds.; Springer: Boston, MA, USA, 1946; Volume 340.
11. Zhuang, Z.; Guo, X.; Liu, B.; Hu, F.; Dai, J.; Zhang, Y.; Li, Y.; Tao, T.; Zhi, T.; Xie, Z.; et al. Great enhancement in the excitonic recombination and light extraction of highly ordered InGaN/GaN elliptic nanorod arrays on a wafer scale. *Nanotechnology* **2015**, *27*, 015301. [CrossRef] [PubMed]
12. Yin, Y.-F.; Lan, W.-Y.; Hsu, Y.-H.; Hsu, Y.-F.; Wu, C.-H.; Huang, J. High-speed modulation from the fast mode extraction of a photonic crystal light-emitting diode. *J. Appl. Phys.* **2016**, *119*, 013103. [CrossRef]
13. Taflove, A.; Hagness, S.C. *Computational Electrodynamics: The Finite-Difference Time-Domain Method*; Artech House: Boston, MA, USA, 1996; pp. 1817–1818.
14. Ryu, H.-Y. Investigation of the purcell effect in GaN-based vertical LED structures using FDTD simulation. *Opt. Quantum Electron.* **2015**, *48*, 6. [CrossRef]
15. Ryu, H.-Y. Large enhancement of light extraction efficiency in AlGaN-based nanorod ultraviolet light-emitting diode structures. *Nanoscale Res. Lett.* **2014**, *9*, 58. [CrossRef] [PubMed]

16. Ryu, H.-Y. Modification of internal quantum efficiency and efficiency droop in GaN-based flip-chip light-emitting diodes via the Purcell effect. *Opt. Express* **2015**, *23*, A1157–A1166. [CrossRef] [PubMed]
17. Lau, E.K.; Lakhani, A.; Tucker, R.S.; Wu, M.C. Enhanced modulation bandwidth of nanocavity light emitting devices. *Opt. Express* **2009**, *17*, 7790–7799. [CrossRef] [PubMed]
18. Nami, M.; Feezell, D. Optical properties of Ag-coated GaN/InGaN axial and core–shell nanowire light-emitting diodes. *J. Opt.* **2015**, *17*, 25004. [CrossRef]
19. David, A.; Grundmann, M.J. Droop in InGaN light-emitting diodes: A differential carrier lifetime analysis. *Appl. Phys. Lett.* **2010**, *96*, 103504. [CrossRef]
20. Yin, Y.-F.; Lin, Y.-C.; Liu, Y.-C.; Shen, Y.-C.; Chiang, H.-P.; Huang, J. Correlation of angular light profiles of light-emitting diodes to spatial spontaneous emissions from photonic crystals. *J. Appl. Phys.* **2013**, *114*, 143104. [CrossRef]
21. Ding, Q.A.; Li, K.; Kong, F.; Zhao, J.; Yue, Q. Improving the vertical light extraction efficiency of gan-based thin-film flip-chip led with double embedded photonic crystals. *IEEE J. Quantum Electron.* **2014**, *51*, 1–9. [CrossRef]
22. Gao, H.; Li, K.; Kong, F.-M.; Chen, X.-L.; Zhang, Z.-M. Improving light extraction efficiency of GaN-based LEDs by $Al_xGa_{1-x}N$ confining layer and embedded photonic crystals. *IEEE J. Selected Top. Quantum Electron.* **2012**, *18*, 1650–1660. [CrossRef]
23. Hangleiter, A.; Netzel, C.; Fuhrmann, D.; Hitzel, F.; Hoffmann, L.; Bremers, H.; Rossow, G.A.; Hinze, P. Anti-localization suppresses non-radiative recombination in GaInN/GaN quantum wells. *Philos. Mag.* **2007**, *87*, 2041–2065. [CrossRef]

micromachines

MDPI

Article

A Designed Broadband Absorber Based on ENZ Mode Incorporating Plasmonic Metasurfaces

Phuc Toan Dang [1], Khai Q. Le [2,3], Ji-Hoon Lee [1,*] and Truong Khang Nguyen [2,3,*]

1 Division of Electronics Engineering, Chonbuk National University, Jeonju 54896, Korea; dangphuctoan@jbnu.ac.kr
2 Division of Computational Physics, Institute for Computational Science, Ton Duc Thang University, Ho Chi Minh City 700000, Vietnam; lequangkhai@tdtu.edu.vn
3 Faculty of Electrical and Electronics Engineering, Ton Duc Thang University, Ho Chi Minh City 700000, Vietnam
* Correspondence: jihoonlee@jbnu.ac.kr (J.-H.L.); nguyentruongkhang@tdtu.edu.vn (T.K.N.)

Received: 6 August 2019; Accepted: 2 October 2019; Published: 4 October 2019

Abstract: In this paper, we present a numerical study of a metamaterial absorber that provides polarization-insensitive absorption over a broad bandwidth of operation over the mid-infrared region. The absorber consists of a periodically patterned metal-dielectric-metal structure integrated with an epsilon-near-zero (ENZ) nanolayer into the insulating dielectric gap region. Such an anomalous broadband absorber is achieved thanks to a couple of resonant modes including plasmon and ENZ modes that are excited under mid-IR light illumination. By adding a 0.06-μm-thick ENZ layer between the patterned gold rectangular grating and the SiO_2 dielectric layer, the absorber captures >95% light over a 1.5 μm bandwidth centered at a near-8-μm wavelength over a wide range of oblique incidence under transverse-magnetic and -electric polarizations. The designed ENZ-based wideband absorber has potential for many practical applications, including sensing, imaging and solar energy harvesting over a wide frequency regime.

Keywords: epsilon-near-zero; wideband absorber; plasmon mode; Brewster mode

1. Introduction

Nowadays, perfect light absorption has attracted much attention for the majority of the modern optoelectronic devices working either in the infrared regime or in the visible spectrum. Among these, perfect wideband absorption has attracted great attention due to its high applicability in practical applications such as surface-enhanced sensing and imaging sensors. Especially, in solar cell applications, high-performance photodetectors all rely on absorption of light. The absorption spectrum can be adjusted in terms of absorption strength and bandwidth and can expand the range of many practical device applications. So, a number of techniques have been proposed over the past years to design the absorption of materials, examples of which include metamaterials, ultrathin semiconductors, coherent absorption and plasmonic perfect absorber [1–12].

Up to now, most of the research related to wideband absorption has mainly been based on plasmonic resonances of nanostructures having delicate structural patterns [13,14], or some studies based on epsilon-near-zero (ENZ) metamaterial structures which can also achieve wideband perfect absorption [15,16]. The complete light absorption in thin films allows perfect wideband carrier collection thanks to the mutual interaction between plasmonic nanostructures and semiconductors absorbers, as shown in previous studies [17–19]. In addition, the omnidirectional broadband absorption can be obtained by nonresonant plasmonic Brewster effects [20].

Recently, combinations of the gap plasmonic mode and the ENZ mode have been emerging as good candidates to optimize broadband absorption [21]. However, the coupling regime of the mode is still a weak and narrow bandwidth. In this work, we present a study of ENZ-enhanced wideband absorption near the ENZ wavelength in the mid-infrared region when an ENZ material is integrated into the gap region of a metal-insulator-metal plasmonic structure. Our study is based on the combination of the gap plasmon mode, ENZ mode, and plasmonic Brewster effects. A mode called "epsilon-near-zero" (ENZ) has been observed on subwavelength film thicknesses. The ENZ mode is also known as a Berreman mode which can appear only as small film ENZ thicknesses [22]. The role of ENZ layers and losses in such materials has been also discussed in previous studies [23–26].

In detail, the requirement for ENZ mode operation is that their thickness must be on the order of or less than the wavelength where the real part of the dielectric constant becomes zero [27]. ENZ modes have shown very large densities of states which make them attractive for enhancing light–matter interactions. Moreover, by varying the doping concentration (i.e., depending on the doping level) or other growth conditions, the ENZ wavelength can be controlled in a definite spectral range in the near- and mid-infrared bands [28,29]. One such ENZ material is InAsSb [27], with an ENZ wavelength that can be generated at around 8 μm. At this wavelength, the electric field in plasmonic subwavelength thin films becomes very strong, and this can lead to extremely large light absorption in the film [10]. Several studies have demonstrated that thin films made of metal, doped semiconductors, or polar materials can support surface plasmon polaritons (SPPs) [30,31]. The gap plasmon resonance localized in the dielectric gap region between two metal layers, which has a strong electric field oriented in the out-of-plane direction [32], can be incorporated efficiently to the ENZ mode of the ENZ layer.

The plasmonic Brewster effects are based on an inherently non-resonant mechanism. At a specific angle, the Brewster angle, light absorption can be achieved over broad bandwidth thanks to the impedance-matching mechanism at the entrance and exit metal surfaces with minimizing reflections through the corrugated metal screen [20]. In this study, we have optimized ENZ mode, plasmon mode, and Brewster effect. These resonances have been controlled to be efficiently coupled together in close proximity. As a result, high and wideband absorption is achieved simultaneously in the wavelength range of interest.

2. Absorber Structure and Simulation Details

The absorber device structure was designed on a 2-μm supporting silicon substrate. As for the silicon dioxide dielectric spacer layer, there are two considerations when calculating thickness. The first consideration is that the spacer layer thickness affects the absorption intensity of the gap plasmon resonance, as has been seen in many previous studies on metal-insulator-metal structures without an integrated ENZ layer [33]. The second consideration is that too thick of a spacer layer makes the coupling of the gap plasmon mode to the ENZ mode weaker because of the spatial overlap of these modes decreasing gradually with the increasing thickness of the spacer layer. Contrariwise, if the spacer layer thickness is too thin, the coupling regime is made robust and can split the two resonances in the hybrid resonance modes [34]. Choosing the exact thickness of the spacer layer depends on the thickness of the ENZ layer.

Figure 1 illustrates the proposed absorber structure with the following specific parameters. The exact thickness of the dielectric layer was 0.75 μm to separate the two metal layers. The repeat period P was 4.55 μm in both x and y directions. A metallic TiN film worked as a ground plane. The gold gratings were placed on the top of the ENZ nanofilm, which has the resonance wavelength of 8 μm. For the device, the absorption band can be broadened significantly in the mid-infrared region around the ENZ resonance wavelength. The thickness of the ENZ layer was 0.06 μm. On the top layers, the height, the lengths of the rectangular grating are designated as t_{Au}, L_x and L_y in the x and y directions with values of 0.9, 1.59 and 1.5 μm, respectively.

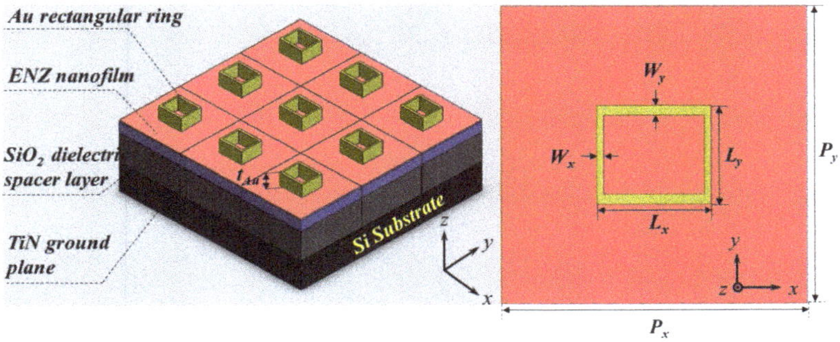

Figure 1. Sketch and top view of the epsilon-near-zero (ENZ)-based absorber with a top rectangular grating.

Similarly, the thicknesses of the rectangular grating's rim W_x and W_y were 0.09 μm and 0.1 μm, respectively. Optical properties of TiN, SiO_2, Si layers were extracted from the material library of the simulation software and verified with experimental data [27], while the gold electric permittivities were taken from Palik's Handbook of Optical Constants [35]. Dispersive properties of the constituent materials of the device absorber were taken into the simulations.

ENZ materials are known to possess some unusual properties when a class of the materials which have their real part of permittivity crossing zero at a certain wavelength (i.e., ENZ wavelength) [27]. Figure 2 shows the resultant spectral dependence of the real and imaginary parts of the ENZ permittivity (ε). In order to demonstrate ENZ-enhanced wideband absorption, we show the coupling of the gap plasmon resonance mode which localized in the gap SiO_2 layer with a strong electric field region and ENZ mode which was excited into the gap region of metal-insulator-metal interfaces. In addition, the effects of angles of incident and other parameters in the structure were also considered.

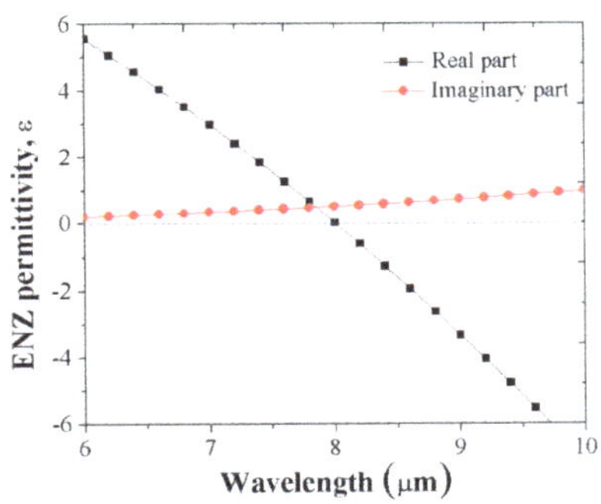

Figure 2. The dielectric constant of the ENZ material.

3. Results

A significant factor in getting broadband absorption is the coupling between the gap plasmon mode and the ENZ mode must be robust, but still be weak coupling. Therefore, the simultaneous combination of these two resonances and the Brewster mode has been investigated in repelling each other to generate wideband absorption. Moreover, the wideband absorption also has an incident angle and therefore it is more suitable for many device applications that require multiple-direction light absorption and energy conversion. We utilized the commercial CST Microwave Studio (CST MWS; Computer Simulation Technology AG., Darmstadt, Germany) software based on a finite element method (FEM) [36] to simulate the device with appropriate boundary conditions. In the device considered here where an optical TiN thick metallic ground plane is used (i.e., transmission is equal to zero), the simple expression of A = 1 − R will show the relationship between the absorption A and reflectivity R. This section will provide a concise and precise description of the simulated results, their mechanism as well as the conclusions that can be drawn.

The absorption spectrum of the integrated ENZ wideband absorber is also shown at the normal incident (black square line) and at the Brewster angle (red circle line) with a wideband absorption of approximately 95% over wavelength ranges of 1.26 μm (from 7.94 μm to 9.2 μm) and 1.46 μm (from 7.77 μm to 9.23 μm), respectively, shown in Figure 3a for transverse-magnetic (TM) polarization. Along with that comparison, in order to reobtain the perfect absorption spectrum in the device without the ENZ layer, herein called a "ENZ-removed" perfect absorber, it was necessary to change the device parameters to a 0.55-μm spacer layer, a 0.4-μm grating thickness, lengths of the grating (L_x and L_y) which have same value of 1.31 μm and thicknesses of grating's rim (W_x and W_y) being 0.11 μm in the x and y directions; all other parameters remained the same. The calculated bandwidth for 95% absorption is 0.58 μm (from 7.94 μm to 8.52 μm) for the "ENZ-removed" perfect absorber (blue triangle line) also shown in Figure 3a for TM mode. A similar comparison is shown in Figure 3b for transverse-electric (TE) mode with various spectral absorption ranges. The angles of incidence play an important role in the optical response of nanostructured surfaces. Degraded performance stems from large oblique angles. To understand more about the performance of the optimized integrated wideband ENZ absorber, we calculated the absorption as a function of the incident angle over a range of 0–90° for both TE and TM modes for the structure. The full structure or the optimized integrated wideband ENZ absorber structure exhibited angular stability up 24° with wideband absorption intensity of over 95% as shown in Figure 3c for TM mode. The absorption gradually decreased with increase in incidence angle. Similarly, the wideband absorption and absorption intensity were reduced when incident angles were over 22° for TE mode as shown Figure 3d. For both TE and TM modes, the absorption intensity on the shorter wavelength side of the absorption bandwidth began to decrease as the incident angle increased above these bandwidths, while the long-wavelength side of the absorption band remained until below 70°. Figure 3e shows the percentage of absorption in the metal layers of the structure where the absorption loss was around 20% in the wavelength range of interest. It can be seen that the absorption bandwidth in the TM mode was wider than that of the TE mode which is due to the temporal and spatial interference between the optical cavity in the SiO_2 spacer and the grating plasmonic modes modifying the angular response. This can be seen in Figure 3a,b where the absorption bandwidths in TM and TE modes are almost same at the normal incident. Though the plasmon lifetime also affects such an absorption bandwidth [5,6], this effect was minor in this proposed design.

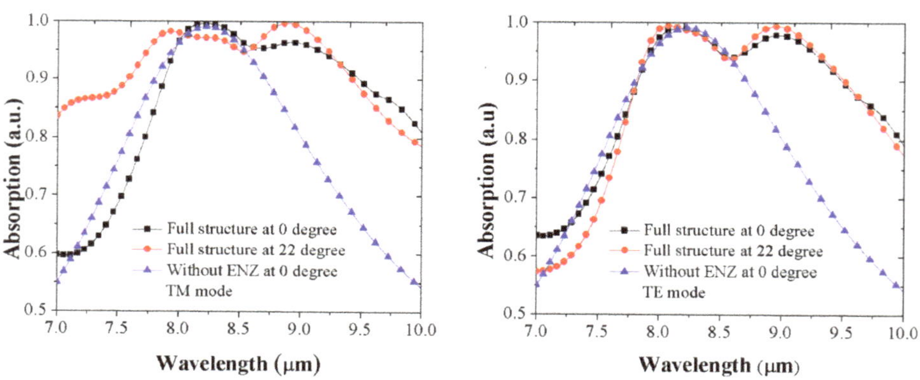

(a) (b)

Figure 3. *Cont.*

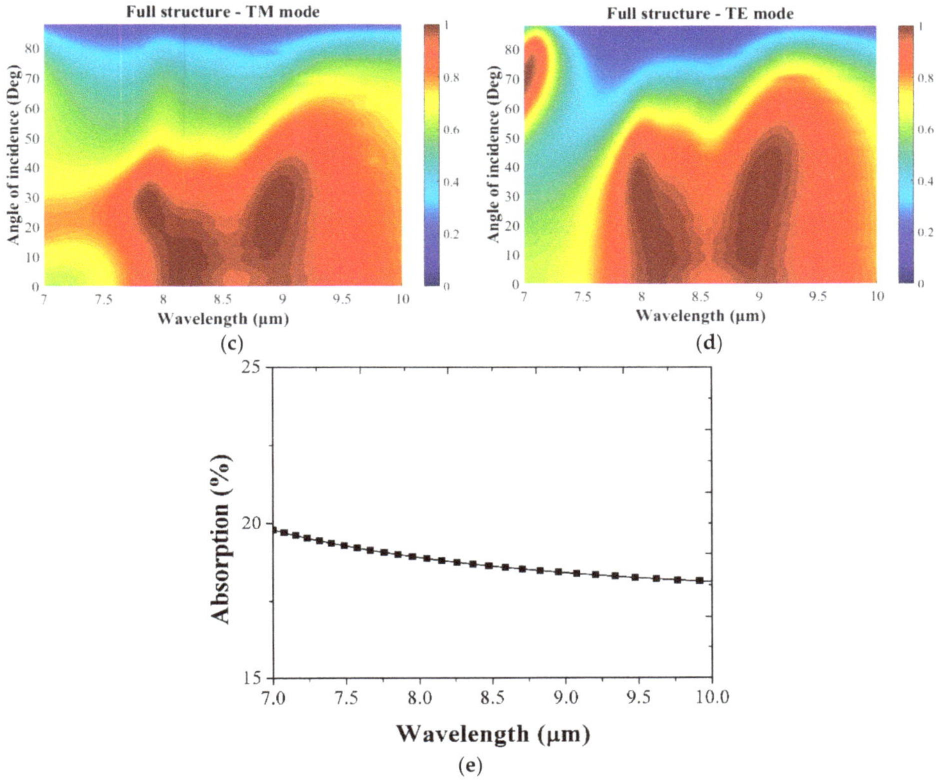

Figure 3. Simulated characteristics of the absorber structures for different incident angles. (**a**,**c**) transverse-magnetic (TM) mode; and (**b**,**d**) transverse-electric (TE) mode; and (**e**) the percentage of absorption only in the metal as a function of wavelength.

The absorption mechanism of the structure can be explained by combination of plasmonic effect, ENZ mode and Brewster mode. These effects cause the strong electromagnetic resonances at the absorption wavelengths. As shown in Figure 2, when the dielectric constant ε becomes zero (i.e., ENZ) at near 8 μm we could still achieve reasonably high field enhancement at the ENZ wavelength. As an effective explanation for the mechanisms of the observed absorption of three peaks of light, Figure 4 shows the electromagnetic field distributions at the specific resonant wavelengths. At 7.9 μm, the first absorption peak, we investigated the distribution of the magnitude of the electric field. As a result, the electric field was excited and confined strongly into the ENZ layer, as shown in Figure 4a. This behavior indicates that the excitation of the ENZ mode has occurred. Figure 4b shows the magnetic field magnitude distribution of the third peak at 8.9 μm. It can be found that the magnetic field enhancement is mainly located in the dielectric gap between the gold grating and the TiN metallic film, indicating that the gap plasmon resonance mode is strongly excited. This effect can be understood to be simply due to the patterned gold grating and the ENZ layer generating electric fields and currents when the structure has the presence of an electromagnetic wave. The high charge concentration at the edges of the gold grating results in the strong enhancement of the field around the edges of the metal structure. In the opposite direction, the conductive TiN ground plane generated the currents and electric fields thanks to the near-field coupling. Then, the antiparallel currents generated magnetic fields in the SiO_2 layer.

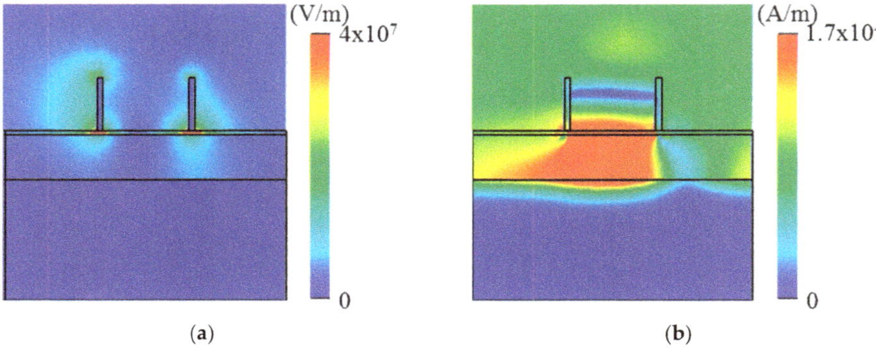

Figure 4. (**a**) E-field magnitude distribution (in units of V/m) indicating ENZ mode at 7.9 μm wavelength and (**b**) H-field magnitude distribution (in units of A/m) indicating the gap plasmon mode at 8.9 μm wavelength within the perfect absorption band.

In the structure, we set up a periodic structure with boundary conditions and the incident waves excited along the z-axis by using Floquet ports at the top and bottom of the structure. When the light propagates as a plane wave on this structure, light will be reflected or absorbed depending on the mismatch of impedance in this structure to impedance in free space. By using numerical analysis, the effective refractive index n and the effective impedance z can be extracted from the relationship with S-parameters expressed in Equations (1)–(4), where S_{11} and S_{21} are the S-parameter results from the two-port model of the aforementioned structure, k_0 is wave number and m is an integer [37].

$$S_{11} = \frac{\Gamma\left(1 - e^{i4nk_0d}\right)}{1 - \Gamma^2 e^{i4nk_0d}} \tag{1}$$

$$S_{21} = \frac{\left(1 - \Gamma^2\right)e^{i4nk_0d}}{1 - \Gamma^2 e^{i4nk_0d}} \tag{2}$$

$$z = \pm\sqrt{\frac{(1 + S_{11})^2 - {S_{21}}^2}{(1 - S_{11})^2 - {S_{21}}^2}} \tag{3}$$

$$n = \frac{1}{2k_0d}\{[\mathrm{Im}(\ln(e^{j2nk_0d}))] + 2m\pi - i\,[\mathrm{Re}(\ln(e^{j2nk_0d}))]\} \tag{4}$$

Similar to the explanation of the mechanism of metamaterial absorber structures in previous studies, the mechanism of the behavior of this device is also explained through impedance matching mechanism [38,39]. In this mechanism, the absorber structure plays a role as a thin slab which made from a homogeneous structure. The mechanism of absorption in the structure is due to the coupling of resonance modes to Brewster mode to generate the strong electromagnetic resonances at the absorption peaks as explained in the above section. Also, we utilized the effective medium theory to explain the absorption mechanism. We calculated the effective parameters of the structure through reflection and transmission coefficients of the structure. The relationship of the thickness of a homogeneous slab (d), free space vector (k_0), S-parameters and the effective parameters of the input impedance (z) and the refractive index (n) as expressed in Equation (3). However, this structure is backed by a metallic layer, therefore, Equation (3) will become:

$$z = \pm\sqrt{\frac{(1 + S_{11})^2}{(1 - S_{11})^2}} = \left|\frac{1 + S_{11}}{1 - S_{11}}\right| \tag{5}$$

Figure 5a shows the effective impedance at resonance wavelengths. Figure 5a indicates that the absorption is maximum when the real part of the impedance of the structure is equal to free space (i.e., equal to 1) and the imaginary part has a value of zero. Otherwise, the absorption will be reduced due to the impedance mismatch between the structure and free space, which results in reflection. Meanwhile, the effective refractive index is given Equation (4). In Figure 5b, at the perfect absorption wavelength of 8.9 μm, the effective refractive index exhibits a large imaginary part, $n'' >> n'$. $n = n' + in'' = 0.12 + 3.16i$ has the large imaginary part (i.e., $n'' = 3.16, n' = 0.12$) with a high ratio contrasting the real part and the imaginary part of up to 26.3, so that the incident electromagnetic wave enters the device without any reflectivity and then rapidly decays to zero inside the structure.

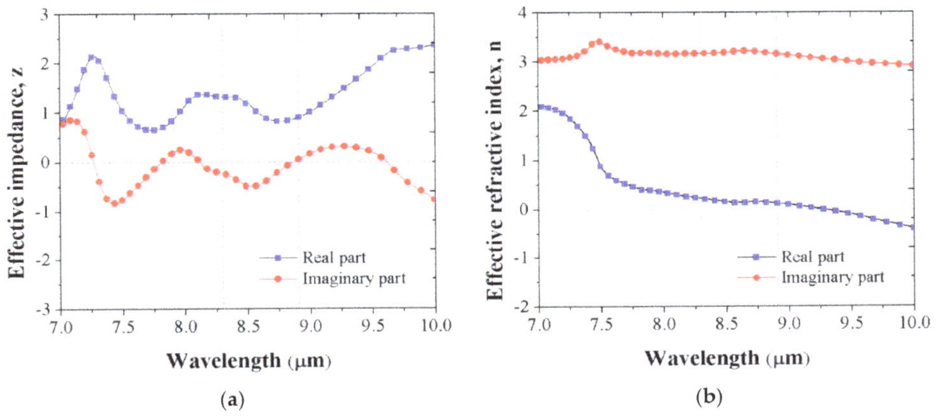

Figure 5. (**a**) Effective impedance and (**b**) effective refractive index for ENZ-based absorber model.

When light propagates between two media of differing refractive indexes, generally some of it is reflected at the boundary. However, light with one particular polarization cannot be reflected at one particular angle of incidence. This angle of incidence is Brewster's angle θ_B which can be determined by:

$$\theta_B = \arctan\left(\frac{n_2}{n_1}\right) \tag{6}$$

This equation is called Brewster's law in which n_1 is the refractive index of the incident medium (usually air) and n_2 is the effective index of the medium in the structure. In this work, we investigated a light wave passing from air (n_1 = 1.00) to the effective medium (n_2 = 0.41) at a wavelength of 7.9 μm; the Brewster angle, θ_B, was calculated to be 22°.

We have also investigated the effects of periodicity (P) on the absorption light in the structure by increasing and decreasing it by 10% compared to the optimized value of P for both TM mode and TE mode. Figure 6 shows the absorption characteristics for various periodicities. In general, the absorption peaks shifted, and absorption intensity was also affected, especially at the absorption peak around 8 μm, as P varied, which was observed both in TM and TE modes. The optimized P was selected regarding to the highest absorption and widest bandwidth.

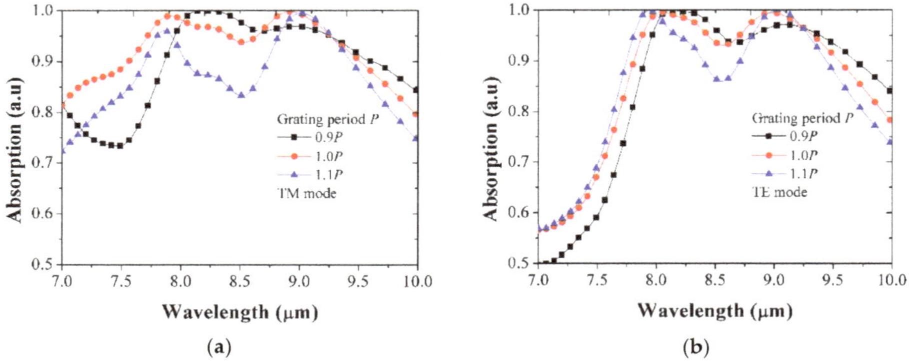

Figure 6. Investigation of the effect of periodicity *P* for absorption light in the structure. (**a**) for TM mode, (**b**) for TE mode.

4. The Absorptivity as a Function of Polarization angle

The insensitivity of polarization is an essential factor in practical applications. In this paper, the polarization independence of the proposed absorber was investigated for polarization angle. Figure 6 shows plots of the absorption spectra of the proposed light absorber to demonstrate the absorption spectra for the different polarizations. Optimization of the integrated ENZ wideband absorber can be observed at the Brewster angle for the TM and TE modes and circular polarizations; such an absorber shows the spectral band with 95% approximate absorption over a 1.46 μm spectral range (from 7.77 μm to 9.23 μm) for TM mode (black square line) and over a 1.4 μm wideband absorption range (from 7.78 μm to 9.27 μm) for TE mode (red circle line). As a comparison, the absorption spectrum of the integrated ENZ wideband absorber is also shown at the Brewster angle (blue triangle line) for circular polarization with a wideband absorption of approximate 95% over a wavelength range of 1.26 μm (from 7.94 μm to 9.2 μm). It can be found from Figure 7 that for both TE and TM polarized waves and circular polarization, the absorption curves are quite similar, which reveals that the proposed absorber is polarization-insensitive.

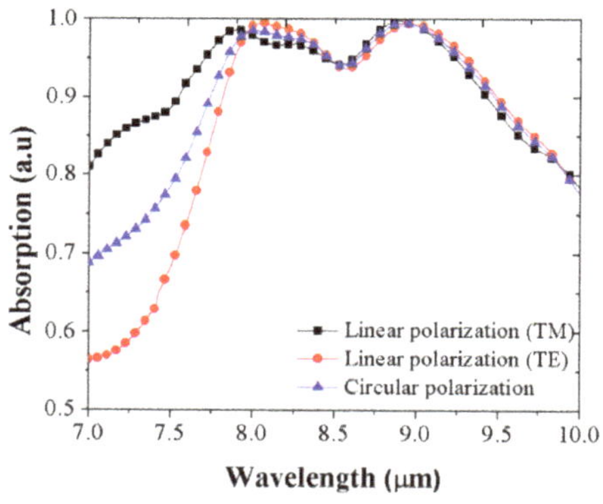

Figure 7. Comparison of TM mode, circular polarization and TE mode for simulated absorption curves from devices of integrated ENZ perfect absorber at Brewster angle.

Next, we also investigated the polarization independence of the proposed absorber for polarization angle. Figure 8 illustrates the absorption spectra with various polarization angles for the both TM and TE modes at normal incident, Brewster angle, and 40° incident. The absorption curves are similar, which indicates that the enhanced absorption is independent of the light polarization, at least up to Brewster angle. However, beyond this angle, TE mode can keep its polarization-insensitivity but the TM mode cannot.

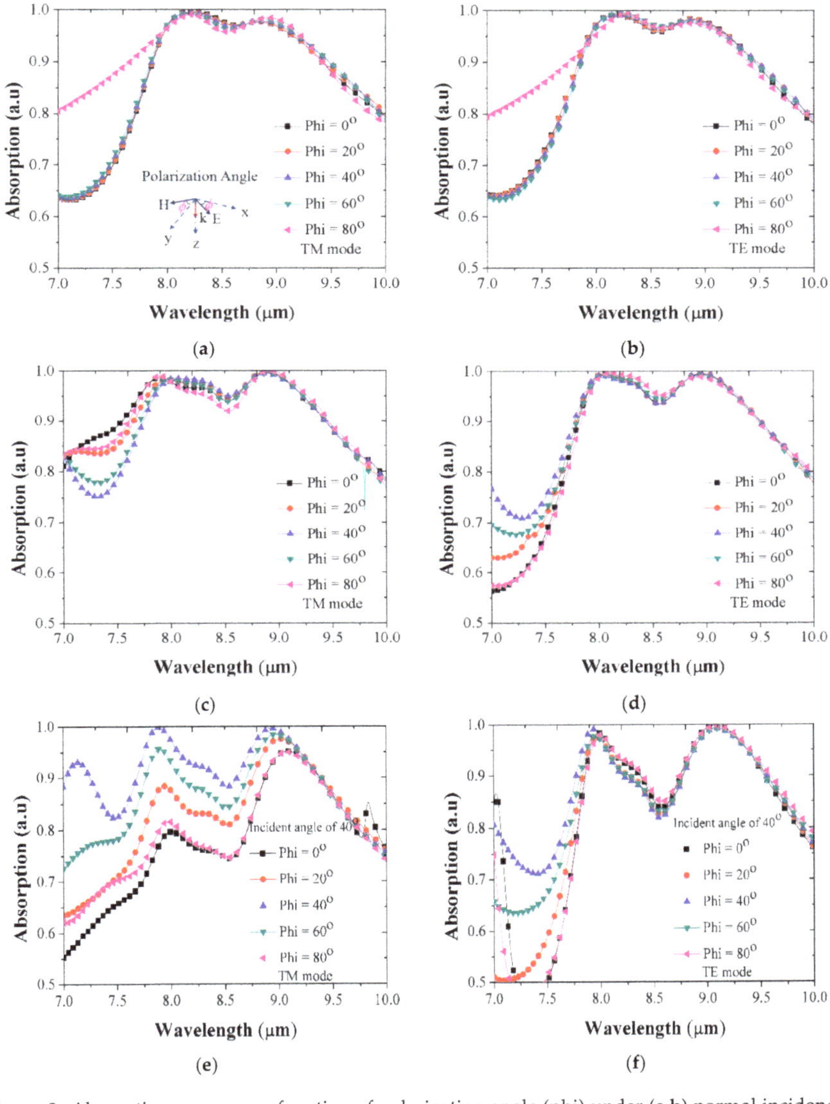

Figure 8. Absorption curves as a function of polarization angle (phi) under (**a**,**b**) normal incidence; (**c**,**d**) Brewster angle; (**e**,**f**) and 40° incidence.

5. Conclusions

We have proposed an absorber based on a combination of ENZ mode and the gap plasmon mode and Brewster effect. The simulated results reveal that this device has a wide range absorption bandwidth of mid-infrared radiation (IR) wavelengths for polarized waves at normal incident and Brewster angle. Furthermore, the ENZ-based absorber is polarization insensitive and keeps high absorption in a broad wavelength range at oblique incidence for TM and TE polarization wave. The proposed design is expected to work at other different wavelength ranges based on the studied mechanisms. The ENZ-based absorber will be a potential candidate for many applications, such as detection, sensing, imaging and defense applications. Also, our device can help improve light-harvesting efficiency with enhanced absorption both in terms of intensity and broadband.

Author Contributions: Investigation, writing—original draft preparation, P.T.D.; data curation, K.Q.L.; writing—review and editing, T.K.N.; project administration, J.-H.L.

Funding: This study was supported by National Research Foundation (NRF, 2019R1A2B5B01069580, 2019R1A6A1A09031717) and Ministry of Education and National Research Foundation (MOE, NRF, Leaders in Industry-University Cooperation + Project 2019). This research is also funded by Vietnam National Foundation for Science and Technology Development (NAFOSTED) under grant number 102.04-2019.04.

Conflicts of Interest: The authors declare no conflict of interest.

References

1. Fu, P.; Liu, F.; Ren, G.J.; Su, F.; Li, D.; Yao, J.Q. A broadband metamaterial absorber based on multi-layer graphene in the terahertz region. *Opt. Commun.* **2018**, *417*, 62–66. [CrossRef]
2. Landy, N.I.; Sajuyigbe, S.; Mock, J.J.; Smith, D.R.; Padilla, W.J. Perfect metamaterial absorber. *Phys. Rev. Lett.* **2008**, *100*, 207402. [CrossRef] [PubMed]
3. Tang, J.; Xiao, Z.; Xu, K.; Liu, D. A polarization insensitive and broadband metamaterial absorber based on three-dimensional structure. *Opt. Commun.* **2016**, *372*, 64–70. [CrossRef]
4. Sturmberg, B.C.P.; Chong, T.K.; Choi, D.-Y.; White, T.P.; Botten, L.C.; Dossou, K.B.; Poulton, C.G.; Catchpole, K.R.; McPhedran, R.C.; De Sterke, C.M. Total absorption of visible light in ultrathin weakly absorbing semiconductor gratings. *Optica* **2016**, *3*, 556–562. [CrossRef]
5. Safaei, A.; Chandra, S.; Leuenberger, M.N.; Chanda, D. Wide Angle Dynamically Tunable Enhanced Infrared Absorption on Large-Area Nanopatterned Graphene. *ACS Nano* **2019**, *13*, 421–428. [CrossRef] [PubMed]
6. Thongrattanasiri, S.; Koppens, F.H.L.; Javier Garcı´a de Abajo, F. Complete Optical Absorption in Periodically Patterned Graphene. *Phys. Rev. Lett.* **2012**, *108*, 047401. [CrossRef]
7. Nguyen, T.K.; Dang, P.T.; Park, I.; Le, K.Q. Broadband THz radiation through tapered semiconductor grating on high-index substrate. *J. Opt. Soc. Am. B* **2017**, *34*, 583–589. [CrossRef]
8. Cleary, J.W.; Nader, N.; Leedy, K.D.; Soref, R. Tunable short- to mid-infrared perfectly absorbing thin films utilizing conductive zinc oxide on metal. *Opt. Mater. Express* **2015**, *9*, 1898–1909. [CrossRef]
9. Zhao, Y.; Huang, Q.; Cai, H.; Lin, X.; Lu, Y. A broadband and switchable VO_2-based perfect absorber at the THz frequency. *Opt. Commun.* **2018**, *426*, 443–449. [CrossRef]
10. Badsha, M.A.; Jun, Y.C.; Hwangbo, C.K. Admittance matching analysis of perfect absorption in unpatterned thin films. *Opt. Commun.* **2014**, *332*, 206–213. [CrossRef]
11. Yang, Y.; Kelley, K.; Sachet, E.; Campione, S.; Luk, T.S.; Maria, J.-P.; Sinclair, M.B.; Brenner, I. Femtosecond optical polarization switching using a cadmium oxide-based perfect absorber. *Nat. Photonics* **2017**, *11*, 390–395. [CrossRef]
12. Rensberg, J.; Zhou, Y.; Richter, S.; Wan, C.; Zhang, S.; Schoppe, P.; Schmidt-Grund, R.; Ramanathan, S.; Capasso, F.; Kats, M.A.; et al. Epsilon-near-zero substrate engineering for ultrathin-film perfect absorbers. *Phys. Rev. Appl.* **2017**, *8*, 014009. [CrossRef]
13. Hendrickson, J.; Guo, J.; Zhang, B.; Buchwald, W.; Soref, R. Wideband perfect light absorber at mid-wave infrared using multiplexed metal structures. *Opt. Lett.* **2012**, *37*, 371–373. [CrossRef] [PubMed]
14. Zhang, B.; Hendrickson, J.; Guo, J. Multi-spectral near perfect metamaterial absorbers using spatially multiplexed plasmon resonance metal square structures. *J. Opt. Soc. Am. B* **2013**, *30*, 656–662. [CrossRef]
15. Dang, P.T.; Pham, T.T.; Le, K.Q.; Nguyen, T.K. Epsilon-near-zero enhanced plasmonic Brewster transmission through subwavelength tapered metallic gratings. *J. Korean Phys. Soc.* **2018**, *72*, 38–44. [CrossRef]
16. Feng, S.; Halterman, K. Coherent perfect absorption in epsilon-near-zero metamaterials. *Phys. Rev. B* **2012**, *86*, 165103. [CrossRef]
17. Wang, Y.; Sun, T.; Paude, T.; Zhang, Y.; Ren, Z.; Kempa, K. Metamaterial-plasmonic absorber structure for high efficiency amorphous silicon solar cells. *Nano Lett.* **2012**, *12*, 440–445. [CrossRef]

18. Nakayama, K.; Tanabe, K.; Atwater, H.A. Plasmonic nanoparticle enhanced light absorption in GaAs solar cells. *Appl. Phys. Lett.* **2008**, *93*, 121904. [CrossRef]
19. Jin, Y.; Feng, J.; Zhang, X.L.; Xu, M.; Bi, Y.G.; Chen, Q.D.; Wang, H.Y.; Sun, H.B. Surface-plasmon enhanced absorption in organic solar cells by employing a periodically corrugated metallic electrode. *Appl. Phys. Lett.* **2012**, *101*, 163303. [CrossRef]
20. Argyropoulos, C.; Le, K.Q.; Mattiucci, N.; D'Aguanno, G.; Alú, A. Broadband absorbers and selective emitters based on plasmonic Brewster metasurfaces. *Phys. Rev. B* **2013**, *87*, 205112. [CrossRef]
21. Hendrickson, J.R.; Vangala, S.; Dass, C.; Gibson, R.; Goldsmith, J.; Leedy, K.; Walker, D.E., Jr.; Cleary, J.W.; Kim, W.; Guo, J. Coupling of epsilon-near-zero mode to gap plasmon mode for flat-top wideband perfect light absorption. *ACS Photonics* **2018**, *5*, 776–781. [CrossRef]
22. Vassant, S.; Hugonin, J.P.; Marquier, F.; Greffet, J.J. Berreman mode and epsilon near zero mode. *Opt. Expr.* **2012**, *20*, 23971–23977. [CrossRef] [PubMed]
23. Campione, S.; de Ceglia, D.; Vincenti, M.A.; Scalora, M.; Capolino, F. Electric field enhancement in ε-near-zero slabs under TM-polarized oblique incidence. *Phys. Rev. B* **2013**, *87*, 035120. [CrossRef]
24. Vincenti, M.A.; de Ceglia, D.; Scalora, M. Nonlinear dynamics in low permittivity media: The impact of losses. *Opt. Exp.* **2013**, *21*, 29949–29954. [CrossRef]
25. de Ceglia, D.; Campione, S.; Vincenti, M.A.; Capolino, F.; Scalora, M. Low-damping epsilon-near-zero slabs: Nonlinear and nonlocal optical properties. *Phys. Rev. B* **2013**, *87*, 155140. [CrossRef]
26. Campione, S.; Brener, I.; Marquier, F. Theory of epsilon-near-zero modes in ultrathin films. *Phys. Rev. B* **2015**, *91*, 121408(R). [CrossRef]
27. Adams, D.C.; Inampudi, S.; Ribaudo, T.; Slocum, D.; Vangala, S.; Kuhta, N.A.; Goodhue, W.D.; Podolskiy, V.A.; Wasserman, D. Funneling light through a subwavelength aperture with epsilon-near-zero materials. *Phys. Rev. Lett.* **2011**, *107*, 133901. [CrossRef]
28. Naik, G.V.; Shalaev, V.M.; Boltasseva, A. Alternative plasmonic materials: Beyond gold and silver. *Adv. Mater.* **2013**, *25*, 3264–3294. [CrossRef]
29. Kinsey, N.; Devault, C.; Kim, J.; Ferrera, M.; Shalaev, V.M.; Boltasseva, A. Epsilon-near-zero Al-doped ZnO for ultrafast switching at telecom wavelengths. *Optica* **2015**, *2*, 616–622. [CrossRef]
30. Kliewer, K.L.; Fuchs, R. Optical modes of vibration in an ionic crystal slab including retardation. II. Radiative region. *Phys. Rev.* **1966**, *150*, 573. [CrossRef]
31. Sarid, D. Long-range surface-plasma waves on very thin metal films. *Phys. Rev. Lett.* **1981**, *47*, 1927–1930. [CrossRef]
32. Chikkaraddy, R.; de Nijs, B.; Benz, F.; Barrow, S.J.; Scherman, O.A.; Rosta, E.; Demetriadou, A.; Fox, P.; Hess, O.; Baumberg, J. Single-molecule strong coupling at room temperature in plasmonic nanocavities. *Nature* **2016**, *535*, 127–130. [CrossRef] [PubMed]
33. Hao, J.; Wang, J.; Liu, X.; Padilla, W.J.; Zhou, L.; Qiu, M. High performance optical absorber based on a plasmonic metamaterial. *Appl. Phys. Lett.* **2010**, *96*, 251104. [CrossRef]
34. Hendrickson, J.R.; Vangala, S.; Dass, C.K.; Gibson, R.; Leedy, K.; Walker, D.; Cleary, J.W.; Luk, T.S.; Guo, J. Experimental Observation of Strong Coupling Between an Epsilon-Near-Zero Mode in a Deep Subwavelength Nanofilm and a Gap Plasmon Mode. Available online: https://arxiv.org/abs/1801.03139 (accessed on 9 January 2018).
35. Palik, E.D. *Handbook of Optical Constants of Solids*, 1st ed.; Academic Press: Cambridge, MA, USA, 1991.
36. CST Microwave Studio, CST GmbH. Available online: http://cst.com (accessed on 1 January 2016).
37. Chen, X.; Grzegorczyk, T.M.; Wu, B.-I.; Pacheco, J., Jr.; Kong, J.A. Robust method to retrieve the constitutive effective parameters of metamaterials. *Phys. Rev. E* **2004**, *70*, 016608. [CrossRef] [PubMed]
38. Tao, H.; Landy, N.I.; Bingham, C.M.; Zhang, X.; Averitt, R.D.; Padilla, W.J. A metamaterial absorber for the terahertz regime: Design, fabrication and characterization. *Opt. Expr.* **2008**, *16*, 7181–7188. [CrossRef] [PubMed]
39. Tittl, A.; Harats, M.G.; Walter, R.; Yin, X.; Schäferling, M.; Liu, N.; Rapaport, R.; Giessen, H. Quantitative angle-resolved small-spot refectance measurements on plasmonic perfect absorbers: Impedance matching and disorder efects. *ACS Nano* **2014**, *8*, 10885–10892. [CrossRef]

micromachines

MDPI

Article

Narrow Linewidth Distributed Bragg Reflectors Based on InGaN/GaN Laser

Wuze Xie [1,2], Junze Li [1,2,*], Mingle Liao [1,2], Zejia Deng [1,2], Wenjie Wang [1,2] and Song Sun [1,2,*]

1 Microsystems and Terahertz Research Center, China Academy of Engineering Physics, Chengdu 610200, China
2 Institute of Electronic Engineering, China Academy of Engineering Physics, Mianyang 621999, China
* Correspondence: lijunze@mtrc.ac.cn (J.L.); sunsong@mtrc.ac.cn (S.S.)

Received: 5 July 2019; Accepted: 9 August 2019; Published: 11 August 2019

Abstract: A variety of emerging technologies, such as visible light communication systems, require narrow linewidths and easy-to-integrate light sources. Such a requirement could be potentially fulfilled with the distributed Bragg reflector (DBR) lasers, which are also promising for the monolithical integration with other optical components. The InGaN/GaN-based surface etched DBR is designed and optimized using the finite-difference-time-domain (FDTD) method to obtain very narrow-band reflectors that can serve as a wavelength filter. The results reveal that the ultimate reflectivity depends on the grating period and duty ratio of the DBR. Based on the design, the DBR lasers with various duty ratios are fabricated, specifically, the 19th, 13th and 3rd order DBR grating with duty ratio set as 50%/75%/95%. The minimum linewidth could be achieved at 0.45 nm from the 19th order grating with a 75% duty ratio. For comparison, the Fabry–Pérot (F–P) based on the same indium gallium nitride/gallium nitride (InGaN/GaN) epitaxial wafer are fabricated. The full width at half maximum (FWHM) of the DBR laser shrank by 65% compared to that of the conventional F–P laser, which might be helpful in the application of the visible light communication system.

Keywords: distributed Bragg reflectors; gratings; GaN-based lasers; linewidth

1. Introduction

The visible light communication (VLC) has been paid great attention in both the academic and engineering fields for its broad range of applications, including Li–Fi, vehicle to vehicle communication, underwater communication, and information display, among the others. However, some common challenges still exist for VLC [1–5], such as the integration of the VLC system and higher modulation bandwidth of the light source. The wide-bandgap-GaN-based solid-state lighting source has incomparable advantages for the high-speed VLC [6–8] due to its small size, excellent beam quality, long lifetime, and good stability.

The integration of various components on a single chip still poses an enormous challenge to enhance the functionality, speed, efficiency, and robustness of the overall system. Recently, the chip integration of modulators, waveguides, photo-detectors with gallium nitride (GaN) light-emitting diodes (LEDs) on various substrates have been demonstrated for VLC [9–11]. Most of the current VLC systems use LEDs, which suffer from low output powers, bad beam convergences, low modulation bandwidths, and wide linewidths. These parasitic drawbacks are due to the light emission nature of the LEDs, which can be effectively overcome with high-performance lasers. Meanwhile, the laser should be integrated easily into the VLC systems.

A promising light source is the Fabry–Pérot (F–P) laser integrated with the surface etched distributed Bragg reflector (DBR) [12,13]. On the one hand, the convolution between the F–P cavity and the DBR grating selects the longitudinal modes, which is capable of obtaining a more stable and narrower laser emission. On the other hand, the DBRs could be simultaneously fabricated with the ridge waveguide, yielding a much simpler fabrication process than the conventional coating process. Altogether, a high performance on-chip integrated optical system could be envisioned with the DBR laser.

Although DBR lasers in the infrared wavelength have been extensively studied [14–20], there are few reports about the GaN-based DBR lasers in the visible spectrum [21–23]. Cho et al. demonstrated the room temperature operation of an electrically injected InGaN/GaN-based DBR laser and reduced the threshold current density [24]. Wang et al. observed a 50% reduction in the threshold pumping intensity by introducing the DBRs at the end of an InGaN/GaN-based multiple quantum well (MQW) laser [12]. The carrier lifetime and optical bistability in the InGaN-based DBR lasers were reported by Dorsaz et al. [25]. The influence of the surface etched DBRs was studied, for instance, the tilt of the sidewall [26,27]. While more attention is focused on the threshold conditions, the characteristics of the emission linewidth of the GaN-based laser with DBRs have not been thoroughly investigated, which is crucial for the high-speed VLC application.

In this work, the etched surface grating is designed and optimized for the InGaN/GaN-based DBR lasers using the finite-difference-time-domain (FDTD) method. Based on the designed structure, the DBR lasers with different-orders and various duty ratios are fabricated and analyzed to obtain a better light source. The uniform grating structures are defined by electron-beam lithography (EBL) and etched by inductively coupled plasma (ICP). In addition, the conventional F–P laser is also fabricated on the same wafer as the reference sample. The slope efficiency and threshold current for the GaN-based DBR lasers and F–P are compared. Furthermore, the linewidth of the GaN-based lasers with and without the DBR structure is illustrated.

2. Simulation and Fabrication

The FDTD simulation is carried out to design the DBR, as shown in Figure 1. To mimic the experimental condition, the grating is assumed to consist of rectangular-shaped grooves. The etched grooves are covered by air ($n_{\text{air}} = 1$) and the grating pair number is fixed at 15, which is sufficient to provide high reflectivity with moderate computing resources. According to the Bragg condition $\lambda_0 = 2n_{eff}\Lambda/m$, for a given etch depth $d = 800$ nm, the lasing wavelength λ_0 varied with Λ and the duty ratio $D = (\Lambda - o)/\Lambda$, where n_{eff} is the longitudinally averaged effective refractive index of the structure, Λ is the grating period, m is the order of the Bragg grating, and o is the width of the etched grooves. In Figure 1, it shows the dependence of reflectivity on Λ and D, respectively. In Figure 1a,b, the duty ratio is selected at $D = 80\%$, the grating period Λ varies from 0.24 μm to 1.6 μm. In Figure 1c,d, the grating period is fixed at $\Lambda = 1.6$ μm, and the duty ratio D varies from 0 to 1.0. For comparison, the result for a period of 1050 nm is also plotted in Figure 1d.

It can be seen from Figure 1a,b that the center wavelength of 400 nm could be obtained for an appropriate pair of Λ and D. A larger maxima reflectivity could be obtained with a shorter grating period, indicating smaller loss. It is noteworthy that the scattering loss at the rough sidewalls is not included in the simulation and it is expected to be more critical for the short period grating. From this point of view, the shorter period DBR does not necessarily have a better reflection performance. On the contrary, the higher-order grating has a narrower reflection bandwidth [28], which is favorable in the communication systems. Combining the above rules, the grating periods of 0.24 μm, 1.05 μm, and 1.55 μm are chosen to obtain various orders of DBR at a specific duty ratio. The desired values are denoted as circles in Figure 1a.

It shows in Figure 1c that the duty ratio has less effect on the maximum reflectivity and the bandwidth of the same order. However, the fabrication tolerance decreases as the grating period increases, since in a long periodic grating, the peak reflectivity has a lower duty ratio tolerance, as shown in Figure 1d. An alternation in the duty ratio actually implies the most common fabrication errors, since a small change in the dose of the e-beam writing process would result in the change in the line/space width, instead of the grating period. The optimized duty ratios of 50%, 75%, and 95% at the period 1.55 μm are set to demonstrate their characteristics. In brief, careful balance between the loss, bandwidth, and fabrication tolerance should be made to achieve the appropriate performances. In the following, the DBR gratings on one end of the F–P cavity with the optimized periods and duty ratios are fabricated and tested to obtain the narrow linewidth lasers.

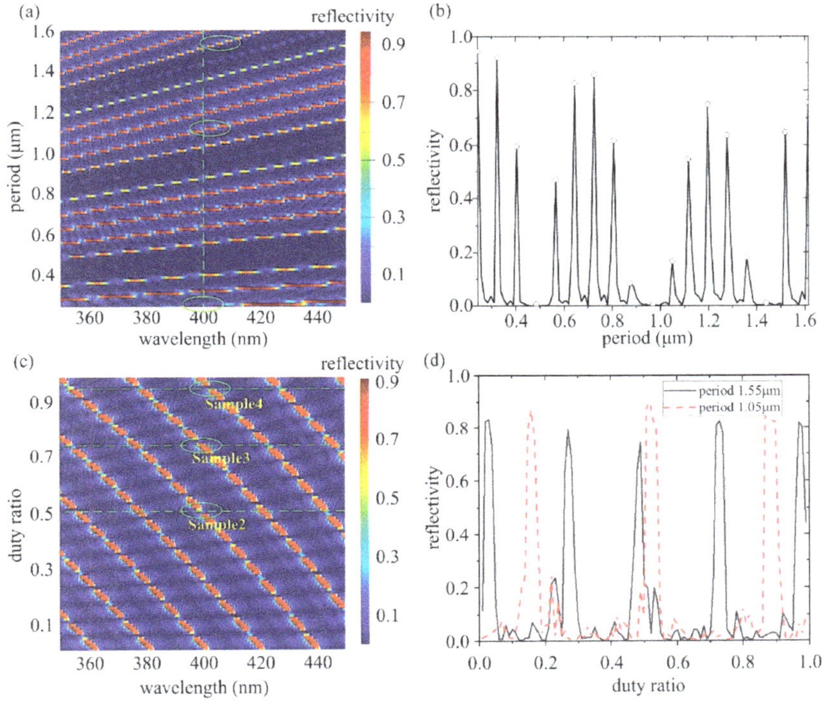

Figure 1. Simulation results of reflection spectra of varied period (**a**) and duty ratio (**c**). (**b**,**d**) are the reflectivity at 400 nm corresponding to (**a**) and (**c**), respectively. For comparison, period of 1050 nm is also plotted in (**d**).

The top and cross-section views of the intensity profile of the optimized DBR are shown in Figure 2, where the grating period is 1.55 μm and the duty ratio is 75%. It could be observed that the optical field is confined in the waveguide, and is gradually decayed along the transmission direction, indicating that the optical field is effectively reflected.

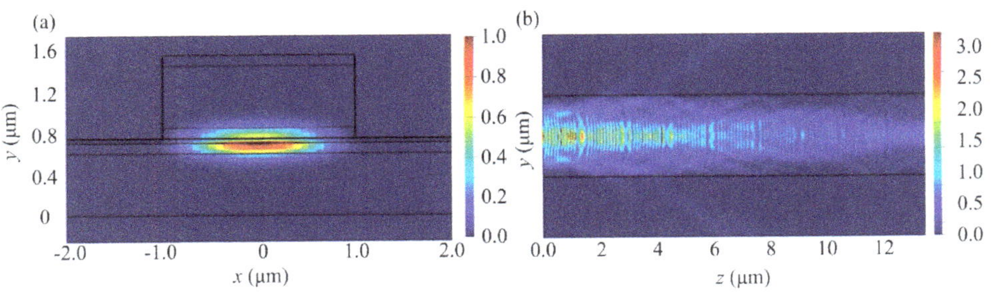

Figure 2. Optical power intensity profile of optimized distributed Bragg reflector (DBR) grating. (**a**) Cross-sectional view; (**b**) top-section view.

The complete structure of the GaN laser diode integrated with the DBR is shown in Figure 3. The sample is first processed into a 10 μm wide ridge structure including an 800 μm length gain section and the DBR section, with the ridge etching stopping at the p-electron block layer by the ion beam etching. Before the deposition of the p-contact metal, a 200 nm thick SiO_2 layer is deposited on the side of the ridge in order to confine the injection section on the top of the ridge and prevent additional optical loss from the metal contact. Aiming to achieve an emission wavelength at 400 nm, a period of 1550 nm is chosen for the 19th order Bragg grating containing 100 pairs of gratings. This period grating is large enough, and could be precisely defined over the pre-etched DBR section via the electron beam lithography. Followed by the inductively coupled plasma (ICP) etch, the grating structure could be simultaneously transferred into the epitaxial layer. A 10/20/30 nm Pd/Ni/Au p-ohmic contact metal is deposited on the top of the ridge with a 15/300 nm Ti/Au contact pad deposited on the top of it. The GaN substrate is thinned to 100 μm thickness by the chemical mechanical polish method. A 50/50/200 nm Ti/Pt/Au n-contact metal is deposited on the back of the GaN substrate. Finally, the sample is cleaved to form the laser facet, which is perpendicular to the ridge structure.

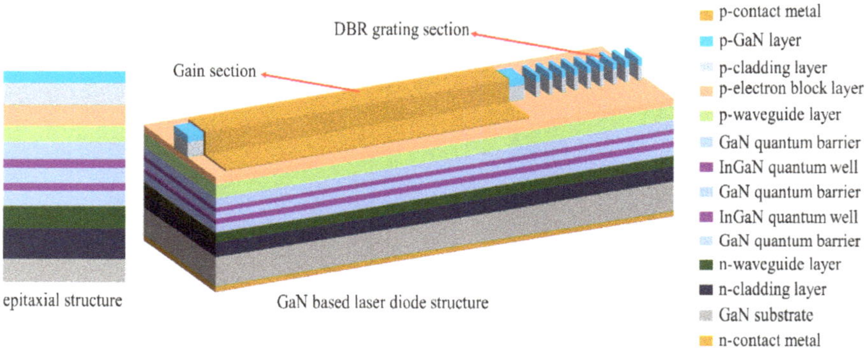

Figure 3. Schematic cross-section of gallium nitride (GaN)-based epitaxial wafer and structure of fabricated DBR laser diode.

The morphologies of the samples are measured by the scanning electron microscope (SEM) with the FEI Nova NanoSEM 450, and the resolution is 1.4 nm at the high-vacuum mode. The power-current-voltage measurement is characterized under the pulse driving condition with a 1 μs pulse width and 10 kHz repetition rate. The spectral characteristics are measured by a fiber spectrometer (BIM6002, Brolight, Hangzhou, China) with a resolution 0.16 nm under the pulse driving condition with a 500 ns pulse width and 1 kHz repetition rate.

3. Results

The DBR grating structures of the fabricated laser diodes are shown in Table 1. An F–P ridge waveguide laser diode is set as the blank reference sample named as Sample 1. The DBR grating laser diodes with duty ratios of 50%, 75%, 95%, are denoted as Sample 2, Sample 3, and Sample 4, respectively. All the samples are processed from the same epitaxial wafer and share the same fabrication procedure. The fabricated structures of the DBR grating show little deviations from the designed structure, which is attributed to the slight alternation in the dose of the e-beam writing process.

Table 1. The structure of the fabricated distributed Bragg reflector (DBR) grating.

Sample Name	Sample 1	Sample 2	Sample 3	Sample 4
Designed period	-	1550 nm	1550 nm	1550 nm
Designed duty ratio	-	50%	75%	95%
Fabricated period	-	1583 nm	1601 nm	1589 nm
Duty ratio	-	44%	75%	92%

Figure 4 shows the scanning electron microscopy (SEM) picture of a typical dry-etched DBR grating of Sample 4. Obviously, the grating is precisely fabricated over the desired DBR section at the end of ridge waveguide of the laser diode, as shown in Figure 4a. The period and duty ratio of DBR grating are shown in Figure 4b. There is a slight deviation in the grating period and duty ratio between the designed structure and the fabricated device. However, such a small deviation wouldn't affect the performance of the device. In order to characterize the depth of the device, the DBR grating is cleaved along the ridge waveguide direction. Figure 4c,d show the sidewall and the corresponding sectional view of the DBR grating. The sidewall of DBR grating structure looks smooth, and the depth of the DBR grating is 414.4 nm, which again proves that the devices are well fabricated.

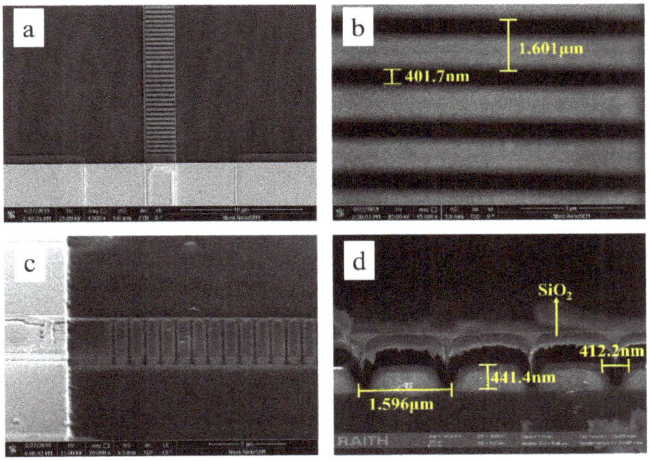

Figure 4. Scanning electron microscope (SEM) imagine of top and side view of 19th order, 75% duty ratio DBR grating fabricated by electron-beam lithography (EBL) and inductively coupled plasma (ICP) dry etch. (**a**) top view; (**b**) top view with high resolution; (**c**) bird view; (**d**) cross section view.

Figure 5 shows the power-current-voltage (P-I-V) characteristics of all the samples. The inset in Figure 5 shows the slope efficiencies of the four samples. The current through the ridge waveguide varies from 10 mA to 850 mA with a step of 50 mA. The threshold current (*I*th) is 490 mA, 540 mA, 530 mA, and 500 mA for the F–P laser diode and three DBR lasers, respectively. All the samples demonstrate similarly high-threshold currents around 500 mA, which is attributed to the imperfect qualities of the epitaxial materials. Another reason could be that the grating is not etched deep enough to provide high reflectivity as desired. The slope efficiency is 0.205 W/A, 0.14 W/A, 0.125 W/A, and 0.156 W/A for the four samples, respectively. It is obvious that the slope efficiency of the F–P laser diode is higher than that of the DBR laser diodes, which is attributed to the diffraction and scattering losses from the gratings. It shares a similar level of the slope efficiency among the DBR laser diodes, while Sample 4 (with a period of 1601 nm of the 19th order gratings DBR laser) shows a relatively higher slope efficiency compared to other counterparts, indicating a smoother sidewall and smaller scattering loss and diffraction loss. It could also be speculated from the current-voltage (I-V) curve that the F–P laser possesses a lower threshold voltage, which is confirmed by the loss of reflectivity in the DBR in the above simulations.

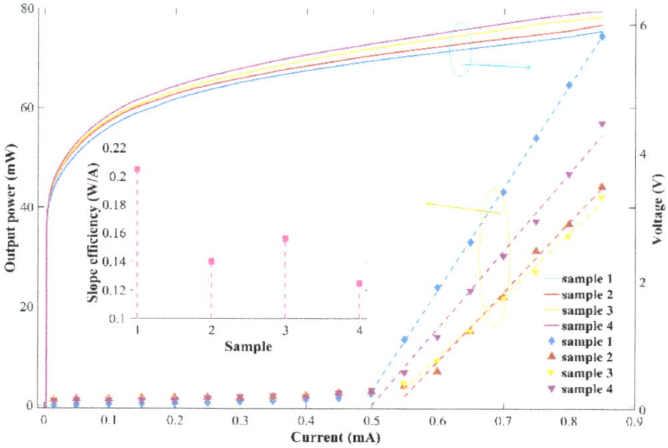

Figure 5. Power-current-voltage (P-I-V) measurement and slope efficiency of Sample 1, Sample 2, Sample 3, and Sample 4 that are characterized under pulse driving conditions, 1 μs pulse width, and 10 kHz repetition rate.

The emission spectra of the F–P laser diode and DBR laser diodes (Sample 1 to Sample 4) are characterized at a driving current of 1.5 times that the threshold current (1.5*I*th) by a fiber optic spectrometer with a spectrum resolution of 0.16 nm, and is shown in Figure 6. The emission wavelengths of all samples show slight fluctuations around 400 nm which could be attributed to the inhomogeneity of the epitaxial structure. The full width at half maxima (FWHM) of samples are 1.29 nm, 0.50 nm, 0.45 nm, and 0.63 nm, respectively. It is obvious that the FWHMs of the DBR laser diodes show more than 65% reductions compared to that of the F–P laser, due to the mode selection realized by the DBR gratings. The DBR laser diode with the 75% duty ratio (Sample 3) shows the narrowest line width of 0.45 nm.

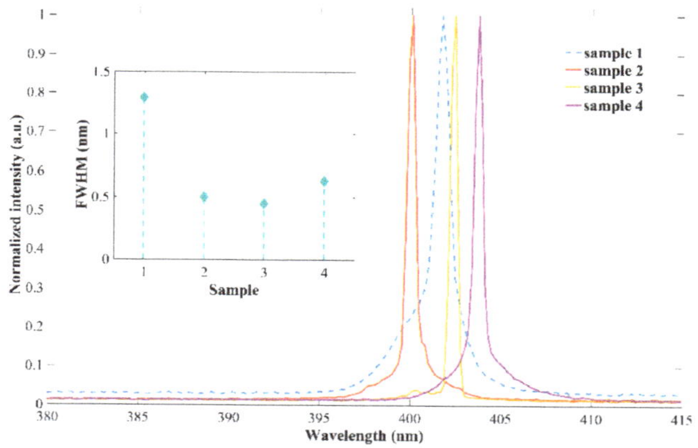

Figure 6. Emission spectra fabricated laser diodes (Sample 1 to Sample 4) under pulse driving conditions with 500 ns pulse width, 1 kHz repetition rate.

The DBR laser diodes with different periods are chosen for the Bragg grating structures to study the influence of the period on the emission spectra. The periods of 1550 nm, 1050 nm, and 240 nm are chosen for the 19th, 9th and 3rd order DBR gratings with the same duty ratio of 75%. Figure 7 shows the emission spectra of the four samples under the pulse driving condition with the injection current of $1.5I_{th}$. It is obvious that the line width of the emission spectrum increases with the decrease of the DBR grating period, which is consistent with the simulation results. In addition, it is easier to fabricate a high-quality DBR grating for a large period, because the structure fluctuation of a large period grating could be easily controlled during the grating manufacture procedure.

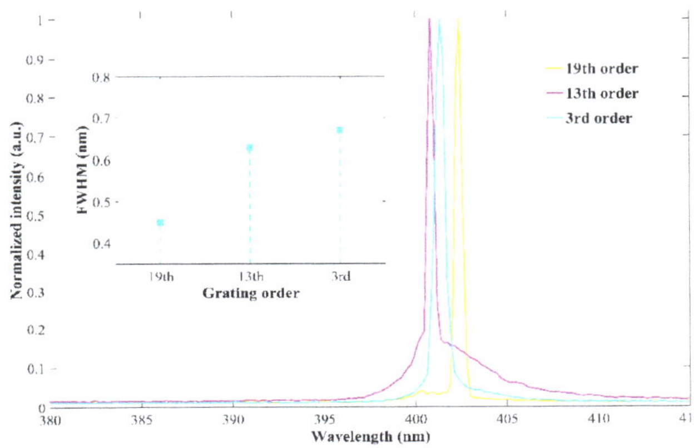

Figure 7. Emission spectrum property of fabricated DBR laser diodes under pulse driving condition with 500 ns pulse width, 1 kHz repetition rate.

4. Conclusions

In summary, the InGaN/GaN-based laser diode integrated with the surface etched DBR grating is designed and optimized using the FDTD method. On top of that, the DBR grating with even and the smooth sidewall is fabricated by the EBL and ICP dry etch, fulfilling the designed structure in the FDTD simulation. The optical and electrical properties are characterized among the DBR laser diodes. About 65% reduction on the emission linewidth can be achieved with a DBR laser diode of 1601 nm period and 75% duty ratio, compared to that of a conventional F–P laser diode. However, the slope efficiency of the DBR laser is lower than the F–P laser because of the scattering losses and absorption loss in the grating. Moreover, the InGaN/GaN-based DBR lasers with 19th, 13th and 3rd, 75% duty ratios are fabricated to investigate the influence of the grating period on the emission linewidth. A minimum linewidth of 0.45 nm could be obtained with a DBR laser diode of 19th order and 75% duty ratio, due to the minimized fabrication tolerance at large scale in the manufacturing procedure.

Author Contributions: Conceptualization—W.X. and J.L.; Data curation—W.X. and M.L.; Methodology—M.L. and W.X.; Project administration—J.L.; Supervision—J.L. and S.S.; Validation—Z.D. and W.W.; Writing original draft—W.X., S.S. and J.L.

Funding: This work was funded by Science Challenge Project (No. TZ2016003-2), National Natural Science Foundation of China (No. 61804140), and National Key R&D Program of China (No. 2017YFB0403103).

Acknowledgments: The authors would like to thank the Nanofabrication facility in Suzhou Institute of Nano-tech and Nano-bionics (CAS) for equipment access.

Conflicts of Interest: The authors declare no conflict of interest.

References

1. Killinger, D. Free space optics for laser communication through the air. *Opt. Photon. News* **2002**, *13*, 36–42. [CrossRef]
2. Watson, S.; Tan, M.; Najda, S.P.; Perlin, P.; Leszczynski, M.; Targowski, G.; Grzanka, S.; Kelly, A.E. Visible light communications using a directly modulated 422 nm GaN laser diode. *Opt. Lett.* **2013**, *38*, 3792–3794. [CrossRef] [PubMed]
3. Yang, Y.; Zhu, B.; Shi, Z.; Wang, J.; Li, X.; Gao, X.; Yuan, J.; Li, Y.; Jiang, Y.; Wang, Y. Multi-dimensional spatial light communication made with on-chip InGaN photonic integration. *Opt. Mater.* **2017**, *66*, 659–663. [CrossRef]
4. Khan, L.U. Visible light communication: Applications, architecture, standardization and research challenges. *Digit. Commun. Netw. J.* **2017**, *3*, 78–88. [CrossRef]
5. Zafar, F.; Bakaul, M.; Parthiban, R. Laser-diode-based visible light communication: Toward gigabit class communication. *IEEE Commun. Mag.* **2017**, *55*, 144–151. [CrossRef]
6. Pimputkar, S.; Speck, J.S.; DenBaars, S.P.; Nakamura, S. Prospects for LED lighting. *Nat. Photonics.* **2009**, *3*, 180. [CrossRef]
7. Kang, C.M.; Kong, D.J.; Shim, J.P.; Kim, S.; Choi, S.B.; Lee, J.Y.; Min, J.H.; Seo, D.J.; Choi, S.Y.; Lee, D.S. Fabrication of a vertically-stacked passive-matrix micro-LED array structure for a dual color display. *Opt. Express* **2017**, *25*, 2489–2495. [CrossRef]
8. Dadabayev, R.; Shabairou, N.; Zalevsky, Z.; Malka, D. A visible light RGB wavelength demultiplexer based on silicon-nitride multicore PCF. *Opt. Laser. Technol.* **2019**, *111*, 411–416. [CrossRef]
9. Feng, M.; Wang, J.; Zhou, R.; Sun, Q.; Gao, H.; Zhou, Y.; Liu, J.; Huang, Y.; Zhang, S.; Ikeda, M.; et al. On-chip integration of GaN-based laser, modulator, and photodetector grown on Si. *IEEE J. Sel. Top. Quant.* **2018**, *24*, 1–5. [CrossRef]
10. Wang, Y.; Wang, X.; Zhu, B.; Shi, Z.; Yuan, J.; Gao, X.; Liu, Y.; Sun, X.; Li, D.; Amano, H. Full-duplex light communication with a monolithic multicomponent system. *Light Sci. Appl.* **2018**, *7*, 83. [CrossRef]
11. Wang, Y.; Wang, X.; Yuan, J.; Gao, X.; Zhu, B. Monolithic III–nitride photonic circuit towards on-chip optical interconnection. *Appl. Phys. Express* **2018**, *11*, 122201. [CrossRef]
12. Wang, H.; Kumagai, M.; Tawara, T.; Nishida, T.; Akasaka, T.; Kobayashi, N.; Saitoh, T. Fabrication of an InGaN multiple-quantum-well laser diode featuring high reflectivity semiconductor/air distributed Bragg reflectors. *Appl. Phys. Lett.* **2002**, *81*, 4703–4705. [CrossRef]
13. Hofling, E.; Schafer, F.; Reithmaier, J.P.; Forchel, A. Edge-emitting GaInAs-AlGaAs microlasers. *IEEE Photon. Tech. Lett.* **1999**, *11*, 943–945. [CrossRef]
14. Oku, S.; Kondo, S.; Noguchi, Y.; Hirono, T.; Nakao, M.; Tamamura, T. Surface-grating distributed Bragg reflector lasers with deeply etched grooves formed by reactive beam etching. *Jpn. J. Appl. Phys.* **1999**, *38*, 1256. [CrossRef]
15. Roh, S.D.; Lee, K.E.; Hughes, J.S.; Coleman, J.J. Single and tunable dual-wavelength operation of an InGaAs-GaAs ridge waveguide distributed Bragg reflector laser. *IEEE Photon. Tech. Lett.* **2000**, *12*, 16–18. [CrossRef]
16. Paschke, K.; Behrendt, J.; Maiwald, M.; Fricke, J.; Wenzel, H.; Erbert, G. High power single mode 980nm DBR tapered diode lasers with integrated sixth order surface gratings based on simplified fabrication process. In Proceedings of the SPIE Photonics Europe, Strasbourg, France, 3–7 April 2006; p. 618401.
17. Bach, L.; Rennon, S.; Reithmaier, J.P.; Forchel, A.; Gentner, J.L.; Goldstein, L. Laterally coupled DBR laser emitting at 1.55 μm fabricated by focused ion beam lithography. *IEEE Photon. Tech. Lett.* **2002**, *14*, 1037–1039. [CrossRef]
18. Hou, L.; Haji, M.; Dylewicz, R.; Qiu, B.; Bryce, A.C. Monolithic 45-GHz mode-locked surface-etched DBR laser using quantum-well intermixing technology. *IEEE Photon. Tech. Lett.* **2010**, *22*, 1039–1041. [CrossRef]
19. Jiang, L.; Achtenhagen, M.; Amarasinghe, N.V.; Young, P.; Evans, G. High-power DBR laser diodes grown in a single epitaxial step. In Proceedings of the SPIE OPTO: Integrated Optoelectronic Devices, San Jose, CA, USA, 24–29 January 2009; p. 72301F.
20. Sumpf, B.; Hasler, K.H.; Adamiec, P.; Bugge, F.; Fricke, J.; Ressel, P.; Wenzel, H.; Erbert, G.; Tränkle, G. 1060 nm DBR tapered lasers with 12 W output power and a nearly diffraction limited beam quality. In Proceedings of the SPIE OPTO: Integrated Optoelectronic Devices, San Jose, CA, USA, 24–29 January 2009; p. 72301E.

21. Li, J.; Liu, W.; Li, Y.; Visovsky, N.; Pikula, D.; Heberle, A.; Brown, G.; Piech, G.; Butler, D.; Zah, C.E. 350 mW green light emission from a directly frequency-doubled DBR laser in a compact package. In Proceedings of the IEEE Photonic Society 24th Annual Meeting, Arlington, VA, USA, 9–13 October 2011; pp. 292–293.
22. Sanz, D.C.; Rorison, J.M.; Yu, S. InGaN/GaN MQW laser diodes with 4th order FIB-etched gratings. In Proceedings of the Quantum Electronics and Laser Science Conference, Baltimore, MD, USA, 22–27 May 2005; p. JTuC82.
23. Dumitru, V.; Schweizer, H.; Grabeldinger, H.; Harle, R.; Bader, S.; Brurderl, G.; Weimar, A.; Lell, A.; Harle, V. InGaN/GaN multi-quantum well distributed Bragg reflector laser diode with second-order gratings. *Electro. Lett.* **2003**, *39*, 372–373. [CrossRef]
24. Cho, J.; Cho, S.; Kim, B.J.; Chae, S.; Sone, C.; Nam, O.H.; Lee, J.W.; Kim, T.I. InGaN/GaN multi-quantum well distributed Bragg reflector laser diode. *Appl. Phys. Lett.* **2000**, *76*, 1489–1491. [CrossRef]
25. Dorsaz, J.; Boïko, D.L.; Sulmoni, L.; Carlin, J.F.; Scheibenzuber, W.G.; Schwarz, U.T.; Grandjean, N. Optical bistability in InGaN-based multisection laser diodes. *Appl. Phys. Lett.* **2011**, *98*, 191115. [CrossRef]
26. Ren, Q.; Zhang, B.; Zhang, B.; Xu, J.; Yang, Z.J.; Hu, X.D. Micro-zone optical measurements on GaN based nitride/air distributed Bragg reflector (DBR) mirrors made by focused ion beam milling. *Phys. Status Solidi C* **2004**, *1*, 2450–2453. [CrossRef]
27. Bao, S.; Song, Q.; Xie, C. The influence of grating shape formation fluctuation on DFB laser diode threshold condition. *Opt. Rev.* **2018**, *25*, 330–335. [CrossRef]
28. Fricke, J.; John, W.; Klehr, A.; Ressel, P.; Weixelbaum, L.; Wenzel, H.; Erbert, G. Properties and fabrication of high-order bragg gratings for wavelength stabilization of diode lasers. *Semicond. Sci. Tech.* **2012**, *27*, 055009. [CrossRef]

micromachines

MDPI

Article

Enhanced Photon Emission Efficiency Using Surface Plasmon Effect of Pt Nanoparticles in Ultra-Violet Emitter

Hee-Jung Choi [1], Sohyeon Kim [1], Eun-Kyung Chu [1], Beom-Rae Noh [1], Won-Seok Lee [1], Soon-Hwan Kwon [1], Semi Oh [2,*] and Kyoung-Kook Kim [1,3,*]

1 Department of Advanced Convergence Technology, and Research Institute of Advanced Convergence Technology, Korea Polytechnic University, 237 Sangidaehak-ro, Siheung-si 15073, Korea
2 Department of Electrical Engineering and Computer Science, University of Michigan, Ann Arbor, MI 48109, USA
3 Department of Nano-optical engineering, Korea Polytechnic University, 237 Sangidaehak-ro, Siheung-si 15073, Korea
* Correspondence: semio@umich.edu (S.O.); kim.kk@kpu.ac.kr (K.-K.K.); Tel.: +82-31-8041-0720 (K.-K.K.)

Received: 29 June 2019; Accepted: 7 August 2019; Published: 9 August 2019

Abstract: We demonstrate the surface plasmon (SP)-enhanced ultraviolet (UV) emitter using Pt nanoparticles (NPs). The UV emitter is hole-patterned on the *p*-AlGaN layer to consider the penetration depth of Pt NPs. The Pt NPs with sizes under 50 nm are required to realize the plasmonic absorption in UV wavelength. In this study, we confirm the average Pt NP sizes of 10 nm, 20 nm, and 25 nm, respectively, at an annealing temperature of 600 °C. The absorption of annealed Pt NPs is covered with the 365-nm wavelength. The electroluminescence intensity of SP-UV is 70% higher than that of reference UV emitter without hole-patterns and Pt NPs. This improvement can be attributed to the increase of spontaneous emission rate through resonance coupling between the excitons in multiple quantum wells and Pt NPs deposited on the *p*-AlGaN layer.

Keywords: ultraviolet (UV) emitter; surface plasmon; Pt nanoparticles; hole-pattern; photon emission efficiency

1. Introduction

Ultraviolet (UV) photon emitters are currently in high demand for applications including lighting, lamps, chemical and biological agent detection, sterilization, and many medical uses [1–3]. However, the UV emitters have more technical problems enhancing the light efficiency than visible emitters because of the shorter wavelength of the UV emitters, such as the light absorption in *p*-AlGaN layer and transverse magnetic mode emission [4–7]. Therefore, it is necessary to improve the total light efficiency in UV emitters.

Recently, surface plasmon (SP) have drawn great attention for their ability to enhance the photon emission efficiency (PEE) of the emitter [8–13]. SPs are the collective oscillations of free electrons at a metal-dielectric interface [14]. The SP coupled between metal and semiconductor creates an alternative emission channel besides the intrinsic emission through carrier recombination. Therefore, it has been reported that the metal nanostructures could play a significant role in improving the PEE of emitters by SP coupling [15,16].

Among these metallic nanostructures, nanoparticles (NPs) are of great interest and have demonstrated their usefulness in PEE enhancement by controlling the energy transfer between multiple quantum wells (MQWs) and SP [13,15,16]. However, the metal NPs buried inside the UV emitters for approaching the MQWs can even deteriorate the epitaxial quality of the MQWs and, consequently, degrade the electrical characteristics of devices. The suppression of emission might be caused by the small size and low height due to strong resonant absorption and poor photon extraction, as reported in [17–20]. Therefore, these NPs are usually formed by the rapid thermal annealing (RTA) of metals, and the challenges lie in accurately controlling their shapes and dimensions. Furthermore, the penetration depth, which effectively causes plasmonic resonance, is determined for each metal material [21]. The closer the metal particles are formed in the MQWs region, the more effectively both the internal quantum efficiency and the PEE of the photon emitter can be increased.

In this study, we demonstrate the enhancement of the photon efficiency of the UV emitter using Pt NPs. For high plasmonic resonance, we fabricated periodic hole-patterns on top of the *p*-AlGaN layer of UV emitter. The finite difference time domain (FDTD) simulations were also conducted to compare the experimental results.

2. Materials and Methods

Figure 1 shows a schematic diagram and the fabrication procedure of SP-UV emitter. The epitaxy layers of UV emitter were grown on a sapphire substrate by metal-organic chemical vapor deposition.

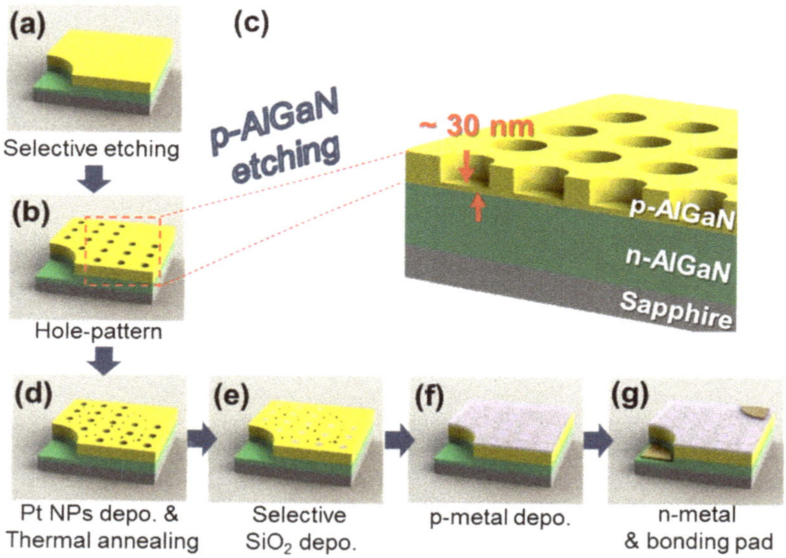

Figure 1. (**a–g**) Schematic diagram of surface plasmon-ultraviolet (SP-UV) emitter with the periodic hole-patterns and Pt nanoparticles (NPs).

The UV emitter (350 × 350 μm^2 in chip size) consisted of a 2 nm-thick *p*-GaN:Mg layer, a 0.1 μm-thick *p*-AlGaN:Mg ($n_a = 5 \times 10^{17}$ cm^{-3}) layer, a 20 nm-thick AlGaN electron blocking layer, a 100 nm-thick MQWs with $In_{0.002}Ga_{0.998}N/Al_{0.04}Ga_{0.96}N$ active layers, a 2.0 μm-thick *n*-$Al_{0.08}Ga_{0.92}N$:Si ($n_d = 5 \times 10^{18}$ cm^{-3}) layer, and a 2.0 μm-thick un-doped GaN layer on the sapphire substrate.

In order to fabricate the device, the epitaxy layer was selectively etched by an inductively coupled plasma (ICP) etching process until the *n*-AlGaN layer was partially exposed. The *p*-AlGaN layer was hole-patterned through the photolithography and ICP etching process. Then, Pt NPs were formed on the hole patterned *p*-AlGaN layer using the RTA process at 600 °C annealing temperature. After Pt NPs deposition, selective SiO_2 was deposited using an electron-beam evaporator. The ITO layer was deposited by an electron-beam evaporator on the *p*-AlGaN layer and annealed at 600 °C in O_2 ambient for 1 min using RTA. The Ti (50 nm)/Al (200 nm) layers were deposited as an n-electrode and annealed at 550 °C in N_2 ambient.

For comparison, a conventional UV (C-UV) emitter (without any pattern and Pt NPs), a periodic hole-patterned UV (HP-UV) emitter, and an SP-UV emitter sample (with HPs and Pt NPs) were simultaneously prepared to compare the light emission property. The penetration depth of Pt NPs is defined by the following formula [21].

$$\text{Penetrationdepth} = \frac{\lambda}{2\pi}\sqrt{\frac{\epsilon_d - \epsilon_m}{\epsilon_{m^2}}} \tag{1}$$

The λ is the wavelength of the emitter, ϵ_d is the dielectric constant of dielectric material, and ϵ_m is the dielectric constant of metal. Using this formula, we can calculate that the penetration depth of the Pt layer is about 45 nm at the 365-nm wavelength [22]. Based on this calculation, the p-AlGaN layer was periodically hole-patterned with the 5 μm diameter and 90-nm thickness (30 nm-thick p-AlGaN layer was remained under the holes).

The Pt layers were deposited on top of the hole-patterned UV emitter by sputtering for 60 s (2 nm-thick), 90 s (3 nm-thick), and 120 s (4 nm-thick) in a vacuum of 0.09 mbar, respectively (thickness data not shown). The deposited Pt layers were annealed at 500 °C, 600 °C, 700 °C, and 800 °C, respectively, to realize the Pt NPs.

Pt NPs were examined by Field Emission scanning electron microscopy (FE-SEM) (Hitachi, S-4100) and atomic force microscopy (AFM) (Park systems, XE-100). The photoluminescence (PL) intensity was measured using the laser excited with DPSS (Diode Pumping Solid State) (λ = 266 nm) and 30 mW power. The PL intensity was detected by Ocean Optics USB 2000. The current–voltage (I–V), Light output intensity–injection current (L–I), and electroluminescence (EL) measurements of fabricated UV emitters were carried out using a parameter analyzer (Keithley-2420 source meter), and Ocean optics-US 4000 spectrometer with the 0.1–10.0-nm full width and half maximum (FWHM) resolution.

3. Results and Discussion

The SEM images of Pt NPs with different deposition thicknesses and annealing temperatures are shown in Figure 2. Figure 2a–l show the Pt NPs with 2 nm (sputtered for 60 s), 3 nm (sputtered for 90 s), and 4 nm-thick Pt layers (sputtered for 120 s) at different annealing temperatures, respectively. As the Pt thicknesses are increased, the Pt NPs sizes are also increased. The average NPs sizes as annealing temperatures are shown in Figure 2m. As shown in this graph, the Pt NPs sizes are decreased with increasing annealing temperatures. Several research groups reported the change of absorption peak positions as Pt NPs sizes [18,19]. On the basis of some experimental results, we fabricated Pt NPs under 50-nm diameters for application to UV emitter. At annealing temperature of 600 °C, we can confirm the average NPs sizes are 10 nm (2 nm deposition), 20 nm (3 nm deposition), and 25 nm (4 nm deposition), respectively. Thus, we measured the electrical and optical properties of UV emitter with Pt NPs annealed at 600 °C.

Figure 2n–p show the AFM images of Pt layers annealed at 600 °C. The average root mean square (RMS) of Pt NPs are shown 0.883, 1.285, and 1.353, respectively. As the thickness of the Pt layers becomes thicker, the RMS of the Pt NPs are also increased. These results show a similar tendency with the surface images of Pt NPs as shown in Figure 2.

The absorption and transmittance of C-UV and SP-UV with the Pt layers annealed at 600 °C are measured. The reflectance of all Pt NPs was only near 0.2%. Therefore, the reflectance is not considered in this experiment. Figure 3 shows that the thickness of the Pt layer are increased, the absorption is also increased. The absorption peaks of Pt NPs are located near the 300-nm wavelength region with the broadband spectrum. Therefore, the absorption tail covers the 365-nm wavelength region. The absorption peak is red-shifted with increasing diameter of NPs due to the aspect-ratio dependence of the localized surface plasmon resonance [18,19,23,24]. In contrast to other materials, such as gold and silver, this linearity holds even for small particle sizes and makes Pt a highly interesting material to apply in the optoelectronic devices [14–16]. The decrease of the absorption intensity of Pt NPs is connected with the fact that the coverage is smaller for smaller particles, which implies a smaller number of electrons undergoing interband transitions, thus yielding a lower absorption intensity [18,19].

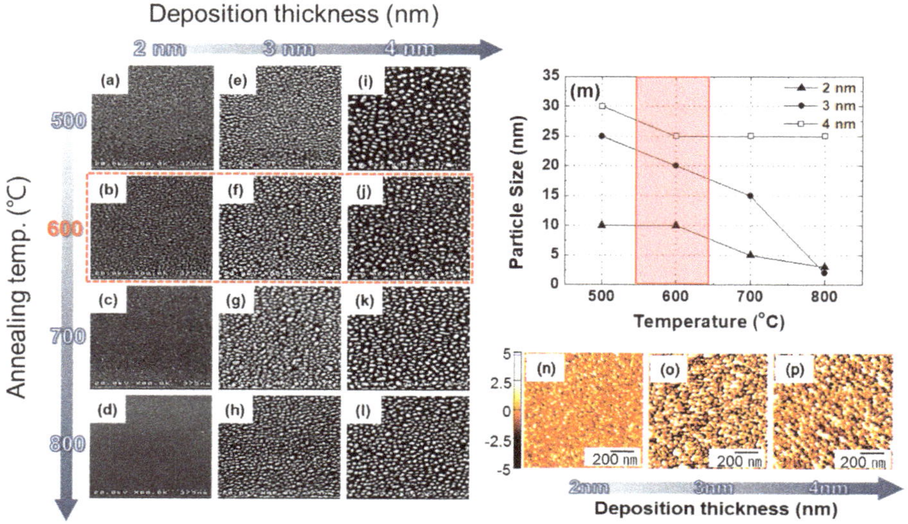

Figure 2. SEM images of Pt NPs with different deposition thicknesses and annealing temperatures. (**a–d**) 2 nm, (**e–h**) 3 nm, and (**i–l**) 4 nm-thick Pt layers with 500–800 °C annealing temperatures; (**m**) Average diameters of Pt NPs with annealing temperatures; (**n–p**) AFM images of annealed 2–4 nm of Pt layers at 600 °C annealing temperature.

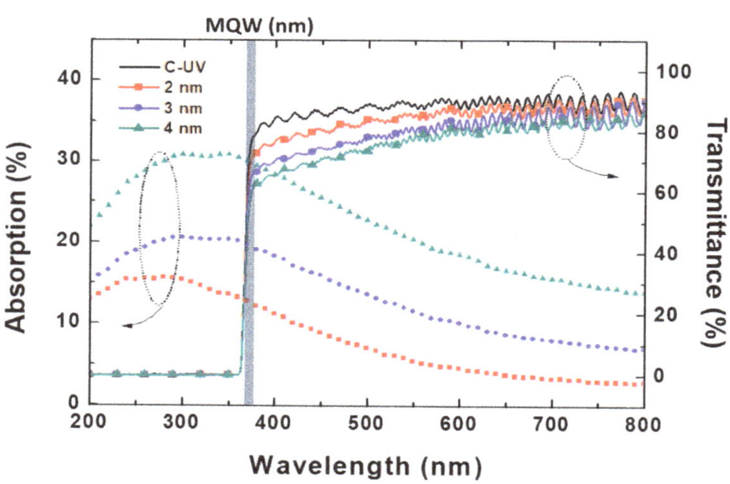

Figure 3. Absorption and transmittance of the Pt NPs annealed at 600 °C.

Figure 4 shows finite-difference time-domain (FDTD) simulations according to the diameters of Pt NPs on UV emitters. The Pt NPs are located on the *p*-AlGaN layer of the UV emitter, and the dipole source is set into the AlGaN layer with the 30-nm distance from air region. The refractive index of Pt NPs follows Palik model [25–28]. The size and distance of the Pt NPs have 10 nm, 20 nm, and 25 nm, respectively, similar to the SEM images of Figure 2. The absorption of Pt NPs on UV emitter was estimated using the transmittance and reflectance from the simulation results of Figure 4a–c. As shown in these simulation results, the absorption is also increased and red-shifted as the diameters of Pt NPs increased. The absorbed peaks simulated by the FDTD are almost the same as the experimental data. The field propagation images are also shown in Figure 4d–f. The UV emitter with Pt NPs (10-nm diameter) has higher light propagation than other UV emitters. It is expected that the light absorption in 10 nm is lower than the other particle sizes and has a higher light scattering center [18–20,23,24].

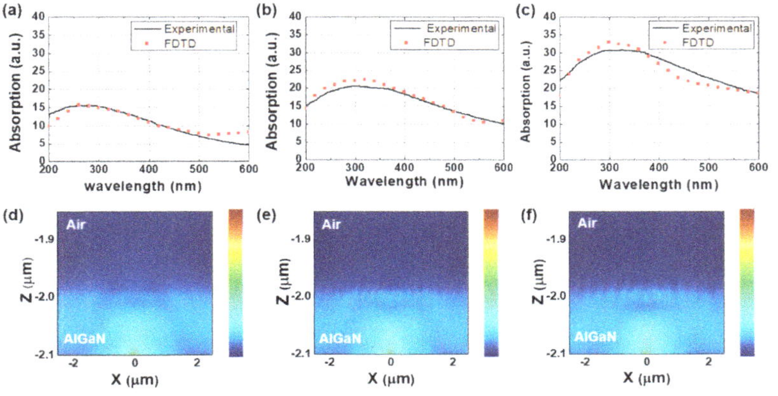

Figure 4. Finite-difference time-domain (FDTD) simulations according to the diameters of Pt NPs in UV emitters. Calculated and experimented absorption and field propagations according to the diameters of Pt NPs (**a**,**d**) 10 nm, (**b**,**e**) 20 nm, and (**c**,**f**) 25 nm, respectively.

Before measurement the electrical and optical properties of emitters, we confirmed the PL properties of UV emitters in Figure 5. The 366-nm peak is strongly shown in all samples, that is, the photon emitted from the active layer in the emitters. The peak measured at 358 nm is inferred to the photon emitted from the barrier located between the QW active layers. The peak observed at 348 nm is considered to the photon emitted from the *n*-AlGaN layer. The UV emitter without Pt NPs shows the lower PL intensity than those with Pt NPs. The PL intensity of the UV emitters with the 2 nm, 3 nm, and 4 nm-thick Pt layers were 2.3, 2, and 1.6 times higher than without Pt NPs. This enhancement can be attributed to the strong coupling by SP between the MQWs and the Pt NPs. The PL intensity is decreased as Pt thickness is increased because the intensity of the laser reaching to the MQWs is reduced as the Pt NPs sizes are increased. The photon emitted by the laser is also reflected and absorbed by the Pt NPs on the surface of UV emitter [29–31].

In order to investigate the electrical and optical properties of C-, HP, and SP-UV emitters, *I–V*, *L–I*, and EL characteristics were measured. The *I–V* curves in Figure 6a show that the voltages of C-UV, HP-UV, and SP-UV are 3.5, 3.7, and 3.6 V at an injection current of 20 mA, respectively. This shows that HP-UV and SP-UV samples with HP have a slightly higher series resistance than C-UV sample because of the surface defect during formation of hole-patterns. In addition, the decrease of electrical property is partly attributed to the resistance increase from the reduction of *p*-AlGaN volume [32]. Figure 6b shows the *L–I* characteristics of the UV emitters. The SP-UV shows the highest light output intensity in every injection current than other UV emitters. The EL intensities for each sample with an injection current of 20 mA are shown in Figure 6c. The EL intensity of HP-UV is 20% lower than that of C-UV and SP-UV is 70% higher than that of C-UV. This is because the HP-UV emitter is reduced of the electrical properties and caused the current crowing due to etching damage by fabrication of hole-patterns. If we realize the more stable current spreading in HP-UV chip, the light intensity of the HP-UV emitter can enhance light emission compared to C-UV. The large light enhancement of SP-UV is due to the SP effect, which occurred between Pt NPs and the photons emitted from UV emitter. Figure 6d–f also shows the emission images of UV emitters with the same injection current. Although the emission image of HP-UV is slightly darker than that of C-UV, the SP-UV chip has a brighter emission image compared with C-UV chip. However, we can confirm that this damage can be overcome by SP effect using Pt nanoparticles, although the SP-UV has also electrical damage during the etching process.

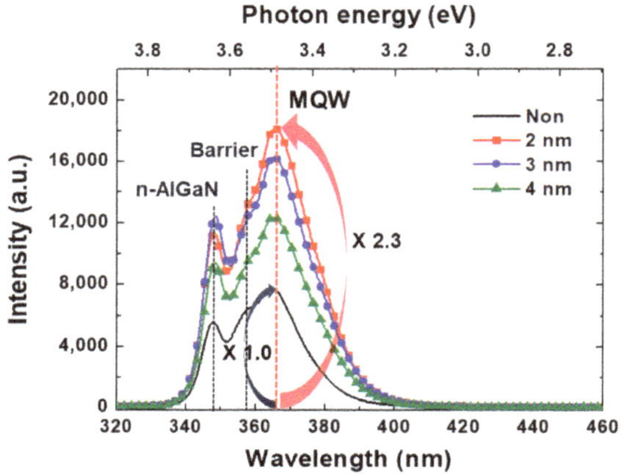

Figure 5. PL results of UV emitters with different sizes Pt NPs.

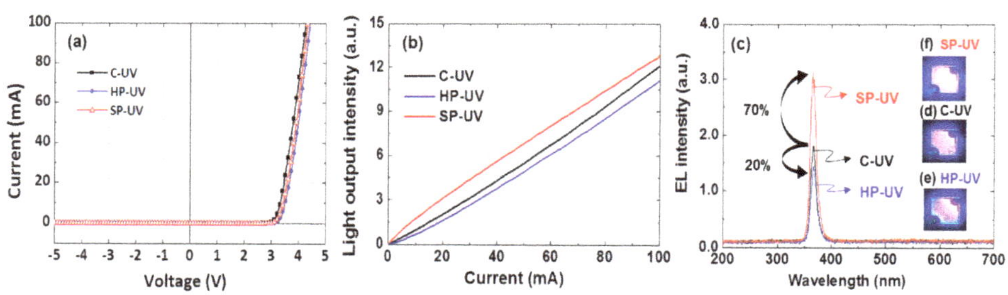

Figure 6. (**a**) *I–V*, (**b**) *L–I*, and (**c**) EL intensity of C-UV, HP-UV, and SP-UV emitters. (**d–f**) Emission images of UV emitters at an injection current of 20 mA.

4. Conclusions

In conclusion, we demonstrate the enhanced PEE of the 365-nm wavelength UV emitter using the Pt NPs. In order to effectively use the SP effect to the emitter, the sizes of NPs are controlled by deposition thickness of Pt layers and the annealing process. At 600 °C annealing temperature, we realized the 10 nm, 20 nm, and 25 nm sizes of Pt NPs. In addition, to consider penetration depth of Pt NPs, we fabricated periodic hole-patterns on top of the *p*-AlGaN layer. Through the absorption data, we can find that all NPs have an absorption intensity near the 300-nm wavelength and have the broadband absorption spectrum covered with the 365-nm wavelength. We confirmed that the PL intensity of UV emitter with the 10-nm sizes of Pt NPs was 2.6 times higher than that without Pt NPs. The small size of Pt NPs (10 nm) has lower light absorption and higher light scattering center compared with the other sizes of Pt NPs (20 nm and 25 nm). Through the *I–V* results, we can confirm the HP- and the SP-UV emitters have a slightly higher series resistance than C-UV emitter because of the etching damage of p-AlGaN layer for periodic hole-patterns. However, the EL intensity shows that the SP-UV emitter has a higher intensity of 70% than that of C-UV emitter. This large enhancement is because of the strong coupling by SP between the MQWs and the Pt NPs.

Author Contributions: H.-J.C., S.K., K.-K.K. conceived and designed the experiments; S.K., E.-K.C., S.O. conducted the simulation; H.-J.C., S.K., B.-R.N. performed the experiments; H.-J.C., S.-H.K., W.-S.L., S.K. analyzed the data; H.-J.C., S.O. and K.-K.K. wrote the paper.

Funding: This research was supported by the Ministry of Science and ICT (MSIT), under the Information Technology Research Center (ITRC) support program (IITP-2019–2018-0-01426) supervised by the Institute for Information & communications Technology Planning & Evaluation (IITP), the Korea Institute for Advancement of Technology (KIAT) grant funded by the Korea Government (MOTIE) (N0001364, the Competency Development Program for Industry Specialist), and the Technology Innovation Program (20002694, Gas sensor) funded By the Ministry of Trade.

Conflicts of Interest: The authors declare no conflict of interest.

References

1. Tsuzuki, H.; Mori, F.; Takeda, K.; Iwaya, M.; Kamiyama, S.; Amano, H.; Isamu, A.; Harumasa, Y.; Masakazu, K.; Yoji, Y.; et al. Novel UV devices on high-quality AlGaN using grooved underlying layer. *J. Cryst. Growth* **2009**, *311*, 2860–2863. [CrossRef]
2. Lei, P.H.; Yang, C.D.; Huang, P.C.; Yeh, S.J. Enhancement of light extraction efficiency for InGaN/GaN light-emitting diodes using silver nanoparticle embedded ZnO thin films. *Micromachines* **2019**, *10*, 239. [CrossRef] [PubMed]
3. Choi, C.H.; Han, J.; Park, J.S.; Seong, T.Y. Near ultraviolet InGaN/AlGaN-based light-emitting diodes with highly reflective tin-doped indium oxide/Al-based reflectors. *Opt. Express* **2013**, *21*, 26774–26779. [CrossRef] [PubMed]
4. Yao, Y.C.; Hwang, J.M.; Yang, Z.P.; Haung, J.Y.; Lin, C.C.; Shen, W.C.; Chou, C.Y.; Wang, M.T.; Haung, C.Y.; Chen, C.Y.; et al. Enhanced external quantum efficiency in GaN-based vertical- type light-emitting diodes by localized surface plasmons. *Sci. Rep.* **2016**, *6*, 22659. [CrossRef] [PubMed]
5. Wierer, J.J.; Krames, M.R.; Epler, J.E.; Gardner, N.F.; Craford, M.G.; Wierer, J.J. InGaN/GaN quantum-well heterostructure light-emitting diodes employing photonic crystal structures. *Appl. Phys. Lett.* **2004**, *84*, 3885. [CrossRef]
6. Zhang, J.; Zhao, H.; Tansu, N. Effect of crystal-field split-off hole and heavy-hole bands crossover on gain characteristics of high Al-content AlGaN quantum well lasers. *Appl. Phys. Lett.* **2011**, *97*, 111105. [CrossRef]
7. Kolbe, T.; Knauer, A.; Chua, C.; Yang, Z.; Kueller, V.; Einfeldt, S.; Vogt, P.; Johnson, N.M.; Weyers, M.; Kneissi, M. Effect of temperature and strain on the optical polarization of (In)(Al) GaN ultraviolet light emitting. *Appl. Phys. Lett.* **2011**, *99*, 261105. [CrossRef]
8. Kwon, B.M.; Kim, J.; Kim, B.; Park, I.; Cho, C.; Byeon, C.C.; Park, S.J. Surface-Plasmon-Enhanced Light-Emitting Diodes. *Adv. Mater.* **2008**, *20*, 1253–1257. [CrossRef]
9. Sung, J.H.; Kim, B.S.; Choi, C.H.; Lee, M.W.; Lee, S.G.; Park, S.G.; Lee, E.H.; Hoan, O.B. Enhanced luminescence of GaN-based light-emitting diode with a localized surface plasmon resonance. *Microelectron. Eng.* **2009**, *86*, 1120–1123. [CrossRef]
10. Cho, C.Y.; Kim, K.S.; Lee, S.J.; Kwon, M.K.; Ko, H.; Kim, S.T.; Jung, G.Y.; Park, S.J. Surface plasmon-enhanced light-emitting diodes with silver nanoparticles and SiO_2 nano-disks embedded in p-GaN. *Appl. Phys. Lett.* **2011**, *99*, 041107. [CrossRef]
11. Cho, C.Y.; Kwon, M.K.; Lee, S.J.; Han, S.H.; Kang, J.W.; Kang, S.E.; Lee, D.Y.; Park, S.J. Surface plasmon-enhanced light-emitting diodes using silver nanoparticles embedded in p-GaN. *Nanotechnology* **2010**, *21*, 205201. [CrossRef] [PubMed]
12. Sung, J.H.; Yang, J.S.; Kim, B.S.; Choi, C.H.; Lee, M.W.; Lee, S.G.; Lee, E.H.; Hoan, O.B. Enhancement of electroluminescence in GaN-based light-emitting diodes by metallic nanoparticles. *Appl. Phys. Lett.* **2011**, *96*, 261105. [CrossRef]
13. Hong, S.H.; Cho, C.Y.; Lee, S.J.; Yim, S.Y.; Lim, W.; Kim, S.T.; Park, S.J. Localized surface plasmon-enhanced near-ultraviolet emission from InGaN/GaN light-emitting diodes using silver and platinum nanoparticles. *Opt. Express* **2013**, *21*, 2822–2825. [CrossRef] [PubMed]

14. Cho, C.Y.; Zhang, Y.; Cicek, E.; Rahnema, B.; Bai, Y.; McClintock, R.; Razeghi, M. Surface plasmon enhanced light emission from AlGaN-based ultraviolet light-emitting diodes grown on Si (111). *Appl. Phys. Lett.* **2013**, *102*, 211110. [CrossRef]
15. Gontijo, I.; Boroditsky, M.; Yablonovitch, E.; Keller, S.; Mishra, U.K.; DenBaars, S.P. Coupling of InGaN quantum-well photoluminescence to silver surface plasmons. *Phys. Rev. B* **1999**, *60*, 564–567. [CrossRef]
16. Neogi, A.; Lee, C.W.; Everitt, H.O.; Kuroda, T.; Tackeuchi, A.; Yablonovitch, E. Enhancement of spontaneous recombination rate in a quantum well by resonant surface plasmon coupling. *Phys. Rev. B* **2002**, *66*, 153305. [CrossRef]
17. Mulvaney, P. Surface Plasmon Spectroscopy of Nanosized Metal Particles. *Langmuir* **1996**, *12*, 788–800. [CrossRef]
18. Bigall, N.C.; Ha, T.; Klose, M.; Simon, P.; Eng, L.M.; Eychmuller, A. Monodisperse Platinum Nanospheres with Adjustable Diameters from 10 to 100 nm: Synthesis and Distinct Optical Properties. *Nano Lett.* **2008**, *8*, 4588–4592. [CrossRef] [PubMed]
19. Langhammer, C.; Yuan, Z.; Zoric, I.; Kasemo, B. Plasmonic Properties of Supported Pt and Pd Nanostructures. *Nano Lett.* **2006**, *4*, 833–838. [CrossRef]
20. Near, R.; Hayden, S.; El-sayed, M. Extinction vs. Absorption: Which Is the Indicator of Plasmonic Field Strength for Silver Nanocubes? *J. Phys. Chem.* **2012**, *116*, 23019–23026. [CrossRef]
21. Okamoto, K.; Niki, I.; Shvartser, A.; Narukawa, Y.; Mukai, T.; Scherer, A. Surface-plasmon-enhanced light emitters based on InGaN quantum wells. *Nat. Mater.* **2004**, *3*, 601–605. [CrossRef] [PubMed]
22. Werner, W.S.M.; Glantschnig, K.; Ambrosch-draxl, C. Optical Constants and Inelastic Electron-Scattering Data for 17 Elemental Metals. *J. Phys. Chem. Ref. Data* **2009**, *38*, 1013–1092. [CrossRef]
23. Rivadulla, J.F.; Vergara, M.C.; Blanco, M.C.; Lopez-Quintela, M.A.; Rivas, J. Optical Properties of Platinum Particles Synthesized in Microemulsions. *J. Phys. Chem. B* **1997**, *45*, 8997–9004. [CrossRef]
24. Luther, J.M.; Jain, P.K.; Ewers, T.; Alivisatos, A.P. Localized surface plasmon resonances arising from free carriers in doped quantum dots. *Nat. Mater.* **2011**, *10*, 361–366. [CrossRef] [PubMed]
25. Road, P.; Design, P. An FDTD method for the simulation of dispersive metallic structures. *Opt. Quantum Electron.* **2006**, *38*, 843–856.
26. Vincenzo, A.; Roberto, P.; Marco, F.; Onofrio, M.M.; Maria Antonia, I. Surface plasmon resonance in gold nanoparticles: A review. *J. Phys. Condens. Matter* **2017**, *29*, 203002. [CrossRef]
27. Lee, K.S.; El-Sayed, M.A. Gold and silver nanoparticles in sensing and imaging: Sensitivity of plasmon response to size, shape, and metal composition. *J. Phys. Chem. B* **2006**, *110*, 19220–19225. [CrossRef] [PubMed]
28. Spinelli, P.; Hebbink, M.; De Waele, R.; Black, L.; Lenzmann, F.; Polman, A. Optical impedance matching using coupled plasmonic nanoparticle arrays. *Nano Lett.* **2011**, *11*, 1760–1765. [CrossRef]
29. Zhang, A.; Zhang, J.; Fang, Y. Photoluminescence from colloidal silver nanoparticles. *J. Lumin.* **2008**, *128*, 1635–1640. [CrossRef]
30. Daniel, M.C.; Astruc, D. Gold Nanoparticles: Assembly, Supramolecular Chemistry, Quantum-Size-Related Properties, and Applications toward Biology, Catalysis, and Nanotechnology. *Chem. Rev.* **2004**, *104*, 293–346. [CrossRef]
31. Ledoux, G.; Guillois, O.; Porterat, D.; Reynaud, C.; Huisken, F.; Kohn, B.; Pailard, V. Photoluminescence properties of silicon nanocrystals as a function of their size. *Phys. Rev.* **2000**, *62*, 942–951. [CrossRef]
32. Cho, H.K.; Jang, J.; Choi, J.H.; Choi, J.; Kim, J.; Lee, J.S.; Lee, B.; Choe, Y.H.; Lee, K.D.; Kim, S.H.; et al. Light extraction enhancement from nanoimprinted photonic crystal GaN-based blue light emitting diodes. *Opt. Express* **2006**, *14*, 8654–8660. [CrossRef] [PubMed]

micromachines

Article

Full-Color InGaN/AlGaN Nanowire Micro Light-Emitting Diodes Grown by Molecular Beam Epitaxy: A Promising Candidate for Next Generation Micro Displays

Ha Quoc Thang Bui [1,2], Ravi Teja Velpula [1], Barsha Jain [1], Omar Hamed Aref [1], Hoang-Duy Nguyen [3,*], Trupti Ranjan Lenka [4] and Hieu Pham Trung Nguyen [1,*]

1 Department of Electrical and Computer Engineering, New Jersey Institute of Technology, Newark, NJ 07102, USA
2 Department of Biomedical Physics, Pham Ngoc Thach University of Medicine, Ho Chi Minh City 700000, Vietnam
3 Institute of Chemical Technology, Vietnam Academy of Science and Technology, 1 Mac Dinh Chi Street, District 1, Ho Chi Minh City 700000, Vietnam
4 Department of Electronics & Communication Engineering, National Institute of Technology Silchar, Assam 788010, India
* Correspondence: nhduy@iams.vast.vn (H.-D.N.); hieu.p.nguyen@njit.edu (H.P.T.N.); Tel.: +1-(973)-596-3523 (H.P.T.N.)

Received: 5 July 2019; Accepted: 22 July 2019; Published: 24 July 2019

Abstract: We have demonstrated full-color and white-color micro light-emitting diodes (µLEDs) using InGaN/AlGaN core-shell nanowire heterostructures, grown on silicon substrate by molecular beam epitaxy. InGaN/AlGaN core-shell nanowire µLED arrays were fabricated with their wavelengths tunable from blue to red by controlling the indium composition in the device active regions. Moreover, our fabricated phosphor-free white-color µLEDs demonstrate strong and highly stable white-light emission with high color rendering index of ~ 94. The µLEDs are in circular shapes with the diameter varying from 30 to 100 µm. Such high-performance µLEDs are perfectly suitable for the next generation of high-resolution micro-display applications.

Keywords: µLED displays; µLEDs; GaN nanowires; core-shell structure

1. Introduction

A display based on inorganic micro light-emitting diodes (µLEDs) has recently been intensively investigated due to its great potential for tech gadgets such as Apple watches, smartphone screens, television screens, billboards, Google glass, and virtual reality devices. Considerable efforts have been poured into this field to bring the novel standard displays to the market [1–5]. The increasing demand for µLED displays in tech screens has received much attention from academia and industry since early last decade. In order to achieve µLEDs displays, it requires several critical stages [6] consisting of making µLEDs, transferring them to a backplane, and precisely controlling each individual LED [7–9]. The first essential step is to have the right type of red, green, and blue (RGB) µLEDs satisfactory for the displays. The µLEDs need a long lifespan, superior brightness, high efficiency, and durability. Besides, different approaches for making RGB and white color µLEDs to form a full-color micro-pixel in the µLED displays [10–13], the monolithic display based on III-nitride nanowire heterostructure µLEDs is promising since it allows more direct control of emission wavelengths of the light-emitting diodes (LEDs). The emission wavelength of nanowire LEDs can be controlled by changing the composition of indium in the InGaN active region. This can be done by adjusting the epitaxial growth conditions including growth temperature and/or In/Ga flux ratios [14–17]. The energy bandgap of InGaN compounds can be varied from 3.4 eV (GaN) to 0.65 eV (InN) [18], covering the full visible region for display applications. Therefore, GaN based μLEDs are a potential candidate for developing novel micro-LED displays [4].

Conventional planar GaN based LEDs have been used in everyday lighting and automotive headlights. However, their poor operating efficiency and efficiency degradation in the green to red spectrums have limited their potential achievements. The presence of polarization fields [19,20], Auger recombination [21,22], poor hole transport [23], defects/dislocations [24,25], and electron leakage and electron overflow [26–28] are the main causes of these above drawbacks. In this regard, nanowire heterostructures have been intensively studied as an alternative candidate for high efficiency light-emitters. Unlike conventional planar structures, nanowires exhibit several distinct advantages, including dramatically reduced strain-induced polarization fields and dislocation densities due to the effective lateral stress relaxation. Moreover, the micron-size nanowire-based LEDs can be much more efficient in heat dissipation due to the reduced current spreading resistance and thereby resulting in increased injection current levels [29–31]. Thus, the performance of the nanowire LED is expected to be better than their thin-film counterparts. As compared to organic and inorganic thin-film devices, the brightness, reliability, energy efficiency and moisture resistance of the nanowire LEDs are predicted to be far superior [32]. Therefore, nanowire μLEDs have emerged as a promising candidate for general lighting and display applications. In this context, InGaN/AlGaN nanowire μLEDs in different sizes have been fabricated and characterized. Such nanowire μLEDs exhibit strong and stable emissions from blue to red wavelengths. Moreover, phosphor-free white-color μLEDs have also been demonstrated with highly stable emission.

2. Experiment

Vertically aligned InGaN/AlGaN core-shell nanowire μLEDs were grown by Veeco Gen II plasma-assisted molecular beam epitaxy (PAMBE) system. Silicon and magnesium dopants were used to grow *n*-GaN and *p*-GaN, respectively. During the epitaxial growth process, the nitrogen flow was kept at 1 sccm and the plasma power was controlled at 350 W. GaN segments were grown at 750 °C, while InGaN in the active region was grown at lower temperatures, in the range of 580–650 °C to enhance the indium incorporation. Figure 1a presents the schematic structure of a single InGaN/AlGaN nanowire on a Si substrate. The nanowire μLEDs consist of GaN:Si grown on a silicon substrate and the GaN:Mg on the top. The ten couples of quantum wells are inserted in the active region. Each quantum well includes a 3 nm InGaN dot and 3 nm AlGaN barrier. During growth of the AlGaN barrier, the AlGaN shell is spontaneously formed, enabling unique core-shell layers [33,34]. The emission color of the μLEDs can be controlled by adjusting the Ga/In flux ratios and the substrate temperature during the MBE growth. For instance, the peak emission wavelength can be shifted from red to blue by gradually increasing the growth temperature of the InGaN active region from 580 °C to 650 °C with ramping rate of 10 °C/min. The nanowire length is controlled by the growth duration. Further information of the core-shell nanowire structures and MBE growth can be found elsewhere [34–38].

Figure 1b shows a scanning electron microscope (SEM) image of InGaN/AlGaN nanowire LEDs taken under a 45° tilted angle. It shows that the nanowires are relatively uniform across the substrate. Figure 1c illustrates the microscopic image of the fabricated μLEDs. The μLED's emissive window has 50 μm in diameter, which is connected with a square electrode pad. The μLEDs were fabricated using standard photolithography, dry etching and contact metallization techniques, which are described elsewhere [17,37,39–41]. During the fabrication process, μLEDs with 30 μm to 100 μm in diameter were defined by standard photolithography. In this paper, the μLEDs with a diameter of 50 μm were chosen for characterization.

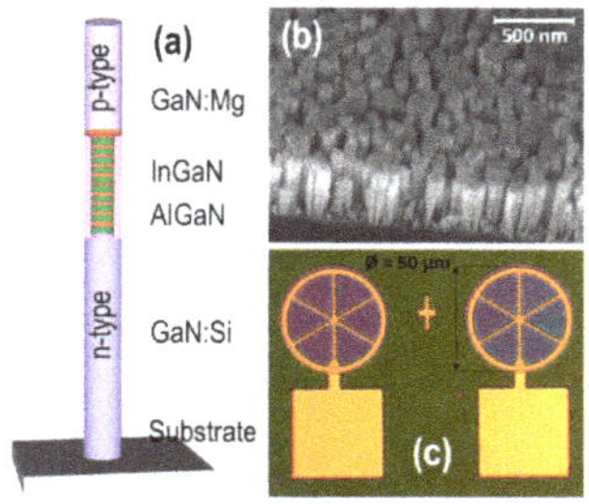

Figure 1. Schematic structure of a nanowire micro light-emitting diodes (μLED) with ten InGaN/AlGaN quantum well heterostructures (**a**); the 45° tilted SEM image of InGaN/AlGaN nanowires on Si substrate (**b**); and optical image of μLEDs and the electrode pads (**c**).

3. Results and Discussion

Figure 2 shows the normalized photoluminescence (PL) spectra of the InGaN/AlGaN core-shell nanowires. The measurement was performed at room temperature with a 405 nm laser excitation source. It clearly shows that strong red, green, and blue emissions were recorded at 645 nm, 550 nm, and 475 nm, respectively. The current-voltage characteristics of RGB μLEDs were characterized. The turn-on voltages increase with the decreasing indium composition in the active region of the nanowires. The less indium composition, the higher energy bandgap, makes the light emissions shift toward the blue region. The higher energy bandgap makes turn-on voltages increase [42], as clearly shown in Figure 3. The turn-on voltages of the red, green and blue μLEDs are approximately 1.6 V, 3.5 V, and 4.6 V, respectively. The I-V characteristics also indicate that the devices with a low resistance and good fabrication processes have been achieved.

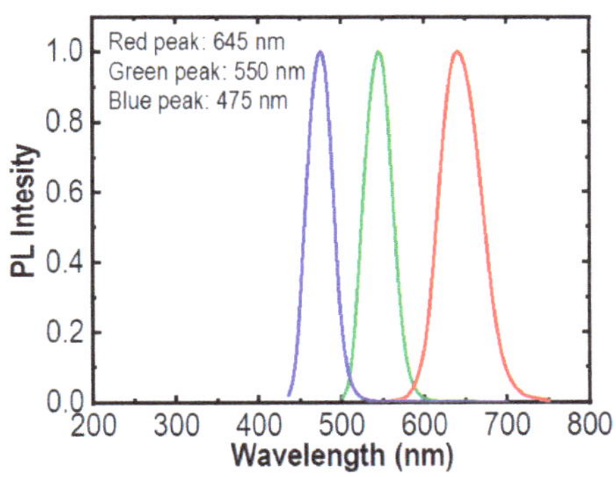

Figure 2. Photoluminescence spectra of the red, green, and blue (RGB) InGaN/AlGaN nanowire μLEDs measured at room temperature.

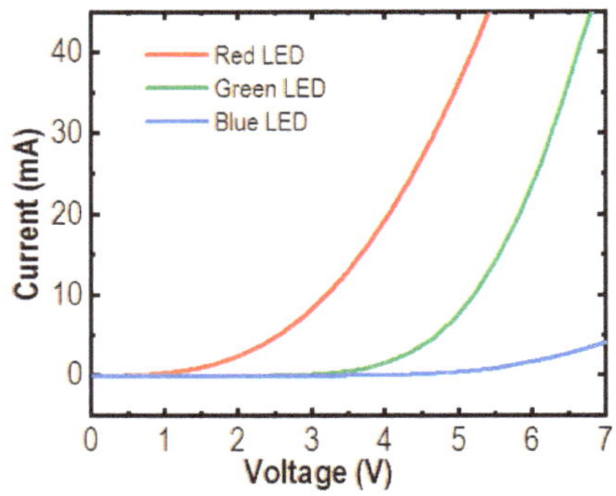

Figure 3. I-V characteristics of the RGB μLEDs are illustrated.

The electroluminescence (EL) spectra of these μLEDs are presented in Figure 4. The measurements were conducted at room temperature using pulse biasing conditions from 50 mA to 350 mA to reduce the heating effect. Strong red, green, and blue emissions were recorded at 475 nm, 550 nm, and 645 nm for blue, green and red μLEDs, respectively. Illustrated in Figure 4a, highly stable emissions with negligible shift in peak wavelengths were measured for blue μLED with an injection current from 50 mA to 350 mA, attributed to the greatly reduced quantum-confined Stark effect (QCSE) [43] by using nanowire structures. Shown in Figure 4b,c, the blue-shifts of about 1.5 nm were measured for the green and 3 nm for the red μLEDs. These values are significantly smaller than those of planar quantum well LEDs operating in the same spectral regime. Summary of peak wavelength variations of the RGB μLEDs are shown in Figure 5.

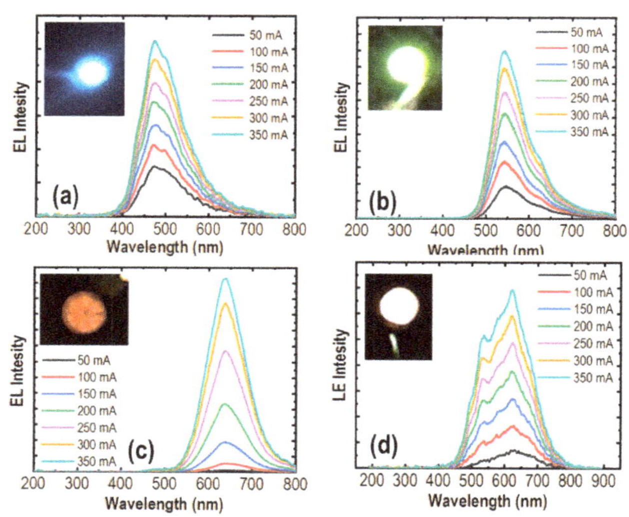

Figure 4. The electroluminescence characteristics of the fabricated blue μLED (**a**), green μLED (**b**), red μLED (**c**), and the white μLED (**d**). The corresponding optical images of these μLEDs are presented in the insets.

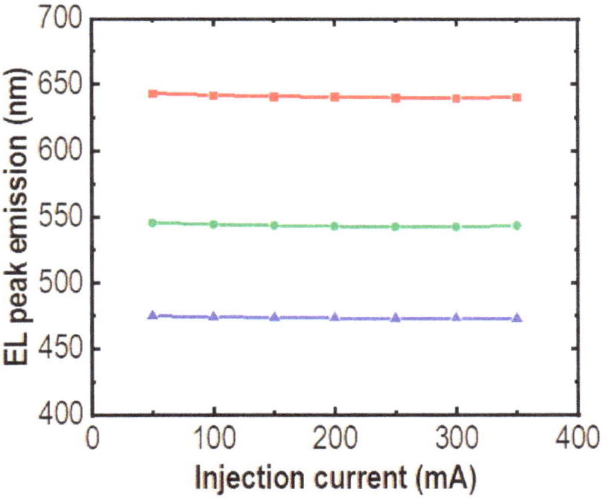

Figure 5. The peak emissions of red, green and blue μLEDs measured under different injection currents from 50 mA to 350 mA.

III-nitride based planar LEDs operating in the green or longer wavelengths are prone to a blue shift with high driving currents due to QCSE [43]. However, using the nanowire structure, the RGB μLEDs have stable peak wavelengths at 645 nm in red, 550 nm in green, and 475 nm in blue due to the negligible QCSE, meaning that the lattice mismatch induced strain and efficiency droop are negligible [44].

We have also demonstrated white-color μLEDs with strong emission by engineering the emission wavelength of the core-shell nanowire structure. The white light emission covers a long range of the visible spectrum, which is from 450 nm to 750 nm. Moreover, the white-color μLED exhibits a stable emission with a small blue-shift of ~4 nm for injection current from 50 mA to 350 mA. The stable emission characteristics of the phosphor-free white-color μLEDs are further illustrated in the 1931 Commission International l'Eclairage chromaticity diagram in Figure 6. The x and y values are derived to be in the ranges of ~0.351-0.362 and 0.391-0.398, respectively. The phosphor-free white-color μLEDs exhibit nearly a neutral white light emission, with correlated color temperature of ~ 4850 K. Additionally, a relatively high color rendering index (CRI) of ~ 94 was measured for this phosphor-free white-color μLED, which is extremely difficult to achieve using planar LED structures. Currently, phosphor-based white LEDs have CRI values in the range of 80–86 [45–48]. With some special design of phosphor converters, the CRI of these types of white LEDs can be increased up to 90–97 [49–53]. However, the fabrication process of these white LEDs is complex and the device reliability is a major concern. To the best of our knowledge, up to this moment, the CRI value for white-color μLEDs have not yet been reported, possibly due to the complexity of the device fabrication for such μLED devices. In this regard, our phosphor-free full-color and white-color μLEDs are highly desirable for the next generation μLED display technology.

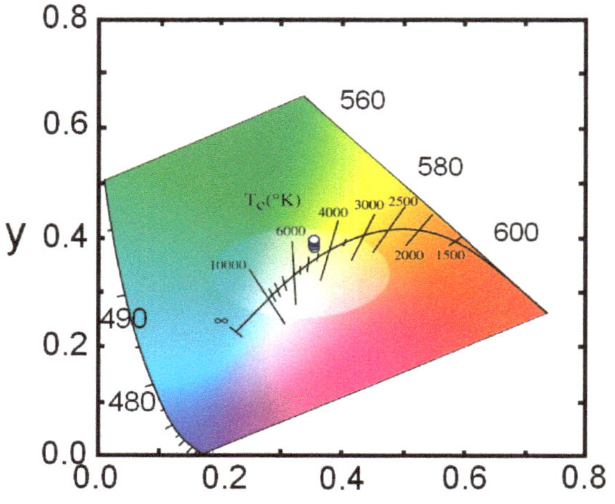

Figure 6. The 1931 Commission International l'Eclairage chromaticity diagram presents stable white light emission characteristics of the phosphor-free white-color InGaN/AlGaN nanowire μLED.

4. Conclusions

We have successfully demonstrated μLEDs using InGaN/AlGaN core-shell nanowire heterostructures grown on silicon substrates by MBE. Strong and stable emission from full-color and white-color were recorded from these μLEDs. The color properties of the μLEDs can be optimized by controlling the spectral distribution of the μLEDs. Using nanowire structures, we have achieved phosphor-free white light with unprecedentedly high CRI of ~ 94. The high performance and stable operation of the red, green and blue μLEDs show promise in monolithic μLED displays. Generally, the high cost of current displays is a bottleneck and slows down the market growth. Due to the progressive demand for electronic devices, it is expected to provide lucrative growth opportunities for the micro-display market. In this regard, using the selective area growth approach, RGB subpixel μLEDs can be integrated on the same chip, eliminating the current pick-and-place process, which requires precisely controlling procedures. Therefore, high efficiency, high color rendering properties, and low power consumption μLEDs using GaN nanowire heterostructures are perfectly suitable as an alternative replacement of current display technologies.

Author Contributions: H.P.T.N. and H.-D.N. designed the experiment. H.Q.T.B. and H.P.T.N. contributed to the MBE growth and device fabrication. H.P.T.N. and H.Q.T.B. performed the IV, EL, and PL measurements. R.T.V. and B.J. contributed to the device design and simulations. H.Q.T.B. and H.P.T.N. wrote the manuscript with contributions from R.T.V., B.J., O.H.A., H.-D.N., and T.R.L.

Funding: This research is supported by New Jersey Institute of Technology, the National Science Foundation (grant numbers EEC-1560131 and EEC-1852375) and funded by Vietnam National Foundation for Science and Technology Development (NAFOSTED) under grant number 103.03-2017.312. Acknowledgment also goes to Visvesvaraya Young Faculty Research Fellowship by MeitY, Govt. of India to T.R. Lenka for his visit to NJIT, US.

Conflicts of Interest: The authors declare no conflict of interest.

References

1. Virey, E. Are microLEDs really the next display revolution? *SID* **2018**, *3*, 22–27. [CrossRef]
2. Choi, M.K.; Yang, J.; Hyeon, T.; Kim, D.H. Flexible quantum dot light-emitting diodes for next-generation displays. *NPJ Flex. Electron.* **2018**, *2*, 1–10. [CrossRef]
3. Sun, Y.; Jiang, Y.; Sun, X.W.; Zhang, S.; Chen, S. Beyond OLED: Efficient Quantum Dot Light-Emitting Diodes for Display and Lighting Application. *Chem. Rec.* **2019**, *19*, 1–25. [CrossRef] [PubMed]

4. Zhang, X.; Li, P.; Zou, X.; Jiang, J.; Yuen, S.H.; Tang, C.W.; Lau, K.M. Active Matrix Monolithic LED Micro-Display Using GaN-on-Si Epilayers. *IEEE Photonics Technol. Lett.* **2019**, *31*, 865–868. [CrossRef]
5. Liu, Z.J.; Chong, W.C.; Wong, K.M.; Lau, K.M. 360 PPI Flip-Chip Mounted Active Matrix Addressable Light Emitting Diode on Silicon (LEDoS) Micro-Displays. *J. Disp. Technol.* **2013**, *9*, 678–682. [CrossRef]
6. Ding, K.; Avrutin, V.; Izyumskaya, N.; Özgür, Ü.; Morkoç, H. Micro-LEDs, a Manufacturability Perspective. *Appl. Sci.* **2019**, *9*, 1206. [CrossRef]
7. Zhang, L.; Ou, F.; Chong, W.C.; Chen, Y.; Li, Q. Wafer-scale monolithic hybrid integration of Si-based IC and III-V epi-layers-A mass manufacturable approach for active matrix μLED micro-displays. *J. Soc. Inf. Disp.* **2018**, *26*, 137–145. [CrossRef]
8. Cok, R.S.; Meitl, M.; Rotzoll, R.; Melnik, G.; Fecioru, A.; Trindade, A.J.; Raymond, B.; Bonafede, S.; Gomez, D.; Moore, T. Inorganic light-emitting diode displays using micro-transfer printing. *J. Soc. Inf. Disp.* **2017**, *25*, 589–609. [CrossRef]
9. Corbett, B.; Loi, R.; Zhou, W.; Liu, D.; Ma, Z. Transfer print techniques for heterogeneous integration of photonic components. *Prog. Quantum Electron.* **2017**, *52*, 1–17. [CrossRef]
10. Chen, K.J.; Chen, H.C.; Tsai, K.A.; Lin, C.C.; Tsai, H.H.; Chien, S.H.; Cheng, B.S.; Hsu, Y.J.; Shih, M.H.; Tsai, C.H. Resonant-Enhanced Full-Color Emission of Quantum-Dot-Based Display Technology Using a Pulsed Spray Method. *Adv. Funct. Mater.* **2012**, *22*, 5138–5143. [CrossRef]
11. Chen, S.W.H.; Shen, C.C.; Wu, T.; Liao, Z.Y.; Chen, L.F.; Zhou, J.R.; Lee, C.F.; Lin, C.H.; Lin, C.C.; Sher, C.W. Full-color monolithic hybrid quantum dot nanoring micro light-emitting diodes with improved efficiency using atomic layer deposition and nonradiative resonant energy transfer. *Photonics Res.* **2019**, *7*, 416–422. [CrossRef]
12. Teng, C.H.; Zhang, L.; Deng, H.; Ku, P.C. Strain-induced red-green-blue wavelength tuning in InGaN quantum wells. *Appl. Phys. Lett.* **2016**, *108*, 071104. [CrossRef]
13. Chung, K.; Sui, J.; Demory, B.; Ku, P.C. Color mixing from monolithically integrated InGaN-based light-emitting diodes by local strain engineering. *Appl. Phys. Lett.* **2017**, *111*, 041101. [CrossRef]
14. Wang, R.; Nguyen, H.P.; Connie, A.T.; Lee, J.; Shih, I.; Mi, Z. Color-tunable, phosphor-free InGaN nanowire light-emitting diode arrays monolithically integrated on silicon. *Opt. Express* **2014**, *22*, A1768–A1775. [CrossRef] [PubMed]
15. Sekiguchi, H.; Kishino, K.; Kikuchi, A. Emission color control from blue to red with nanocolumn diameter of InGaN/GaN nanocolumn arrays grown on same substrate. *Appl. Phys. Lett.* **2010**, *96*, 231104. [CrossRef]
16. Ra, Y.H.; Wang, R.; Woo, S.Y.; Djavid, M.; Sadaf, S.M.; Lee, J.; Botton, G.A.; Mi, Z. Full-Color Single Nanowire Pixels for Projection Displays. *Nano Lett.* **2016**, *16*, 4608–4615. [CrossRef] [PubMed]
17. Philip, M.R.; Choudhary, D.D.; Djavid, M.; Bhuyian, M.N.; Piao, J.; Pham, T.T.; Misra, D.; Nguyen, H.P.T. Controlling color emission of InGaN/AlGaN nanowire light-emitting diodes grown by molecular beam epitaxy. *J. Vac. Sci. Technol. B* **2017**, *35*, 02B108. [CrossRef]
18. Arnaudov, B.; Paskova, T.; Paskov, P.P.; Magnusson, B.; Valcheva, E.; Monemar, B.; Lu, H.; Schaff, W.J.; Amano, H.; Akasaki, I. Energy position of near-band-edge emission spectra of InN epitaxial layers with different doping levels. *Phys. Rev. B* **2004**, *69*, 115216. [CrossRef]
19. Kim, M.H.; Schubert, M.F.; Dai, Q.; Kim, J.K.; Schubert, E.F.; Piprek, J.; Park, Y. Origin of efficiency droop in GaN-based light-emitting diodes. *Appl. Phys. Lett.* **2007**, *91*, 183507. [CrossRef]
20. Schubert, M.F.; Xu, J.; Kim, J.K.; Schubert, E.F.; Kim, M.H.; Yoon, S.; Lee, S.M.; Sone, C.; Sakong, T.; Park, Y. Polarization-matched GaInN/AlGaInN multi-quantum-well light-emitting diodes with reduced efficiency droop. *Appl. Phys. Lett.* **2008**, *93*, 041102. [CrossRef]
21. Shen, Y.C.; Mueller, G.O.; Watanabe, S.; Gardner, N.F.; Munkholm, A.; Krames, M.R. Auger recombination in InGaN measured by photoluminescence. *Appl. Phys. Lett.* **2007**, *91*, 141101. [CrossRef]
22. Zhang, M.; Bhattacharya, P.; Singh, J.; Hinckley, J. Direct measurement of auger recombination in In0.1Ga0.9N/GaN quantum wells and its impact on the efficiency of In0.1Ga0.9N/GaN multiple quantum well light emitting diodes. *Appl. Phys. Lett.* **2009**, *95*, 201108. [CrossRef]
23. Xie, J.Q.; Ni, X.F.; Fan, Q.; Shimada, R.; Ozgur, U.; Morkoc, H. On the efficiency droop in InGaN multiple quantum well blue light emitting diodes and its reduction with p-doped quantum well barriers. *Appl. Phys. Lett.* **2008**, *93*, 121107. [CrossRef]
24. Monemar, B.; Sernelius, B.E. Defect related issues in the "current roll-off" in InGaN based light emitting diodes. *Appl. Phys. Lett.* **2007**, *91*, 181103. [CrossRef]

25. Yang, Y.; Cao, X.A.; Yan, C.H. Investigation of the nonthermal mechanism of efficiency rolloff in InGaN light-emitting diodes. *IEEE Trans. Electron Devices* **2008**, *55*, 1771–1775. [CrossRef]
26. Ozgur, U.; Ni, X.; Li, X.; Lee, J.; Liu, S.; Okur, S.; Avrutin, V.; Matulionis, A.; Morkoc, H. Ballistic transport in InGaN-based LEDs: Impact on efficiency. *Semicond. Sci. Tech.* **2011**, *26*, 014022. [CrossRef]
27. Ni, X.; Li, X.; Lee, J.; Liu, S.; Avrutin, V.; Ozgur, U.; Morkoc, H.; Matulionis, A.; Paskova, T.; Mulholland, G. InGaN staircase electron injector for reduction of electron overflow in InGaN light emitting diodes. *Appl. Phys. Lett.* **2010**, *97*, 031110. [CrossRef]
28. Vampola, K.J.; Iza, M.; Keller, S.; DenBaars, S.P.; Nakamura, S. Measurement of electron overflow in 450 nm InGaN light-emitting diode structures. *Appl. Phys. Lett.* **2009**, *94*, 061116. [CrossRef]
29. Gong, Z.; Jin, S.; Chen, Y.; McKendry, J.; Massoubre, D.; Watson, I.M.; Gu, E.; Dawson, M.D. Size-dependent light output, spectral shift, and self-heating of 400 nm InGaN light-emitting diodes. *J. Appl. Phys.* **2010**, *107*, 013103. [CrossRef]
30. Kim, T.; Jung, Y.H.; Song, J.; Kim, D.; Li, Y.; Kim, H.S.; Song, I.S.; Wierer, J.J.; Pao, H.A.; Huang, Y. High-Efficiency, Microscale GaN Light-Emitting Diodes and Their Thermal Properties on Unusual Substrates. *Small* **2012**, *8*, 1643–1649. [CrossRef]
31. Meyaard, D.S.; Shan, Q.; Cho, J.; Schubert, E.F.; Han, S.H.; Kim, M.H.; Sone, C.; Oh, S.J.; Kim, J.K. Temperature dependent efficiency droop in GaInN light-emitting diodes with different current densities. *Appl. Phys. Lett.* **2012**, *100*, 081106. [CrossRef]
32. Zhao, C.; Ng, T.K.; ElAfandy, R.T.; Prabaswara, A.; Consiglio, G.B.; Ajia, I.A.; Roqan, I.S.; Janjua, B.; Shen, C.; Eid, J.; et al. Droop-Free, Reliable, and High-Power InGaN/GaN Nanowire Light-Emitting Diodes for Monolithic Metal-Optoelectronics. *Nano Lett.* **2016**, *16*, 4616–4623. [CrossRef] [PubMed]
33. Nguyen, H.P.T.; Djavid, M.; Woo, S.Y.; Liu, X.; Connie, A.T.; Sadaf, S.; Wang, Q.; Botton, G.A.; Shih, I.; Mi, Z. Engineering the Carrier Dynamics of InGaN Nanowire White Light-Emitting Diodes by Distributed p-AlGaN Electron Blocking Layers. *Sci. Rep.* **2015**, *5*, 7744. [CrossRef] [PubMed]
34. Tchernycheva, M.; Neplokh, V.; Zhang, H.; Lavenus, P.; Rigutti, L.; Bayle, F.; Julien, F.H.; Babichev, A.; Jacopin, G.; Largeau, L. Core-shell InGaN/GaN nanowire light emitting diodes analyzed by electron beam induced current microscopy and cathodoluminescence mapping. *Nanoscale* **2015**, *7*, 11692–11701. [CrossRef] [PubMed]
35. Nguyen, H.P.T.; Djavid, M.; Woo, S.Y.; Liu, X.; Wang, Q.; Botton, G.A.; Mi, Z. High-power phosphor-free InGaN/AlGaN dot-in-a-wire core-shell white light-emitting diodes. In Proceedings of the SPIE Light-Emitting Diodes: Materials, Devices, and Applications for Solid State Lighting XIX, San Francisco, CA, USA, 9 March 2015.
36. Mi, Z.; Nguyen, H.; Djavid, M.; Zhang, S.; Connie, A.; Sadaf, S.; Wang, Q.; Zhao, S.; Shih, I. High Power phosphor-free InGaNGaNAlGaN core-shell nanowire white light emitting diodes on Si substrates. *ECS Trans.* **2014**, *61*, 9–15. [CrossRef]
37. Philip, M.R.; Choudhary, D.D.; Djavid, M.; Le, K.Q.; Piao, J.; Nguyen, H.P.T. High efficiency green/yellow and red InGaN/AlGaN nanowire light-emitting diodes grown by molecular beam epitaxy. *J. Sci. Adv. Mater. Devices* **2017**, *2*, 150–155. [CrossRef]
38. Philip, M.R.; Bui, T.H.Q.; Djavid, D.D.C.M.; Phuoc, V.; Pham, T.T.; Nguyen, H.P.T. Molecular Beam Epitaxial Growth and Device Characterization of AlGaN UV-B Nanowire Light-Emitting Diodes. *J. Adv. Opt. Photonics* **2018**, *1*, 3–11.
39. Nguyen, H.P.T.; Cui, K.; Zhang, S.; Djavid, M.; Korinek, A.; Botton, G.A.; Mi, Z. Controlling Electron Overflow in Phosphor-Free InGaN/GaN Nanowire White Light-Emitting Diodes. *Nano Lett.* **2012**, *12*, 1317–1323. [CrossRef] [PubMed]
40. Nguyen, H.P.T.; Djavid, M.; Mi, Z. Nonradiative Recombination Mechanism in Phosphor-Free GaN-Based Nanowire White Light Emitting Diodes and the effect of Ammonium Sulfide Surface Passivation. *ECS Trans.* **2013**, *53*, 93–100. [CrossRef]
41. Nguyen, H.P.T.; Zhang, S.; Connie, A.T.; Kibria, M.G.; Wang, Q.; Shih, I.; Mi, Z. Breaking the Carrier Injection Bottleneck of Phosphor-Free Nanowire White Light-Emitting Diodes. *Nano Lett.* **2013**, *13*, 5437–5442. [CrossRef] [PubMed]
42. Schubert, E.F. *Light-Emitting Diodes*, 2nd ed.; Cambridge University Press: New York, NY, USA, 2006; pp. 61–63.

43. Hwang, J.I.; Hashimoto, R.; Saito, S.; Nunoue, S. Development of InGaN-based red LED grown on (0001) polar surface. *Appl. Phys. Express* **2014**, *7*, 071003. [CrossRef]
44. Piprek, J. Efficiency droop in nitride-based light-emitting diodes. *Phys. Status Solidi* **2010**, *207*, 2217–2225. [CrossRef]
45. Wang, X.F.; Yang, Q.; Wang, G.G.; Wang, X.Z.; Han, J.C. A new single-component $KCaY(PO_4)_2$: Dy^{3+}, Eu^{3+} nanosized phosphor with high color-rendering index and excellent thermal resistance for warm-white NUV-LED. *RSC Adv.* **2016**, *6*, 96263–96274. [CrossRef]
46. Kuo, T.W.; Liu, W.R.; Chen, T.M. High color rendering white light-emitting-diode illuminator using the red-emitting Eu^{2+}-activated CaZnOS phosphors excited by blue LED. *Opt. Express* **2010**, *18*, 8187–8192. [CrossRef] [PubMed]
47. Quoc Anh, N.D.; Lee, H.Y.; Thanh Phuong, T.; Khanh Nhan, N.H.; Quang Minh, T.H.; Huu Ly, T. Y_2O_3:Eu^{3+} phosphor: A novel solution for an increase in color rendering index of multi-chip white LED packages. *J. Chin. Inst. Eng.* **2017**, *40*, 228–234. [CrossRef]
48. Nizamoglu, Z.; Demir, G. Color-converting combinations of nanocrystal emitters for warm-white light generation with high color rendering index. *Appl. Phys. Lett.* **2008**, *92*, 031102. [CrossRef]
49. Chen, L.Y.; Chang, J.K.; Cheng, W.C.; Huang, J.C.; Huang, Y.C.; Cheng, W.H. Chromaticity tailorable glass-based phosphor-converted white light-emitting diodes with high color rendering index. *Opt. Express* **2015**, *23*, A1024–A1029. [CrossRef]
50. Huang, C.H.; Chen, T.M. Novel yellow-emitting $Sr_8MgLn(PO_4)_7$: Eu^{2+} (Ln = Y, La) phosphors for applications in white LEDs with excellent color rendering index. *Inorg. Chem.* **2011**, *50*, 5725–5730. [CrossRef]
51. Alvi, N.H.; Ali, S.U.; Hussain, S.; Nur, O.; Willander, M. Fabrication and comparative optical characterization of n-ZnO nanostructures (nanowalls, nanorods, nanoflowers and nanotubes)/p-GaN white-light-emitting diodes. *Scr. Mater.* **2011**, *64*, 697–700. [CrossRef]
52. Fang, M.H.; Ni, C.; Zhang, X.; Tsai, Y.T.; Mahlik, S.; Lazarowska, A.; Grinberg, M.; Sheu, H.S.; Lee, J.F.; Cheng, B.M.; et al. Enhance Color Rendering Index via Full Spectrum Employing the Important Key of Cyan Phosphor. *ACS Appl. Mater. Interfaces* **2016**, *8*, 30677–30682. [CrossRef]
53. Wang, Z.; Yuan, F.; Li, X.; Li, Y.; Zhong, H.; Fan, L.; Yang, S. 53% Efficient Red Emissive Carbon Quantum Dots for High Color Rendering and Stable Warm White-Light-Emitting Diodes. *Adv. Mater.* **2017**, *29*, 1702910. [CrossRef] [PubMed]

micromachines

MDPI

Article

Improved Efficiency of Perovskite Light-Emitting Diodes Using a Three-Step Spin-Coated $CH_3NH_3PbBr_3$ Emitter and a PEDOT:PSS/MoO_3-Ammonia Composite Hole Transport Layer

Yuanming Zhou [1,*], Sijiong Mei [1], Dongwei Sun [1], Neng Liu [1], Wuxing Shi [1], Jiahuan Feng [1], Fei Mei [1,*], Jinxia Xu [1], Yan Jiang [1] and Xianan Cao [2]

1 Hubei Key Laboratory for High-efficiency Use of Solar Energy and Operation Control of Energy Storage System, Hubei University of Technology, Wuhan 430068, China
2 Department of Computer Science and Electrical Engineering, West Virginia University, Morgantown, WV 26506, USA
* Correspondence: zhouym@mail.hbut.edu.cn (Y.Z.); meifei777@163.com (F.M.); Tel.: +86-27-5975-0430 (Y.Z. & F.M.)

Received: 5 June 2019; Accepted: 5 July 2019; Published: 7 July 2019

Abstract: High efficiency perovskite light-emitting diodes (PeLEDs) using PEDOT:PSS/MoO_3-ammonia composite hole transport layers (HTLs) with different MoO_3-ammonia ratios were prepared and characterized. For PeLEDs with one-step spin-coated $CH_3NH_3PbBr_3$ emitter, an optimal MoO_3-ammonia volume ratio (0.02) in PEDOT:PSS/MoO_3-ammonia composite HTL presented a maximum luminance of 1082 cd/m^2 and maximum current efficiency of 0.7 cd/A, which are 82% and 94% higher than those of the control device using pure PEDOT:PSS HTL respectively. It can be explained by that the optimized amount of MoO_3-ammonia in the composite HTLs cannot only facilitate hole injection into $CH_3NH_3PbBr_3$ through reducing the contact barrier, but also suppress the exciton quenching at the HTL/$CH_3NH_3PbBr_3$ interface. Three-step spin coating method was further used to obtain uniform and dense $CH_3NH_3PbBr_3$ films, which lead to a maximum luminance of 5044 cd/m^2 and maximum current efficiency of 3.12 cd/A, showing enhancement of 750% and 767% compared with the control device respectively. The significantly improved efficiency of PeLEDs using three-step spin-coated $CH_3NH_3PbBr_3$ film and an optimum PEDOT:PSS/MoO_3-ammonia composite HTL can be explained by the enhanced carrier recombination through better hole injection and film morphology optimization, as well as the reduced exciton quenching at HTL/$CH_3NH_3PbBr_3$ interface. These results present a promising strategy for the device engineering of high efficiency PeLEDs.

Keywords: perovskite light-emitting diodes; three-step spin coating; hole transport layer; PEDOT:PSS/MoO_3-ammonia composite

1. Introduction

Taking advantage of high photoluminescence quantum yield (PLQY), excellent color purity, high carrier mobility and low-temperature solution-processing, organometal halide perovskites have been studied extensively for their applications in solution-processed light-emitting diodes (LEDs) [1,2]. Since Friend's group reported the first demonstration about room-temperature infrared and green light emission observed in LEDs with $CH_3NH_3PbX_3$ (X is I^-, Br^- or Cl^-) perovskite emission layers (EML) in 2014 [3], organic-inorganic perovskite light-emitting diodes (PeLEDs) have attracted much attention and their external quantum efficiency (EQE) of PeLEDs increased rapidly from 0.1% to exceeding 20% [3–7]. Although possessing the excellent EQE comparable with quantum dot light-emitting diodes (QLEDs) and organic light-emitting diodes (OLEDs), PeLEDs still face challenges for the commercial application, such as device performance and stability.

In PeLEDs, poly(styrenesulfonate)-doped poly(3,4-ethylenedioxythiophene) (PEDOT:PSS) is the most frequently selected hole transport material (HTM), which can reduce the surface roughness of indium tin oxide (ITO) and the energy barrier between ITO and perovskite materials. However, it can erode the ITO substrate because of its acidic nature and finally affect the performance and reliability of devices [8]. Although PEDOT:PSS possesses lowest unoccupied molecular orbital (LUMO) of 5.2 eV, which are beneficial for the hole injection and transport, exciton quenching usually takes place at PEDOT:PSS/perovskite interface [9]. Therefore, further modification of PEDOT:PSS is still required for the improved performance of PeLEDs.

Different methods have been used to modify the PEDOT:PSS layer and prevent exciton quenching at PEDOT:PSS/perovskite interface. Cho et al. constructed a PEDOT:PSS/perfluorinated ionomer (PFI) composite layer to adjust the work function of HTL in PeLEDs, which leads to a reduced hole injection barrier and balanced injection of charge carriers. Meanwhile, the exciton quenching at the PEDOT:PSS/$CH_3NH_3PbBr_3$ interface could be suppressed by increasing the PFI quantity in HTL [4,5]. Besides, transition metal oxides (TMOs) have attracted much attention because of their excellent properties such as high transparency, tunable morphology, and good electrical conductivities. TMOs, such as MoO_3 [10,11], WO_3 [12], and V_2O_5 [13], are stable p-type semiconductor materials, which is promising to substitute or modify PEDOT:PSS. MoO_3, which has high work function, is one of the most studied oxide HTMs used in both LEDs and organic solar cells [14,15]. In detail, MoO_3 has been used as interlayers to enhance the hole injection or transport in OLEDs [14–16], organic solar cells (OPVs) [17–21], and perovskite solar cells [22,23]. However, it has been less employed to modify HTL such as PEDOT:PSS to obtain the improved performance of OLEDs and PeLEDs, not only through balanced carrier transport, but also by suppressing the exciton quenching.

Several groups have tried to employ the MoO_3 doped HTLs in organic-inorganic PeLEDs and other LEDs in order to obtain better performance of devices [24–26]. Kim et al. [25] reported the enhanced performance of $CH_3NH_3PbBr_3$ PeLEDs caused by a solution-processed MoO_3 and PEDOT:PSS (PEDOT:MoO_3) composite HTL with the MoO_3 concentration in the range of 0.1~0.7 wt.%. The hole injection was improved by doping MoO_3 in PEDOT:PSS through a reduction in the contact barrier at HTL/$CH_3NH_3PbBr_3$ interface and enhanced crystallinity of perovskite film. It is noted that the MoO_3 concentration in the composite is small, and the electron transport material (SPW-111) is not usually used in PeLEDs. Besides, Zheng et al. developed a composite hole injection layer (HIL) of MoO_x-doped GO in tris(8-hydroxy-quinolinato)aluminum (Alq_3)-based OLEDs [24], and Meng et al. modified the PEDOT:PSS HTL by doping a MoO_3 ammonia solution with largely adjusted volume ratio of (0~0.8):1 in all inorganic $CsPbBr_3$ PeLEDs [26]. According to these results, the doping of MoO_3 in HTLs is promising to reduce the contact barrier and luminescent quenching at PEDOT:PSS/EML interface. Although a few results have been reported as regards PeLEDs with MoO_3 doped HTLs, there is still great room for an improvement in terms of device's efficiency and stability, and the elucidation of corresponding physical mechanism.

In this paper, PEDOT:PSS/MoO_3-ammonia composite HTLs with different MoO_3-ammonia ratios were introduced in organic-inorganic PeLEDs with a simple structure of ITO/HTL/$CH_3NH_3PbBr_3$/TPBi/LiF/Al in order to suppress the exciton quenching and reduce the contact barrier at HTL/EML interface, facilitating the balanced transport of carriers. Three-step spin coating method was also employed to obtain uniform and dense $CH_3NH_3PbBr_3$ films, which lead to a maximum luminance of 5044 cd/m^2 and maximum current efficiency of 3.12 cd/A, showing enhancement of 750% and 767% compared with the control device respectively.

2. Experimental Section

2.1. Materials

PEDOT:PSS (Clevios P AI4083), Molybdenum trioxide (MoO_3 powder) and Ammonium hydroxide aqueous solution were purchased from Heraeus Materials Technology Co., Ltd. (Hanau, Germany), Shanghai Aladdin Bio-Chem Technology Co., Ltd. (Shanghai, China), and Sinopharm Chemical Reagent Co., Ltd. (Shanghai, China) respectively. Methylammonium bromine (CH_3NH_3Br) was purchased from Xi'an Polymer Light Technology Co., Ltd. (Xi'an, China). Lead Bromide ($PbBr_2$, 99.999%) and N,N-Dimethylformamide (DMF, 99.9%) were purchased from Sigma-Aldrich Co., Ltd. (St. Louis, MO, USA). 2,2',2'-(1,3,5-benzinetriyl)-tris(1-phenyl-1-H-benzimidazole) (TPBi) and Aluminum slug (Al, 99.999%) were purchased from Jilin Optical and Electronic Materials Co., Ltd. (Changchun, China) and Alfa Aesar (Ward Hill, MA, USA) respectively. PEDOT:PSS was filtered through a 0.45 μm PTFE filter before use, while other materials and solvents were used directly without further purification. The perovskite precursor solution was prepared by dissolving MABr and $PbBr_2$ with a molar ratio of 2:1 in DMF solvent to obtain a fixed concentration of 5 wt.%. PEDOT:PSS/MoO_3-ammonia composite HTLs with different MoO_3-ammonia ratios were obtained by using the method referred to the literature [26]. After dissolving MoO_3 powder in ammonium hydroxide aqueous solution to obtain 5 mg/mL MoO_3-Ammonia mixed solution, PEDOT:PSS and MoO_3-Ammonia solution with different volume ratios (1:0.01, 1:0.02, 1:0.03) were mixed to prepare PEDOT:PSS/MoO_3-ammonia composite solution, which should be stirred for 1 h before use.

2.2. Device Fabrication

Our PeLEDs were prepared on pre-patterned ITO-coated glass substrates with the sheet resistance of ~15 Ω/m^2. The basic device structure is ITO/composite HTL/$CH_3NH_3PbBr_3$/TPBi/LiF/Al. Typically, the substrates were cleaned ultrasonically in acetone, methanol and deionized water for 5 min sequentially. After drying with a nitrogen gun, the substrates were treated by oxygen plasma for 5 min in order to modify the work function of ITO effectively.

Next, these substrates were moved into a glovebox to spin-coat pure PEDOT:PSS, PEDOT:PSS/MoO_3-ammonia composite and $CH_3NH_3PbBr_3$ layers. The PEDOT:PSS and PEDOT:PSS/MoO_3-ammonia composite layers were spin-coated onto the substrates at 8000 rpm for 30 s, and then annealed at 150 °C for 15 min in a nitrogen atmosphere. For the one-step spin coating, the perovskite precursor solution was spin-coated at 8000 rpm for 30 s, and then annealed at 80 °C for 10 min. While for the three-step spin coating, the precursor was spin-coated by three times with sequential speeds of 2000, 4000 and 6000 rpm for 30 s, followed by annealing at 80 °C for 10 min for each-step spin coating. No anti-solvent and other additives were used in the spin coating of $CH_3NH_3PbBr_3$ layers.

Finally, the substrates were transferred to a physical vapor thermal evaporation system, in which a 30 nm TPBi, a 0.5 nm LiF and a 100 nm Al were deposited sequentially for electron transport layer (ETL), electron injection layer (EIL) and cathode in a base pressure of ~3×10^{-7} Torr respectively. Each substrate contains four devices with the active area of 0.1 cm^2. All PeLEDs were encapsulated simply with cover glass slides in the glovebox and then tested immediately in ambient air.

2.3. Device Characterization

The thickness of PEDOT:PSS, PEDOT:PSS/MoO_3-ammonia composite and perovskite films were recorded by an Alpha-Step D-600 stylus profiler (KLA Corporation, Milpitas, CA, USA). The absorption spectra, transmittance spectra and photoluminescence (PL) spectra were carried out with a HITACHI U-3900 ultraviolte/visible spectrophotometer and a HITACHI F-4600 luminescence spectrometer (Japan), respectively. The surface morphology of perovskite films were observed with a scanning electron microscopy (SEM, FEI Sirion FEG, FEI Corporation, Eindhoven, Netherlands). X-ray diffraction (XRD) patterns were measured with a PANalytical Empyrean X-ray diffractometer (PANalytical B. V.,

Almelo, Netherlands). The luminance-current density-voltage (L-J-V) characteristics of PeLEDs were tested using a Keithley 2400 source meter and a Keithley 2000 multimeter (Tektronix, Inc., Beaverton, OR, USA) coupled with a calibrated silicon photodetector (1 cm in diameter), which capture and convert photons emitted from the glass side. The electroluminescence (EL) spectra of the devices were monitored by an Ocean Optics fiber-optic spectrometer (Ocean Optics, Inc., Largo, FL, USA).

3. Results and Discussion

Figure 1a shows the schematic structure of our PeLEDs, which consist of ITO as a transparent anode, the PEDOT:PSS/MoO_3-ammonia composite as a HTL, $CH_3NH_3PbBr_3$ as an EML, TPBi as an ETL, LiF as an EIL, and Al as a cathode, respectively. To obtain good surface morphology of $CH_3NH_3PbBr_3$ films, they were prepared by one-step and three-step spin coating of a $CH_3NH_3PbBr_3$ precursor respectively. Figure 1b shows energy level diagrams of the PeLEDs with a pure PEDOT:PSS HTL. It is noted that the energy barrier between the PEDOT:PSS and $CH_3NH_3PbBr_3$ layers is ~0.5 eV, which may result in low device efficiency. The doping of MoO_3-ammonia in the PEDOT:PSS is expected to increase the work function of the PEDOT:PSS HTL and correspondingly reduce the contact barrier between the HTL and the $CH_3NH_3PbBr_3$ EML for efficient hole injection [25,26].

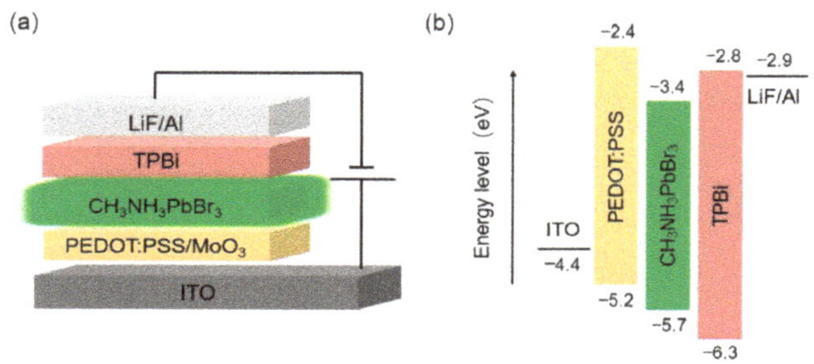

Figure 1. (**a**) The schematic structure of our PeLEDs. (**b**) The energy level diagram of the PeLEDs with a pure PEDOT:PSS HTL.

Figure 2 shows the SEM images of $CH_3NH_3PbBr_3$ films prepared on pure PEDOT:PSS and PEDOT:PSS/MoO_3-ammonia composite HTLs [glass/ITO/composite HTL (~40 nm)/$CH_3NH_3PbBr_3$ (~30 nm for one-step spin coating, ~55 nm for three-step spin coating)], respectively. The effect of the small MoO_3-ammonia amount on the surface morphology of $CH_3NH_3PbBr_3$ film is not evident. As reported in the literature, multi-step spin coating is expected to improve the surface morphology of perovskite film [27,28]. As shown in Figure 2, uniform and compact perovskite films with enhanced crystallinity formed on increasing the coating times from one to three.

Figure 3 shows XRD patterns of one-step and three-step coated $CH_3NH_3PbBr_3$ films (glass/composite HTL (~40 nm)/$CH_3NH_3PbBr_3$ (~30 nm for one-step spin coating, ~55 nm for three-step spin coating)). All the XRD patterns show two characteristic peaks at 15° and 30°, assigned to (100) and (200) crystal planes respectively, suggesting the crystal growth orientation along (100) planes. As shown in Figure 3a, the intensity of diffraction peaks was enhanced in three-step spin-coated $CH_3NH_3PbBr_3$ film compared with one-step spin-coated $CH_3NH_3PbBr_3$ film on pure PEDOT:PSS film, suggesting a better crystallization on increasing the coating time from one to three. As shown in Figure 3b, on increasing the ratio of MoO_3-ammonia from 0 to 0.03, the intensity of diffraction peaks of three-step spin-coated $CH_3NH_3PbBr_3$ film increases monotonically. The similar trend is also found in the one-step spin-coated $CH_3NH_3PbBr_3$ films with different MoO_3-ammonia ratios. This result is well consistent with previously reported results [25,26], which may be explained by that the MoO_3 particles can act as crystal nuclei for the growth of spin-coated perovskite film.

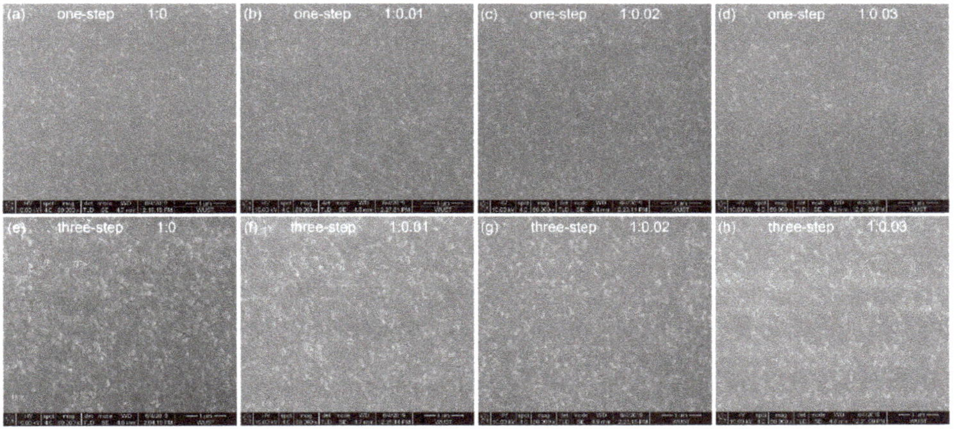

Figure 2. SEM images of the one-step spin-coated $CH_3NH_3PbBr_3$ film samples on (**a**) PEDOT:PSS, (**b**) PEDOT:PSS/MoO_3-ammonia (1:0.01), (**c**) PEDOT:PSS/MoO_3-ammonia (1:0.02), (**d**) PEDOT:PSS/MoO_3-ammonia (1:0.03) composite HTLs, and three-step spin-coated $CH_3NH_3PbBr_3$ film samples on (**e**) PEDOT:PSS, (**f**) PEDOT:PSS/MoO_3-ammonia (1:0.01), (**g**) PEDOT:PSS/MoO_3-ammonia (1:0.02), (**h**) PEDOT:PSS/MoO_3-ammonia (1:0.03) composite HTLs.

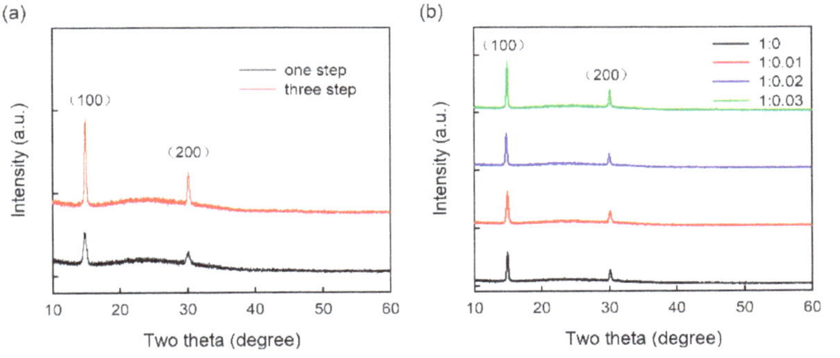

Figure 3. XRD patterns of (**a**) one-step and three-step coated $CH_3NH_3PbBr_3$ films on the pure PEDOT:PSS film, (**b**) three-step coated $CH_3NH_3PbBr_3$ films on the pure PEDOT:PSS film and PEDOT:PSS/MoO_3-ammonia composite HTLs with different MoO_3 ratios. The XRD patters were shifted vertically for clarity.

Figure 4 shows steady-state PL spectra of one-step and three-step coated $CH_3NH_3PbBr_3$ films [glass/ITO/composite HTL (~40 nm)/$CH_3NH_3PbBr_3$ (~30 nm for one-step spin coating, ~55 nm for three-step spin coating)] conducted by using a luminescence spectrometer with an excitation wavelength of 315 nm. All the PL spectra show a well-defined peak at ~528 nm. As shown in Figure 4a, PL intensity was enhanced in three-step spin-coated $CH_3NH_3PbBr_3$ film compared with one-step spin-coated $CH_3NH_3PbBr_3$ film on pure PEDOT:PSS film. It can be explained by that the amount and morphology of perovskite material affect the PL intensity, namely more excitons will be generated by increasing the amount of perovskite particles, leading to the enhancement of the PL intensity when the thickness of perovskite film increases from ~30 nm (one-step coating) to ~55 nm (three-step coating) shown in Figure 4a. As shown in Figure 4b, on increasing the ratio of MoO_3-ammonia from 0 to 0.02, the PL intensity of three-step spin-coated $CH_3NH_3PbBr_3$ film gradually increases, while the further increase of the amount of MoO_3-ammonia leads to the decrease of the PL intensity. The similar trend is also found in the one-step spin-coated $CH_3NH_3PbBr_3$ films with different

MoO_3-ammonia ratios. It is suggested that the optimal MoO_3-ammonia ratio is beneficial for blocking the exciton quenching at the HTL/$CH_3NH_3PbBr_3$ interface, while the excessive MoO_3-ammonia ratio is unfavorable. These results may be due to the increase of MoO_3 on top of the HIL separating excitons generated in the $CH_3NH_3PbBr_3$ EML from the quenching of PEDOT:PSS. However, on increasing the MoO_3-ammonia amount, dopant aggregation or trap states may also occur at the HTL/EML interface, leading to the decay of photoluminescence. Figure 4c shows the transmittance spectra of PEDOT:PSS and PEDOT:PSS/MoO_3-ammonia composite films with different MoO_3-ammonia ratios (glass/ITO/composite HTL (~40 nm)). As shown in transmittance spectra, a small amount of MoO_3 has little effect on the transmittance of PEDOT:PSS/MoO_3-ammonia composite films in the visible range. The transmittances of four samples are near-identical, indicating that the doping of MoO_3-ammonia in PEDOT:PSS HTL cannot impede the light passing through the HTL in this work. Figure 4d shows the absorption spectra of one-step and three-step spin-coated perovskite films on pure PEDOT:PSS HTL. Both two absorption spectra show a well-defined peak at ~526 nm. Furthermore, the absorption intensity was enhanced on increasing the coating time from one to three, which can be attributed to the increase of the thickness of perovskite film from ~30 nm (one-step coating) to ~55 nm (three-step coating).

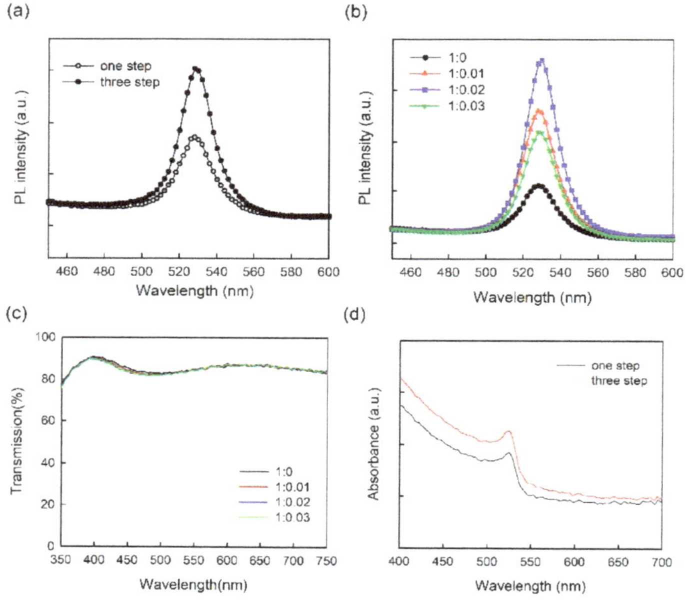

Figure 4. PL spectra of (**a**) one-step and three-step coated $CH_3NH_3PbBr_3$ films on the pure PEDOT:PSS film, (**b**) three-step coated $CH_3NH_3PbBr_3$ films on the pure PEDOT:PSS film and PEDOT:PSS/MoO_3-ammonia composite HTLs with different MoO_3-ammonia ratios. (**c**) The transmittance spectra of the pure PEDOT:PSS film and PEDOT:PSS/MoO_3-ammonia composite HTLs with different MoO_3-ammonia ratios. (**d**) The absorption spectra of one-step and three-step spin-coated perovskite films on pure PEDOT:PSS HTL.

Figure 5 shows (a) the current density vs. voltage (J-V), (b) the luminance vs. current density (L-J), (c) the current efficiency vs. current density (CE-J), and (d) the EQE vs. current density (EQE-J) curves for the single-step $CH_3NH_3PbBr_3$ PeLEDs with pure PEDOT:PSS and PEDOT:PSS/MoO_3-ammonia (1:0.01, 1:0.02, 1:0.03) composite HTLs. These four devices are labelled as S1, S2, S3, and S4 for clarity respectively. The detailed device parameters of the PeLEDs (S1, S2, S3, S4) are summarized in Table 1.

As shown in Figure 5a, on increasing the MoO_3-ammonia ratio from 0 to 0.03, the turn-on voltage, which are defined as the driving voltage at ~1 mA/cm^2, decreases monotonically from 4.3 V to 4.14 V. Besides, the current density of the PeLEDs increases on increasing the MoO_3-ammonia amount, suggesting a reduced energy barrier at HTL/EML interface, inducing more efficient hole injection into $CH_3NH_3PbBr_3$ layer [25,26]. As described in the L-J, CE-J, and EQE-J characteristics, a maximum luminance of 1082 cd/m^2, a maximum CE of 0.7 cd/A and a maximum EQE of 0.11% were observed in the device with the MoO_3-ammonia ratio of 0.02 (device S3), indicating the optimal volume ratio, while for the control device with pure PEDOT:PSS HTL (device S1), the maximum luminance of 593 cd/m^2 and maximum CE of 0.36 cd/A were obtained. Therefore, the optimized device shows a 82% enhancement in the maximum luminance and 94% enhancement in the maximum CE respectively.

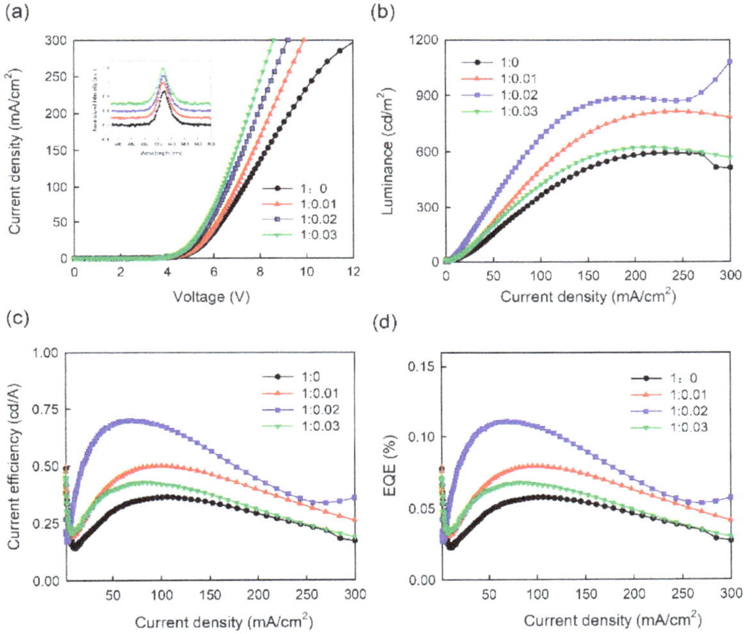

Figure 5. (**a**) J-V, (**b**) L-J, (**c**) CE-J, (**d**) EQE-J curves of PeLEDs with a one-step spin-coated emitter and a PEDOT:PSS/MoO_3-ammonia (1:0, 1:0.01, 1:0.02, 1:0.03) composite HTL (device S1, S2, S3, S4). The inset is normalized EL spectra of the PeLED devices at 20 mA/cm^2, which were shifted vertically for clarity.

Table 1. Summary of the device parameters of PeLEDs (device S1, S2, S3, S4) with a single-step spin-coated emitter and a PEDOT:PSS/MoO_3-ammonia (1:0, 1:0.01, 1:0,02, 1:0.03) composite HTL.

Volume Ratio of PEDOT:PSS/MoO_3-Ammonia	L_{max} (cd/m^2)	CE_{max} (cd/A)	EQE_{max} (%)	Turn-on Voltage (V)
1:0 (S1)	593	0.36	0.057	4.3
1:0.01 (S2)	816	0.5	0.079	4.22
1:0.02 (S3)	1082	0.7	0.11	4.18
1:0.03 (S4)	624	0.42	0.068	4.14

These results suggest that the hole injection in the PeLEDs with the PEDOT:PSS/MoO_3-ammonia composite layer can be improved by reducing the contact barrier [25,26] and blocking the exciton quenching at the HTL/$CH_3NH_3PbBr_3$ interface [26]. However, the device efficiency decreases when an excessive MoO_3-ammonia amount was doped in the PEDOT:PSS/MoO_3-ammonia composite HTL

(0.03), possibly due to the trap states formed at HTL/EML interface after the doping of the excessive MoO_3-ammonia. The inset in Figure 5a shows the normalized electroluminescence (EL) spectra of $CH_3NH_3PbBr_3$ PeLEDs using PEDOT:PSS/MoO_3-ammonia composite HTLs with different amounts of MoO_3-ammonia, indicating that the EL spectra of PeLEDs with composite HTLs (S2, S3, S4) are nearly identical to that with pure PEDOT:PSS HTL (S1). It is suggested that the MoO_3-ammonia doping cannot modify the emission profiles of $CH_3NH_3PbBr_3$ PeLEDs, which have an EL peak at ~528 nm.

Figure 6 shows J-V, L-J, CE-J, and EQE-J curves for the three-step $CH_3NH_3PbBr_3$ PeLEDs with pure PEDOT:PSS and PEDOT:PSS/MoO_3-ammonia (1:0.01, 1:0.02, 1:0.03) composite HTLs, which are labelled as T1, T2, T3, and T4 for clarity respectively. The detailed device parameters of the PeLEDs (T1, T2, T3, T4) are summarized in Table 2. As shown in Figure 6a, on increasing the MoO_3-ammonia ratio from 0 to 0.03, the turn-on voltage decreases from 4.08 V to 3.68 V. Besides, the current density of the PeLEDs increases on increasing the MoO_3-ammonia amount, which is similar to the trend observed in one-step devices (S1, S2, S3, S4). As shown in Figure 6c,d, the CE and EQE increase at low current densities because of the rapidly increased luminance (or the number of excitons). On further increasing the current density, the luminance increases more slowly or decreases, leading to the decreased CE and EQE, namely the efficiency roll-off. As reported, the efficiency roll-off of OLEDs is mainly caused by charge imbalance and quenching processes [10,29,30]. Similarly, in this work, higher CE found in the device with the MoO_3-ammonia ratio of 0.02 can be explained by the balance of electrons and holes in the EM, as well as the reduced exciton quenching. From the L-J, CE-J, and EQE-J characteristics, a maximum luminance of 5044 cd/m^2, a maximum CE of 3.12 cd/A and a maximum EQE of 0.5% were also observed in the device with the optimal MoO_3-ammonia ratio of 0.02 (device T3), while for the device with pure PEDOT:PSS HTL (device T1), the maximum luminance of 2309 cd/m^2 and maximum CE of 1.47 cd/A were obtained. Thus, the optimized device shows a 118% enhancement in the maximum luminance and 112% enhancement in the maximum CE respectively. Compared with the control device with pure PEDOT:PSS HTL and one-step spin-coated $CH_3NH_3PbBr_3$ film (device S1), a 750% enhancement in the maximum luminance and 767% enhancement in the maximum CE were obtained for the optimized device (device T3). The inset in Figure 6a shows the normalized EL spectra of three-step $CH_3NH_3PbBr_3$ PeLEDs using PEDOT:PSS/MoO_3-ammonia composite HTLs with different amounts of MoO_3-ammonia, in which all PeLEDs have an EL peak at ~528 nm. Figure 7 shows the EL curves measured at different current densities. The results indicate that all PeLEDs have an EL peak at ~528 nm, suggesting the color stability of our devices.

These results indicate that the hole injection in the PeLEDs with the PEDOT:PSS/MoO_3-ammonia composite layer can be improved by reducing the contact barrier [25,26] and suppressing the exciton quenching at the HTL/$CH_3NH_3PbBr_3$ interface [26]. Besides, three-step spin coating method can improve the surface morphology of the $CH_3NH_3PbBr_3$ perovskite film shown in Figure 2. Furthermore, the as-obtained perovskite layer exhibited a stronger PL intensity shown in Figure 4. These factors induce the significant improvement on luminous performance of our PeLEDs. Therefore, the significantly improved efficiency of PeLEDs using three-step spin-coated $CH_3NH_3PbBr_3$ film and an optimum PEDOT:PSS/MoO_3-ammonia composite HTL can be explained by the enhanced carrier recombination through better hole injection and film morphology optimization, as well as the reduced exciton quenching at HTL/$CH_3NH_3PbBr_3$ interface.

Table 2. Summary of the device parameters of PeLEDs (device T1, T2, T3, T4) with a three-step spin-coated emitter and a PEDOT:PSS/MoO_3-ammonia (1:0, 1:0.01, 1:0,02, 1:0.03) composite HTL.

Volume Ratio of PEDOT:PSS/MoO_3-Ammonia	L_{max} (cd/m^2)	CE_{max} (cd/A)	EQE_{max} (%)	Turn-on Voltage (V)
1:0 (T1)	2309	1.47	0.23	4.08
1:0.01 (T2)	4215	2.84	0.45	3.84
1:0.02 (T3)	5044	3.12	0.5	3.75
1:0.03 (T4)	3055	2.08	0.33	3.68

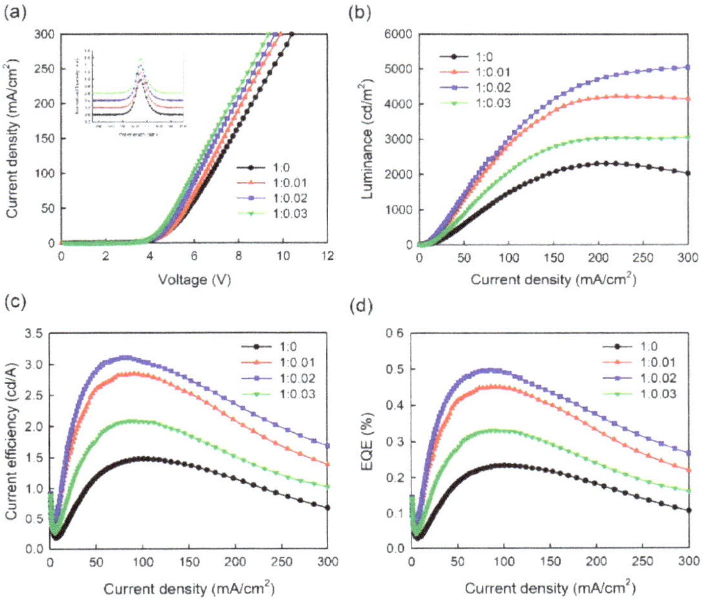

Figure 6. (**a**) J-V, (**b**) L-J, (**c**) CE-J, (**d**) EQE-J curves of PeLEDs with a three-step spin-coated emitter and a PEDOT:PSS/MoO_3-ammonia (1:0, 1:0.01, 1:0.02, 1:0.03) composite HTL (device T1, T2, T3, T4). The inset is normalized EL spectra of the PeLED devices at 20 mA/cm^2, which were shifted vertically for clarity.

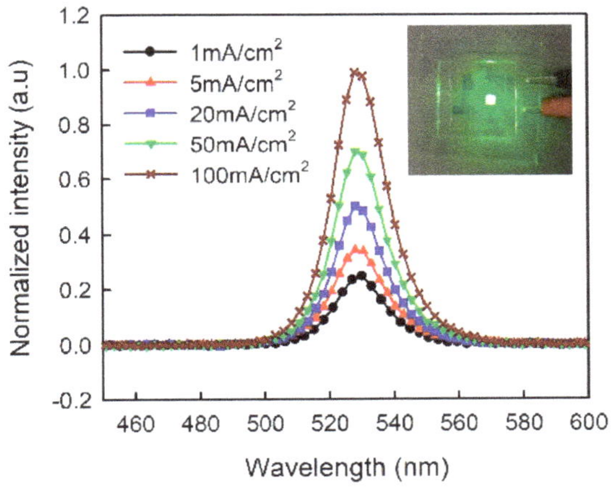

Figure 7. EL spectra of the PeLED devices with a three-step spin-coated emitter and a PEDOT:PSS/MoO_3-ammonia (1:0.02) composite HTL at different current densities. The inset is a luminescence image of the device at 50 mA/cm^2.

4. Conclusions

In summary, we demonstrated improved performance of PeLEDs using a PEDOT:PSS/MoO_3-ammonia composite HTL by reducing the energy barrier and blocking the exciton quenching at the HTL/$CH_3NH_3PbBr_3$ interface. For PeLEDs with one-step spin-coated $CH_3NH_3PbBr_3$ film, an enhancement of 82% in the maximum luminance and 94% in the maximum CE was found in

PeLED with an optimal MoO_3-ammonia volume ratio (0.02) in PEDOT:PSS/MoO_3-ammonia composite HTL compared with the control device with pure PEDOT:PSS HTL respectively. Three-step spin coating method was further used to obtain uniform and dense $CH_3NH_3PbBr_3$ films, which lead to a maximum luminance of 5044 cd/m^2 and maximum CE of 3.12 cd/A, which are 750% and 767% larger than those of the control device respectively. The significantly improved efficiency of PeLEDs using three-step spin-coated $CH_3NH_3PbBr_3$ film and an optimum PEDOT:PSS/MoO_3-ammonia composite HTL can be originated from the enhanced carrier recombination through better hole injection and film morphology optimization, as well as the reduced exciton quenching at HTL/$CH_3NH_3PbBr_3$ interface. These results suggest a promising clue for the device engineering of high efficiency PeLEDs.

Author Contributions: Conceptualization, Y.Z. and F.M.; Funding acquisition, Y.Z. and F.M.; Investigation, S.M., D.S. and N.L.; Methodology, S.M., W.S., J.F., J.X., Y.J. and X.C.; Writing—original draft, S.M.; Writing—review & editing, Y.Z.

Funding: This work was supported by the Foundation of Hubei Provincial Science and Technology Department (Grant No. 2016BKJ005), the Leading Plan of Green Industry of Hubei University of Technology (Grant No. YXQN2016003), the International Science and Technology Cooperation Program of China (Grant No. 2016YFE0124300), the Open Foundation of Hubei Key Laboratory for High-efficiency Use of Solar Energy and Operation Control of Energy Storage System (Grant Nos. HBSEES201801, HBSEES201705), and the National Natural Science Foundation of China (Grant Nos. 51602099, 51371079, 11304092, and 11305056).

Acknowledgments: We thank Lixia Fan in Wuhan University of Science and Technology for technical assistance.

Conflicts of Interest: The authors declare no conflict of interest.

References

1. Era, M.; Morimoto, S.; Tsutsui, T.; Saito, S. Organic-inorganic heterostructure electroluminescent device using a layered perovskite semiconductor $(C_6H_5C_2H_4NH_3)_2PbI_4$. *Appl. Phys. Lett.* **1994**, *65*, 676. [CrossRef]
2. Koutselas, I.; Bampoulis, P.; Maratou, E.; Evagelinou, T.; Pagona, G.; Papavassiliou, G.C. Some unconventional organic–inorganic hybrid low-dimensional semiconductors and related light-emitting devices. *J. Phys. Chem. C* **2011**, *115*, 8475–8483. [CrossRef]
3. Tan, Z.K.; Moghaddam, R.S.; Lai, M.L.; Docampo, P.; Higler, R.; Deschler, F.; Price, M.; Sadhanala, A.; Pazos, L.M.; Credgington, D.; et al. Bright light-emitting diodes based on organometal halide perovskite. *Nat. Nanotechnol.* **2014**, *9*, 687–692. [CrossRef] [PubMed]
4. Cho, H.C.; Jeong, S.H.; Park, M.H.; Kim, Y.H.; Wolf, C.; Lee, C.L.; Heo, J.H.; Sadhanala, A.; Myoung, N.; Yoo, S.; et al. Overcoming the electroluminescence efficiency limitations of perovskite light-emitting diodes. *Science* **2015**, *350*, 1222–1225. [CrossRef] [PubMed]
5. Kim, Y.H.; Cho, H.; Heo, J.H.; Kim, T.S.; Myoung, N.; Lee, C.L.; Im, S.H.; Lee, T.W. Multicolored organic/inorganic hybrid perovskite light-emitting diodes. *Adv. Mater.* **2015**, *27*, 1248–1254. [CrossRef] [PubMed]
6. Lin, K.B.; Xing, J.; Quan, L.N.; de Arquer, F.P.G.; Gong, X.W.; Lu, J.X.; Xie, L.Q.; Zhao, W.J.; Zhang, D.; Yan, C.Z.; et al. Perovskite light-emitting diodes with external quantum efficiency exceeding 20 per cent. *Nature* **2018**, *562*, 245–248. [CrossRef] [PubMed]
7. Cao, Y.; Wang, N.N.; Tian, H.; Guo, J.S.; Wei, Y.Q.; Chen, H.; Miao, Y.F.; Zou, W.; Pan, K.; He, Y.R.; et al. Perovskite light-emitting diodes based on spontaneously formed submicrometre-scale structures. *Nature* **2018**, *562*, 249–253. [CrossRef]
8. Kim, Y.H.; Lee, S.H.; Noh, J.; Han, S.H. Performance and stability of electroluminescent device with self-assembled layers of poly(3,4-ethylenedioxythiophene)-poly(styrenesulfonate) and polyelectrolytes. *Thin Solid Films* **2006**, *510*, 305–310. [CrossRef]
9. Kim, J.S.; Friend, R.H.; Grizzi, I.; Burroughes, J.H. Spin-cast thin semiconducting polymer interlayer for improving device efficiency of polymer light-emitting diodes. *Appl. Phys. Lett.* **2005**, *87*, 023506. [CrossRef]
10. Liu, N.; Mei, S.J.; Sun, D.W.; Shi, W.X.; Feng, J.H.; Zhou, Y.M.; Mei, F.; Xu, J.X.; Jiang, Y.; Cao, X.A. Effects of charge transport materials on blue fluorescent organic light-emitting diodes with a host-dopant system. *Micromachines* **2019**, *10*, 344. [CrossRef]
11. You, H.; Dai, Y.F.; Zhang, Z.Q.; Ma, D.G. Improved performances of organic light-emitting diodes with metal oxide as anode buffer. *J. Appl. Phys.* **2007**, *101*, 026105. [CrossRef]

12. Li, J.Z.; Yahiro, M.; Ishida, K.; Yamada, H.; Matsushige, K. Enhanced performance of organic light emitting device by insertion of conducting/insulating WO_3 anodic buffer layer. *Synth. Met.* **2005**, *151*, 141. [CrossRef]
13. Meyer, J.; Zilberberg, K.; Riedl, T.; Kahn, A. Electronic structure of Vanadium pentoxide: An efficient hole injector for organic electronic materials. *J. Appl. Phys.* **2011**, *110*, 033710. [CrossRef]
14. Shrotriya, V.; Li, G.; Yao, Y.; Chu, C.W.; Yang, Y. Transition metal oxides as the buffer layer for polymer photovoltaic cells. *Appl. Phys. Lett.* **2006**, *88*, 073508. [CrossRef]
15. Shi, W.X.; Liu, N.; Zhou, Y.M.; Cao, X.A. Effects of postannealing on the characteristics and reliability of polyfluorene organic light-emitting diodes. *IEEE Trans. Electron Devices* **2019**, *66*, 1057–1062. [CrossRef]
16. Zhao, Y.B.; Chen, J.S.; Chen, W.; Ma, D.G. Poly(3,4-ethylenedioxythiophene): Poly(styrenesulfonate)/MoO_3 composite layer for efficient and stable hole injection in organic semiconductors. *J. Appl. Phys.* **2012**, *111*, 043716. [CrossRef]
17. Meyer, J.; Khalandovsky, R.; Gorrn, P.; Kahn, A. MoO_3 films spin-coated from a nanoparticle suspension for efficient hole-injection in organic electronics. *Adv. Mater.* **2011**, *23*, 70–73. [CrossRef]
18. Wong, K.H.; Ananthanarayanan, K.; Luther, J.; Balaya, P. Origin of hole selectivity and the role of defects in low-temperature solution-processed molybdenum oxide interfacial layer for organic solar cells. *J. Phys. Chem. C* **2012**, *116*, 16346–16351. [CrossRef]
19. Murase, S.; Yang, Y. Solution processed MoO_3 interfacial layer for organic photovoltaics prepared by a facile synthesis method. *Adv. Mater.* **2012**, *24*, 2459–2462. [CrossRef]
20. Sun, J.Y.; Tseng, W.H.; Lan, S.; Lin, S.H.; Yang, P.C.; Wu, C.I.; Lin, C.F. Performance enhancement in inverted polymer photovoltaics with solution-processed MoO_x and air-plasma treatment for anode modification. *Sol. Energy Mater. Sol. Cells* **2013**, *109*, 178–184. [CrossRef]
21. Liu, F.M.; Shao, S.Y.; Guo, X.Y.; Zhao, Y.; Xie, Z.Y. Efficient polymer photovoltaic cells using solution-processed MoO_3 as anode buffer layer. *Sol. Energy Mater. Sol. Cells* **2010**, *94*, 842–845. [CrossRef]
22. Hou, F.H.; Su, Z.S.; Jin, F.M.; Yan, X.W.; Wang, L.D.; Zhao, H.F.; Zhu, J.Z.; Chu, B.; Li, W.L. Efficient and stable planar heterojunction perovskite solar cells with an MoO_3/PEDOT:PSS hole transporting layer. *Nanoscale* **2015**, *7*, 9427–9432. [CrossRef] [PubMed]
23. Sung, H.; Ahn, N.; Jang, M.S.; Lee, J.K.; Yoon, H.; Park, N.G.; Choi, M. Transparent Conductive Oxide-Free Graphene-Based Perovskite Solar Cells with over 17% Efficiency. *Adv. Energy Mater.* **2016**, *6*, 1501873. [CrossRef]
24. Zheng, Q.H.; Li, W.S.; Zhang, Y.; Xu, K.; Xu, J.W.; Wang, H.; Xiong, J.; Zhang, X.Y.; Zhang, X.W. Solution-processed composite interfacial layer of MoO_x-doped graphene oxide for robust hole injection in organic light-emitting diode. *Phys. Status Solidi RRL* **2018**, *12*, 1700434. [CrossRef]
25. Kim, D.B.; Yu, J.C.; Nam, Y.S.; Kim, D.W.; Jung, E.D.; Lee, S.Y.; Lee, S.; Park, J.H.; Lee, A.Y.; Lee, B.R.; et al. Improved performance of perovskite light-emitting diodes using a PEDOT:PSS and MoO_3 composite layer. *J. Mater. Chem. C* **2016**, *4*, 8161–8165. [CrossRef]
26. Meng, Y.; Ahmadi, M.; Wu, X.Y.; Xu, T.F.; Xu, L.; Xiong, Z.H.; Chen, P. High performance and stable all-inorganic perovskite light emitting diodes by reducing luminescence quenching at PEDOT:PSS/perovskites interface. *Org. Electron.* **2019**, *64*, 47–53. [CrossRef]
27. Feng, Z.Q.; Wang, L.; Yu, H.T.; Ma, X.Q.; Zhang, Q.; Chen, S.F.; Liu, L.H.; Huang, W. Towards efficient perovskite light-emitting diodes: A multi-step spin-coating method for a dense and uniform perovskite film. *Org. Electron.* **2018**, *61*, 18–24. [CrossRef]
28. Qasim, K.; Wang, B.P.; Zhang, Y.P.; Li, P.F.; Wang, Y.S.; Li, S.J.; Lee, S.T.; Liao, L.S.; Lei, W.; Bao, Q.L. Solution-processed extremely efficient multicolor perovskite light-emitting diodes utilizing doped electron transport layer. *Adv. Funct. Mater.* **2017**, *27*, 1606874. [CrossRef]
29. Giebink, N.C.; Forrest, S.R. Quantum efficiency roll-off at high brightness in fluorescent and phosphorescent organic light emitting diodes. *Phys. Rev. B* **2008**, *77*, 235215. [CrossRef]
30. Murawski, C.; Leo, K.; Gather, M.C. Efficiency roll-off in organic light-emitting diodes. *Adv. Mater.* **2013**, *25*, 6801–6827. [CrossRef]

micromachines

MDPI

Communication

The Composition-Dependent Photoluminescence Properties of Non-Stoichiometric $Zn_xAg_yInS_{1.5+x+0.5y}$ Nanocrystals

Jian Feng [1,2], Xiaosheng Yang [1], Rong Li [2], Xianjiong Yang [2] and Guangwei Feng [2,*]

1 State Key Laboratory of Functions and Applications of Medicinal Plants, Guizhou Medical University, Guiyang 550014, Guizhou, China

2 Department of Chemistry, School of Basic Medical Science, Guizhou Medical University, 9 Beijing Road, Guiyang 550004, Guizhou, China

* Correspondence: fengguangwei@gmc.edu.cn

Received: 24 May 2019; Accepted: 24 June 2019; Published: 1 July 2019

Abstract: A facile hot injection approach to synthesize high-quality non-stoichiometric $Zn_xAg_yInS_{1.5+x+0.5y}$ nanocrystals (NCs) in the size range of 2.8–3.1 nm was presented. The fluorescence spectra had single band gap features, and indicated the formation of alloy states rather than simple composite structures. The chemical compositions, photoluminescence (PL) emission wavelengths, and quantum yields of $Zn_xAg_yInS_{1.5+x+0.5y}$ nanocrystals were significantly influenced by the concentration of an organic capping agent. The appropriate proportion of 1-dodecanthiol in the precursor prevented the precipitation, increased the fluorescence quantum yield, and improved their optical properties. The proper ratio of capping agent allowed Zn, Ag, and In to form a better crystallinity and compositional homogeneity of $Zn_xAg_yInS_{1.5+x+0.5y}$ nanocrystals. The photoluminescence was tunable from blue to red in the range of 450–700 nm as the Ag content changed independently. The PL and absorption spectra of $Zn_xAg_yInS_{1.5+x+0.5y}$ nanocrystals showed a significant blue shift with the decrease of Ag content in the precursor. As there were no obvious differences on the average particle sizes of $Zn_xAg_yInS_{1.5+x+0.5y}$ samples, these results fully revealed the composition-dependent photoluminescence properties of $Zn_xAg_yInS_{1.5+x+0.5y}$ nanocrystals. The relative quantum yield reached 35%. The fluorescence lifetimes (τ_1=115–148 ns and τ_2=455–483 ns) were analogous to those of $AgInS_2$ and $(AgIn)_xZn_{2(1-x)}S_2$.

Keywords: nanoparticles; luminescence; non-stoichiometric $Zn_xAg_yInS_{1.5+x+0.5y}$ nanocrystals; photoluminescence properties; tunable fluorescence emission; one-pot approach

1. Introduction

Ternary I-III-VI_2 nanocrystals (NCs), including $AgInS_2$ (AIS) and $CuInS_2$ (CIS), have been developed recently to replace the toxic Cd- and Pb-contained NCs as fluorescence label and probes for biomedical and biological investigations [1–4]. These semiconductor nanomaterials have many advantages, such as high absorption coefficient, low toxicity, high energy conversion efficiency, and stability to solar radiation [5]. Therefore, they have great potential applications in photoelectric conversion and photocatalysis. These NCs have also been used in light-emitting diodes [6] and solar cells [7]. I-III-VI_2-based ternary and quaternary semiconductors, such as $CuInSe_2$ [8], $CuGaSe_2$ [9], $CuGa_xIn_{2-x}S_{3.5}$ [10], and $CuGa_{1-x}In_xSe_2$ [11], have been synthesized. In recent years, ternary AIS has become one of the research focuses in quantum dot synthesis and application investigation. The AIS nanoparticles possess a tunable band gap from 1.87 to 2.03 eV, which matches the solar spectrum very well and does not contain elemental Se. It can be used as a photoelectric conversion material to fabricate high-efficiency solar cells [12]. In addition, the fluorescence emission peak of AIS is also adjustable from visible to near infrared, and it possesses good photostability and large absorption coefficient, making it suitable for lighting and bio-imaging in vivo and in vitro [13,14].

AIS nanocrystals have been synthesized by several methods, such as thermal decomposition of sulfur-containing complexes of different metals or direct reaction of sulfur with metal ions in the presence of coating agents [15]. In the synthetic reaction, metal ions usually form metal complexprecursors with amines or dodecyl mercaptan to ensure controlled growth during the synthesis process. Sulfur powder, mercaptan, dithiocarbamate, or carbon disulfide was used as sulfur sources. Metal mercaptan and dithiocarbamate were the single precursors to provide both metal ions and sulfur elements. The optical properties of bulk $AgInS_2$ were studied in 1976 [16]. Redjai et al. [17] investigated the donor–acceptor pair transition in bulk $AgInS_2$ by time-resolved spectroscopy, and obtained the thermal quenching curve of its emission. The effect of defects in bulk and $AgInS_2$ film, such as vacancies and interstitial atoms, has also been carefully tested [18–20]. Recently, it has been observed that $AgInS_2$ nanocrystals have wide fluorescence emission peaks and relatively large Stokes shifts. It has been suggested that its fluorescence emission was caused by carrier recombination captured by energy levels in the band gap formed by structural defects. It was the characteristic of donor–acceptor pair transition [21].

However, the AIS core has relatively lower photoluminescence (PL) quantum yields (QYs) [22]. Over the last decade, the quantum yields of AIS have been enhanced by two approaches. The first method is the synthesis of core–shell structure coated with a ZnS shell [23–25]. More reports demonstrated the improvement of QYs for core–shell AIS NCs with smaller cores synthesized at lower reaction temperatures. However, smaller cores decreased the crystallinity and increased surface defects, leading to a cumbersome experimental procedure [26]. The second strategy for increasing QYs and tuning photoluminescence (PL) of AIS NCs involved alloyed AIS with ZnS to form Zn-Ag-In-S solid solutions at a higher temperature [26–28]. A series of $(AgIn)_xZn_{2(1-x)}S_2$ solid solution materials with different compositions have also been reported. Unlike the $AgInS_2$/ZnS core–shell structure, in solid solution nanocrystals, Zn was doped into the crystal structure and participated in the formation of conduction bands. The density functional theory (DFT) results showed that the 4d orbital of Ag and the 3p orbital of S formed valence bands of $(AgIn)_xZn_{2(1-x)}S_2$ [29]. The 5s5p orbital of In and 4s4p orbital of Zn participated in the formation of conduction bands. The highest occupied molecular orbital (HOMO) and the lowest unoccupied molecular orbital (LUMO) orbits of solid solution crystals are located between ZnS and $AgInS_2$. By changing the ratio of $AgInS_2$ to ZnS in solid solution, the valence band and conduction band can be changed continuously. Increasing the amount of Zn in AIS directly resulted in a large blue shift of the PL emissions [30], which is ascribed to the increase of the band gap caused by the zinc being incorporated into the AIS core. The alloyed layer abated the lattice mismatch between AIS and ZnS, reduced surface defects, and improved PL properties. Therefore, the band gap and PL emission of AIS-based NCs can be conveniently adjusted by changing their composition and size.

So far, a large number of reports have focused on the preparation and the application of Zn-Ag-In-S NCs. Most of the synthetic methods concentrated on the thermolysis of $(AgIn)_xZn_{2(1-x)}(S_2CN(C_2H_5)_2)_4$ precursors to form $(AgIn)_xZn_{2(1-x)}S_2$ NCs [31–34]. In these reports, the molar ratio of (Zn + Ag) to In was varied to achieve the different PL emissions, in which the relative dose of Zn and Ag remained constant. In our previous work [35], high-quality non-stoichiometric $Zn_xCu_yInS_{1.5+x+0.5y}$ NCs were synthesized using a one-pot strategy, which exhibited excellent PL QYs and tunable PL in 450–640 nm by holding Zn (mole% = 20–35%), In, and S constant and changing the mole percentage of Cu in the precursor. For the prepared $(ZnCu)_xInS_2$ nanoparticles in the same reaction condition, the PL emissions were in the range of 560–640 nm and the PL intensities were lower than those of the Zn constant (25%) samples. Here, we extend this one-pot strategy to fabricate the non-stoichiometric $Zn_xAg_yInS_{1.5+x+0.5y}$ nanocrystals (ZAIS NCs). This is the first exploration of the synthesis of ZAIS NCs with Zn constant (25%) and a different molar ratio of Ag to In, which exhibits composition-dependent photoluminescence properties with high PL QYs and tunable PL wavelength.

2. Materials and Methods

$In(Ac)_3$ (99.99%), sulfur powder (S, 99.99%), $Zn(Ac)_2$ (99.99%), Ag(Ac) (99.99%), oleic acid (OA, 90%), octadecene (ODE, 90%), and 1-dodecanthiol (DDT, 99.9%) were purchased from Sigma-Aldrich (St. Louis, MO, USA). Nitric acid (70%), methanol (AR), and chloroform (AR) were purchased from Beijing Chemical Reagent Ltd., Beijing, China. All chemicals were used without further purification.

$Zn_xAg_yInS_{1.5+x+0.5y}$ (ZAIS) NCs were prepared using our previous methods [35] with a slight modification. For a typical synthetic reaction, 0.01 mmol Ag(Ac), 0.025 mmol $Zn(Ac)_2$, 0.1 mmol $In(Ac)_3$, 1 mmol DDT, 0.4 mmol OA, and 4 g of ODE were loaded in a round bottom flask. The mixture was heated at 180 °C for 30 min to form a clear solution and then kept at 260 °C for 10 min. A sulfur solution of ODE (0.3 mmol)was injected into the three-neck flask and heated for 10 min at 260 °C, followed by cooling to room temperature. The precipitation was obtained by centrifugation at 13,000 rpm for 15 min after adding 8-fold volume of methanol to the reaction solution and then the precipitation was redispersed in chloroform. The purification approach ofthe ZAIS NCs was carried out by repeating the above process several times. Changing the Ag dosage in the precursor, a series of ZAIS NCs with different Ag content (including 0.015, 0.01, 0.005, 0.0025, 0.00125, 0.000625, and 0 mmol, and denoted ZAIS-1 to ZAIS-7, respectively) was obtained. The chemical compositions of ZAIS-1 to ZAIS-7 were carried by inductively coupled plasma optical emission spectrometry/mass spectrometry (ICP-OES/MS, Thermo Jarrell-Ash Corporation, Franklin, MA, USA). The purified ZAIS NCs were first digested by nitric acid. The concentrations of In^{3+}, Ag^{+}, and Zn^{2+} ions in the clear digestion solutions were determined directly by ICP-OES/MS after being diluted with distilled water. The relative PL quantum yield of the ZAIS NCs was determined by comparing the integrated emission of the samples to that of Rhodamine 6G (Rhodamine 6G in ethanol, QY = 95%) solutions with the same optical density at the excitation wavelength and similar fluorescence wavelength.

Transmission electron microscope (TEM, Hillsboro, OR, USA) and high-resolution transmission electron microscope (HRTEM, Hillsboro, OR, USA) images were examined on an FEI Tecnai G2 F20 using 200 kV acceleration voltage. ZAIS NCs were dispersed in chloroform and then dropped on carbon-coated copper grids. X-ray powder diffraction (XRD) patterns were obtained from a Rigaku D/max-2500 X-ray diffractometer (Rigaku Corporation, Tokyo, Japan) with Cu Kα radiation (λ = 1.5418 Å). UV-vis absorption spectra were performed on a Cary 50 spectrophotometer (Varian, Salt Lake City, UT, USA). Fluorescence spectra were recorded on a LS55 luminescence spectrometer (PerkinElmer, Waltham, MA, USA). A Lecroy Wave Runner 6100 digital oscilloscope (Teledyne LeCroy, Chestnut Ridge, NY, USA) was employed to obtain the fluorescence decay curve. A Continuum Sunlite optical parametric oscillator (Teledyne LeCroy, Chestnut Ridge, NY, USA) was the excitation source with tunable laser (pulse width = 4 ns, gate = 50 ns).

3. Results and Discussion

The XRD patterns of ZAIS-1, ZAIS-2, and ZAIS-3 areshown in Figure 1a. The broad peaks of the samples reveal the small-size nature of the NCs. Three diffraction peaks at 28.3, 47.2, and 55.2° areobserved between the peaks of $AgInS_2$ and ZnS, and are practically located at the same diffraction angles of bulk ZnS. These diffraction peaks were attributed to the (111), (220), and (311) planes [28], respectively. This indicated that the XRD diffraction peaks did not originate from the mixture of $AgInS_2$ and ZnS. These results demonstrated that the ZAIS NCs were not a simple mixture of $AgInS_2$ and ZnS but the compositional homogeneous $Zn_xAg_yInS_{1.5+x+0.5y}$ solid solution [35]. Moreover, ZAIS-1, ZAIS-2, and ZAIS-3 presented similar diffraction patterns even though the Ag content was different in the precursor. These results indicated that the ZAIS NCs had monocrystalline structures which coincided with the HRTEM measurements (Figure 1c). One of the high-resolution TEM images of ZAIS-2 is shown in Figure 1c. It presents the continuous lattice fringes, which indicate the high crystallinity of the ZAIS-2 NCs. The TEM photographs (Figure 1b) of ZAIS-2 show that the nanoparticles are approximately spherical in shape and comparatively monodisperse. Average particle sizes of the ZAIS samples were 2.8–3.1 nm. The chemical compositions of the ZAIS NCs were determined by ICP-MS analysis. The analysis results areshown in Table 1. They indicate that the compositions of purified ZAIS NCs can be precisely controlled, which conforms particularly well to the cationic concentration in the precursor.

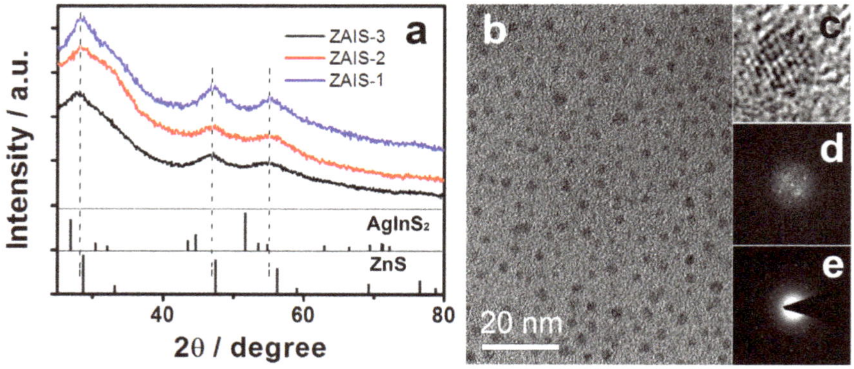

Figure 1. (**a**) X-ray powder diffraction (XRD) patterns of $Zn_xAg_yInS_{1.5+x+0.5y}$ nanocrystals (ZAIS NCs) (ZAIS-1, ZAIS-2, and ZAIS-3), (**b**) transmission electron microscope (TEM), (**c**) high-resolution transmission electron microscope (HRTEM) micrograph, (**d**) corresponding inverse fast fourier transform (FFT) pattern, and (**e**) energy dispersive X-ray spectrometry (EDX) analysis of ZAIS-2. The XRD patterns of $AgInS_2$ (JCPDS 25-1330) and ZnS (JCPDS 65-1691) are shown as reference.

The absorption onsets of ZAIS (Figure 2a) were blue shifted from ZAIS-1 to ZAIS-7, indicating thatthe band gap energy (E_g) of the ZAIS NCs was augmented with the decreasing Ag content. In the serial ZAIS NCs, while keeping In, S, and Zn (mole% = 20–35%) constant and changing the mole ratios of Ag in the precursor, the PL emission wavelengths were tunable from 450 to 700 nm. The PL wavelengths and corresponding chemical compositions of the ZAIS samples are shown in Table 1. The PL and absorption spectra of the ZAIS NCs showed a significant blue shift with the decrease of Ag content. As there were no obvious differences inthe average particle sizes of the ZAIS samples, these results fully revealed the composition-dependent photoluminescence properties of the ZAIS NCs. The onset wavelength varied from 620 to 450 nm. These results agreed well with the phenomena reported in our previous work [30]. The PL peaks of the ZAIS NCs exhibited full width at half maximum around 150 nm. They were remarkably broadened as compared to CdSe but identical to AIS-based NCs. The size-selective precipitation approach [22] was adopted to reduce the particle size distribution of the ZAIS NCs. However, the broadened features of the PL peaks were totally maintained, similar to $CuInS_2$ NCs. This appeared to result from the donor–acceptor pair radiative recombination of charge carriers generated by surface defects and vacancies. The relative PL QYs of the ZAIS NCs are also shown in Table 1. The highest QY reached 35% for ZAIS-4. Most QYs of the ZAIS NCs were higher than those of AIS NCs reported in the literature.

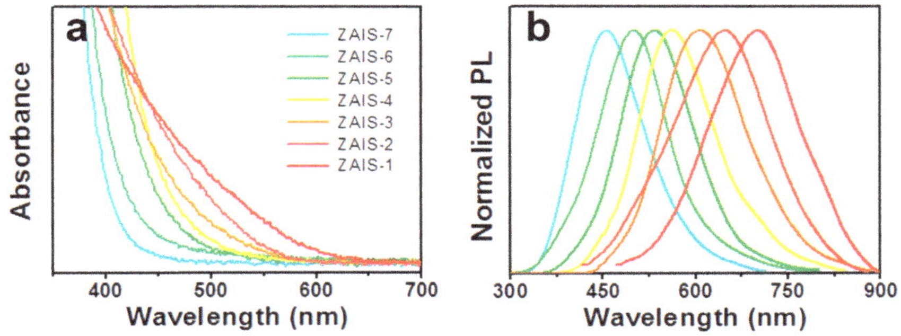

Figure 2. (**a**) UV-vis absorption spectra and (**b**) PL spectra of the ZAIS NCs (ZAIS-1 to ZAIS-7) in chloroform.

Table 1. Chemical compositions of the ZAIS NCs (ZAIS-1 to ZAIS-7) determined by inductively coupled plasma mass spectrometry (ICP-MS) analysis (normalized according to indium ion concentrations).

Sample	PL Emission Wavelength (nm)	Relative Quantum Yield (%)	Silver Dosage in the Precursor y (%)	Composition of the Nanoparticle	
				Ag (%)	Zn (%)
ZAIS-1	700.5	8	15.0	14.1	22.9
ZAIS-2	647.5	15	10.0	9.12	23.1
ZAIS-3	607.0	30	5.00	4.47	24.9
ZAIS-4	561.0	35	2.50	2.33	24.1
ZAIS-5	533.5	23	1.25	1.15	23.5
ZAIS-6	500.5	11	0.625	0.564	24.1
ZAIS-7	456.0	5	-	-	24.4

I-III-VI_2-based ternary and quaternary semiconductor NCs, including $AgInS_2$, $CuInS_2$, and Cu(In,Ga)Se_2 reported previously, had no sharp exciton band. They all exhibited a broad and large absorption band in the visible to near-IR wavelength region. The displayed broad absorption spectra were also afeature of the ZAIS NCs, regardless of their particle sizes and chemical compositions [23]. The broad and large absorption band of the ZAIS NCs signified a high light absorption coefficient in the solar spectrum and highly efficient light energy conversion. Meanwhile, the large full width at half maximum resulted from the donor–acceptor pair radiative recombination of charge carriers generated by surface defects and vacancies. The precise adjustment of the ZAIS NCs could be advantageous to the synthetic control of surface defects, vacancies, and electronic energy structure and the further control of exciton behavior [36]. By precisely controlling the chemical composition of the ZAIS NCs, the effective transfer of photogenerated electrons and holes was induced, and the PL efficiency of the ZAIS NCs was improved.

The chemical compositions, PL emission wavelength, and QY of ZAIS-4 NCs were significantly influenced by the concentration of an organic capping agent in the precursor. The effects of different DDT dosages on the chemical composition and PL emission properties of the solid solution were investigated and are displayed in Table 2. DDT has been documented as an effective capping agent for adjusting the reactivity of metal ions [22]. The chemical composition of the purified ZAIS-4 NCs was decided by the dosage of DDT. The Ag content in the ZAIS-4 NCs was increased with the increase of the DDT dosage in the precursor (shown in Table 2). When DDT, in amounts greater or less than 1.0 mmol, was added to the reaction solution, a white precipitate was produced. This suggested that an appropriate DDT concentration was conducive to the formation of a better crystallinity and compositional homogeneity of the ZAIS NCs. The PL emission wavelength of the ZAIS-4 NCs was a blue shift with the decrease of DDT dosage in the precursor, which maybe related to the decrease of Ag content in the ZAIS-4 NCs. As shown in Table 2, the QY of the ZAIS-4 NCs was also significantly influenced by the dosage of DDT, which was the key factor in the synthesis of the ZAIS solid solution. The appropriate proportion of DDT in the precursor could prevent the precipitation in the reaction system, increase the fluorescence quantum yield, and improve their optical properties.

Table 2. Chemical compositions and photoluminescence (PL) emission properties of ZAIS-4 NCs with different 1-dodecanthiol (DDT) dosage in the precursor (normalized according to indium ion concentrations).

Sample	PL Emission Wavelength (nm)	Relative Quantum Yield (%)	DDT Dosage in the Precursor (mmol)	Composition of the Nanoparticle	
				Ag (%)	Zn (%)
Sample1	542.5	4	0.25	1.57	21.5
Sample2	550.0	12	0.5	2.04	23.7
Sample3	561.0	35	1.0	2.33	24.1
Sample4	572.0	18	1.5	2.55	26.2
Sample5	573.5	7	2.0	2.47	25.6

The excitation spectra of $AgInS_2$ and the ZAIS NCs are shown in Figure 3a. Three excitation peaks of ZAIS at about 329 nm, 391 nm, and 459 nm were observed. This coincides with the three peaks of $AgInS_2$, and a new excitation peak appeared between 226 and 319 nm. It was shown that Zn and Ag in the crystals recombined with the conduction band and valence band of the NCs, forming a new structure of the conduction band and valence band energy of the ZAIS NCs. This was similar to that of Zn-Cu-In-S crystals. Meanwhile, the fluorescence spectra of the ZAIS NCs had single band gap characteristics, and these evidences indicated the formation of alloy states rather than simple composite structures of In-S, Ag-S, and Ag-In-S.

Figure 3b shows the PL decay curves for the ZAIS-1, ZAIS-2, and ZAIS-3 NCs. These curves were well fitted by a biexponential function as the form $I(t) = a\exp(-t/\tau_1) + b\exp(-t/\tau_2)$. Based on the curve-fitting method, the lifetimes of 115–148 ns (τ_1) and 455–483 ns (τ_2) were determined (see Table 3). The average lifetimes calculated according to the formula in the literature [37] were 194, 234, and 237 ns for Ag content of 0.015, 0.01, and 0.005 mmol in the precursors. The fluorescence lifetimes (τ_1 and τ_2) were analogous to those of AIS and $(AgIn)_xZn_{2(1-x)}S_2$ [38]. The components of 115–148 ns can be assigned to donor–acceptor pair recombination related to surface defects and vacancies. The components of 455–483 ns can be allocated to transitions of intrinsic donor–acceptor pairrecombination [34].

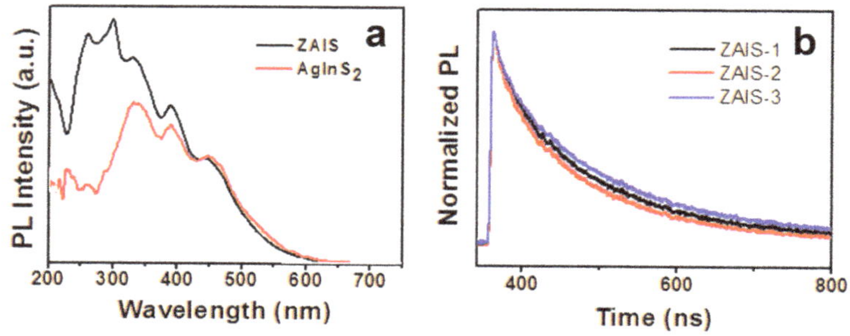

Figure 3. (**a**) PL excitation spectra and (**b**) PL decay curves of ZAIS-1, ZAIS-2, and ZAIS-3.

Table 3. Decay times and amplitude constant ratios of ZAIS-1, ZAIS-2, and ZAIS-3.

Sample	ZAIS-1	ZAIS-2	ZAIS-3	$(AgIn)_xZn_{2(1-x)}S_2$
τ_1/ns	115	146	148	127–131
a	0.423	0.562	0.577	-
τ_2/ns	455	481	483	538–655
b	0.0324	0.0615	0.0641	-
average	194	234	237	-

4. Conclusions

A facile one-pot hot injection approach to successfully synthesize high-quality non-stoichiometric ZAIS NCs in the size range of 2.8–3.1 nm was presented. XRD, HRTEM, and fluorescence spectra showed that Zn, Ag, In, and S formed the alloyed non-stoichiometric solid solution. The fluorescence spectra had single band gap features, and indicated the formation of alloy states rather than simple composite structures. The effects of different reaction conditions on the composition and PL emission properties of the solid solution were investigated. It was found that the appropriate ratio of DDT was beneficial to the formation of alloy state ZAISNCs, which was the key factor in the synthesis of the ZAIS solid solution. The appropriate proportion of DDT can prevent the precipitation in the reaction system, increase the fluorescence quantum yield, reduce their half-peak width, and improve their luminescence properties. The photoluminescence wastunable from blue to red in the range of 450–700 nm since the Zn and Ag contents in the precursor changed. The PL and absorption spectra of the ZAIS NCs showed a significant blue shift with the decrease of Ag content in the precursor. As there were no obvious differences in the average particle sizes of the ZAIS NCs amples, these results fully revealed the composition-dependent photoluminescence properties of the ZAIS NCs. The relative quantum yield reached 35%. The fluorescence lifetimes (τ_1 = 115–148 ns and τ_2 = 455–483 ns) were analogous to those of AIS and $(AgIn)_xZn_{2(1-x)}S_2$. The components of 115–148 ns can be assigned to donor–acceptor pairrecombination related to surface defects and vacancies. The components of 455–483 ns can be allocated to transitions of intrinsic donor–acceptor pair recombination.

Author Contributions: G.F. and R.L. conceived and designed the experiments; J.F. performed the experiments; X.Y. and X.Y. analyzed the data; J.F. wrote the paper. All authors discussed the results and contributed to the manuscript.

Acknowledgments: The authors gratefully acknowledge the financial support of the First-Class Discipline Construction Project of Guizhou Province - Public Health and Preventive Medicine (No. 2017[85]) and the Science and Technology Foundation of Guizhou, China (No. [2016]7366).

Conflicts of Interest: The authors declare no conflict of interest.

References

1. Guo, W.S.; Chen, N.; Dong, C.H.; Tu, Y.; Chang, J.; Zhang, B.B. One-pot synthesis of hydrophilic ZnCuInS/ZnS quantum dots for in vivo imaging. *RSC Adv.* **2013**, *3*, 9470–9475. [CrossRef]
2. Zhang, B.T.; Wang, Y.C.; Yang, C.B.; Hu, S.Y.; Gao, Y.; Zhang, Y.P.; Wang, Y.; Demir, H.V.; Liu, L.W.; Yong, K.T. The composition effect on the optical properties of aqueous synthesized Cu-In-S and Zn-Cu-In-S quantum dot nanocrystals. *Phys. Chem. Chem. Phys.* **2015**, *17*, 25133–25141. [CrossRef] [PubMed]
3. Guo, W.S.; Chen, N.; Tu, Y.; Dong, C.H.; Zhang, B.B.; Hu, C.H.; Chang, J. Synthesis of Zn-Cu-In-S/ZnS core/shell quantum dots with inhibited blue-shift photoluminescence and applications for tumor targeted bioimaging. *Theranostics* **2013**, *3*, 99–108. [CrossRef] [PubMed]
4. Nakamura, H.; Kato, W.; Uehara, M.; Nose, K.; Omata, T.; Otsuka-Yao-Matsuo, S.; Miyazaki, M.; Maeda, H. Tunable photoluminescence wavelength of chalcopyrite $CuInS_2$-based semiconductor nanocrystals synthesized in a colloidal system. *Chem. Mater.* **2006**, *18*, 3330–3335. [CrossRef]
5. Lewerenz, H.J. Development of copperindiumdlsulfide into a solar material. *Sol. Energy Mater. Sol. Cells* **2004**, *83*, 395–407. [CrossRef]

6. Yoon, H.C.; Oh, J.H.; Ko, M.; Yoo, H.; Do, Y.R. Synthesis and characterization of green Zn-Ag-In-S and red Zn-Cu-In-S quantum dots for ultrahigh color quality of down-converted white LEDs. *ACS Appl. Mater. Interfaces* **2015**, *7*, 7342–7350. [CrossRef]
7. Pan, Z.X.; Mora-Sero, I.; Shen, Q.; Zhang, H.; Li, Y.; Zhao, K.; Wang, J.; Zhong, X.H.; Bisquert, J. High-efficiency "green" quantum dot solar cells. *J. Am. Chem. Soc.* **2014**, *136*, 9203–9210. [CrossRef]
8. Allen, P.M.; Bawendi, M.G. Ternary I-III-VI quantum dots luminescent in the red to near infrared. *J. Am. Chem. Soc.* **2008**, *130*, 9240–9241. [CrossRef]
9. Tang, J.; Hinds, S.; Kelley, S.O.; Sargent, E.H. Synthesis of colloidal $CuGaSe_2$, $CuInSe_2$, and $Cu(InGa)Se_2$ nanoparticles. *Chem. Mater.* **2008**, *20*, 6906–6910. [CrossRef]
10. Pan, D.C.; Wang, X.L.; Zhou, Z.H.; Chen, W.; Xu, C.L.; Lu, Y.F. Synthesis of quaternary semiconductor nanocrystals with tunable band gaps. *Chem. Mater.* **2009**, *21*, 2489–2493. [CrossRef]
11. Sun, C.; Gardner, J.S.; Long, G.; Bajracharya, C.; Thurber, A.; Punnoose, A.; Rodriguez, R.G.; Pak, J.J. Controlled stoichiometry for quaternary $CuIn_xGa_{1-x}S_2$ chalcopyrite nanoparticles from single-source precursors via microwave irradiation. *Chem. Mater.* **2010**, *22*, 2699–2701. [CrossRef]
12. Kobosko, S.M.; Jara, D.H.; Kamat, P.V. $AgInS_2$-ZnS quantum dots: Excited state interactions with TiO_2 and photovoltaic performance. *ACS Appl. Mater. Interfaces* **2017**, *9*, 33379–33388. [CrossRef] [PubMed]
13. Shinchi, H.; Wakao, M.; Nagata, N.; Sakamoto, M.; Mochizuki, E.; Uematsu, T.; Kuwabata, S.; Suda, Y. Cadmium-free sugar-chain-immobilized fluorescent nanoparticles containing low-toxicity ZnS-$AgInS_2$ cores for probing lectin and cells. *Bioconjugate Chem.* **2014**, *25*, 286–295. [CrossRef] [PubMed]
14. Deng, D.W.; Cao, J.; Qu, L.Z.; Achilefu, S.; Gu, Y.Q. Highly luminescent water-soluble quaternary Zn-Ag-In-S quantum dots for tumor cell-targeted imaging. *Phys. Chem. Chem. Phys.* **2013**, *15*, 5078–5083. [CrossRef] [PubMed]
15. Tian, L.; Vittal, J.J. Synthesis and characterization of ternary $AgInS_2$ nanocrystals by dual and multiple-source methods. *New J. Chem.* **2007**, *31*, 2083–2087. [CrossRef]
16. Okamoto, K.; Kinoshita, K. Electrical and optical-properties of $AgInS_2$. *Solid-State Electron.* **1976**, *19*, 31. [CrossRef]
17. Redjai, E.; Masse, G. Donor-Acceptor Pair Transitions in $AgInS_2$. *Phys. Status Solidi B* **1985**, *131*, K157–K159. [CrossRef]
18. Uematsu, T.; Doko, A.; Torimoto, T.; Oohora, K.; Hayashi, T.; Kuwabata, S. Photoinduced electron transfer of ZnS-$AgInS_2$ solid-solution semiconductor nanoparticles: emission quenching and photocatalytic reactions controlled by electrostatic forces. *J. Phys. Chem. C* **2013**, *117*, 15667–15676. [CrossRef]
19. Masse, G.; Redjai, E. S-Vacancy energy-levels in $AgInS_2$. *J. Appl. Phys.* **1986**, *59*, 1544–1547. [CrossRef]
20. Krustok, J.; Raudoja, J.; Krunks, M.; Mändar, H.; Collan, H. Nature of the native deep localized defect recombination centers in the chalcopyrite and orthorhombic $AgInS_2$. *J. Appl. Phys.* **2000**, *88*, 205–209. [CrossRef]
21. Hamanaka, Y.; Ogawa, T.; Tsuzuki, M.; Kuzuya, T. Photoluminescence properties and its origin of $AgInS_2$ quantum dots with chalcopyrite structure. *J. Phys. Chem. C* **2011**, *115*, 1786–1792. [CrossRef]
22. Xie, R.; Rutherford, M.; Peng, X. Formation of high-quality I-III-VI semiconductor nanocrystals by tuning relative reactivity of cationic precursors. *J. Am. Chem. Soc.* **2009**, *131*, 5691–5697. [CrossRef] [PubMed]
23. Xiang, W.; Xie, C.; Wang, J.; Zhong, J.; Liang, X.; Yang, H.; Luo, L.; Chen, Z. Studies on highly luminescent $AgInS_2$ and Ag-Zn-In-S quantum dots. *J. Alloy. Compd.* **2014**, *588*, 114–121. [CrossRef]
24. Hong, S.P.; Park, H.K.; Oh, J.H.; Yang, H.; Do, Y.R. Comparisons of the structural and optical properties of o-$AgInS_2$, t-$AgInS_2$, and c-$AgIn_5S_8$ nanocrystals and their solid-solution nanocrystals with ZnS. *J. Mater. Chem.* **2012**, *22*, 18939–18949. [CrossRef]
25. Song, J.; Jiang, T.; Guo, T.; Liu, L.; Wang, H.; Xia, T.; Zhang, W.; Ye, X.; Yang, M.; Zhu, L. Facile synthesis of water-soluble Zn-Doped $AgIn_5S_8$/ZnS core/shell fluorescent nanocrystals and their biological application. *Inorg. Chem.* **2015**, *54*, 1627–1633. [CrossRef]
26. Ko, M.; Yoon, H.C.; Yoo, H.; Oh, J.H.; Yang, H.; Do, Y.R. Highly efficient green Zn-Ag-In-S/Zn-In-S/ZnS QDs by a strong exothermic reaction for down-converted green and tripackage white LEDs. *Adv. Funct. Mater.* **2017**, *27*, 1602638. [CrossRef]
27. Liu, X.; Chen, H.; Wang, R.; Shang, Y.; Zhang, Q.; Li, W.; Zhang, G.; Su, J.; Dinh, C.T.; Li, J. 0D-2D quantum dot: Metal dichalcogenide nanocomposite photocatalyst achieves efficient hydrogen generation. *Adv. Mater.* **2017**, *29*, 1605646. [CrossRef]

28. Kameyama, T.; Takahashi, T.; Machida, T.; Kamiya, Y.; Yamamoto, T.; Kuwabata, S.; Torimoto, T. Controlling the electronic energy structure of ZnS-AgInS_2 solid solution nanocrystals for photoluminescence and photocatalytic hydrogen evolution. *J. Phys. Chem. C* **2015**, *119*, 24740–24749. [CrossRef]
29. Tsuji, I.; Kato, H.; Kobayashi, H.; Kudo, A. photocatalytic H_2 evolution under visible-light irradiation over band-structure-controlled $(CuIn)_xZn_{2(1-x)}S_2$ solid solutions. *J. Phys. Chem. B* **2005**, *109*, 7323–7329. [CrossRef]
30. Park, J.Y.; Kim, S.W. CuInS_2/ZnS core/shell quantum dots by cation exchange and their blue shifted photoluminescence. *J. Mater. Chem.* **2011**, *21*, 3745–3750. [CrossRef]
31. Torimoto, T.; Adachi, T.; Okazaki, K.; Sakuraoka, M.; Shibayama, T.; Ohtani, B.; Kudo, A.; Kuwabata, S. Facile Synthesis of ZnS-AgInS_2 Solid solution nanoparticles for a color adjustable luminophore. *J. Am. Chem. Soc.* **2007**, *129*, 12388–12389. [CrossRef] [PubMed]
32. Ogihara, Y.; Yukawa, H.; Kameyama, T.; Nishi, H.; Onoshima, D.; Ishikawa, T.; Torimoto, T.; Baba, Y. Labeling and in vivo visualization of transplanted adipose tissue-derived stem cells with safe cadmium-free aqueous ZnS coating of ZnS-AgInS_2 nanoparticles. *Sci. Rep.* **2017**, *7*, 40047. [CrossRef] [PubMed]
33. Chevallier, T.; Blevennec, G.L.; Chandezon, F. Photoluminescence properties of AgInS_2-ZnS nanocrystals: The critical role of the surface. *Nanoscale* **2016**, *8*, 7612–7620. [CrossRef] [PubMed]
34. Sharma, D.K.; Hirata, S.; Bujak, L.; Biju, V.; Kameyama, T.; Kishi, M.; Torimotoc, T.; Vacha, M. Influence of Zn on the photoluminescence of colloidal $(AgIn)_xZn_{2(1-x)}S_2$ nanocrystals. *Phys. Chem. Chem. Phys.* **2017**, *19*, 3963–3969. [CrossRef] [PubMed]
35. Feng, J.; Sun, M.; Yang, F.; Yang, X.R. A facile approach to synthesize high-quality $Zn_xCu_yInS_{1.5+x+0.5y}$ nanocrystal emitters. *Chem. Commun.* **2011**, *47*, 6422–6424. [CrossRef] [PubMed]
36. Pu, C.; Qin, H.; Gao, Y.; Zhou, J.; Wang, P.; Peng, X. Syntheticcontrolofexcitonbehaviorincolloidalquantumdots. *J. Am. Chem. Soc.* **2017**, *139*, 3302–3311. [CrossRef] [PubMed]
37. Torimoto, T.; Kamiya, Y.; Kameyama, T.; Nishi, H.; Uematsu, T.; Kuwabata, S.; Shibayama, T. controlling shape anisotropy of ZnS-AgInS_2 solid solution nanoparticles for improving photocatalytic activity. *ACS Appl. Mater. Interfaces* **2016**, *8*, 27151–27161. [CrossRef] [PubMed]
38. You, S.H.; Hong, K.J.; Youn, C.J.; Jeong, T.S.; Moon, J.D.; Kim, H.S.; Park, J.S. Origin of point defects in AgInS_2/GaAs epilayer obtained from photoluminescence measurement. *J. Appl. Phys.* **2001**, *90*, 3894–3898. [CrossRef]

micromachines

MDPI

Article

Effects of Charge Transport Materials on Blue Fluorescent Organic Light-Emitting Diodes with a Host-Dopant System

Neng Liu [1], Sijiong Mei [1], Dongwei Sun [1], Wuxing Shi [1], Jiahuan Feng [1], Yuanming Zhou [1,*], Fei Mei [1,*], Jinxia Xu [1], Yan Jiang [1] and Xianan Cao [2]

1 Hubei Key Laboratory for High-efficiency Utilization of Solar Energy and Operation Control of Energy Storage System, Hubei University of Technology, Wuhan 430068, China; 13618655764@163.com (N.L.); meisijiong@163.com (S.M.); sdwwant@163.com (D.S.); 18971141948@163.com (W.S.); 17764283212@163.com (J.F.); xujx@mail.hbut.edu.cn (J.X.); yanjiang5909@126.com (Y.J.)

2 Department of Computer Science and Electrical Engineering, West Virginia University, Morgantown, WV 26506, USA; xacao@mail.hbut.edu.cn

* Correspondence: zhouym@mail.hbut.edu.cn (Y.Z.); meifei777@163.com (F.M.); Tel.: +86-27-5975-0430 (Y.Z. & F.M.)

Received: 19 April 2019; Accepted: 23 May 2019; Published: 25 May 2019

Abstract: High efficiency blue fluorescent organic light-emitting diodes (OLEDs), based on 1,3-bis(carbazol-9-yl)benzene (mCP) doped with 4,4′-bis(9-ethyl-3-carbazovinylene)-1,1′-biphenyl (BCzVBi), were fabricated using four different hole transport layers (HTLs) and two different electron transport layers (ETLs). Fixing the electron transport material TPBi, four hole transport materials, including 1,1-Bis[(di-4-tolylamino)phenyl]cyclohexane (TAPC), N,N′-Di(1-naphthyl)-N,N′-diphenyl-(1,1′-biphenyl)-4′-diamine(NPB), 4,4′-Bis(N-carbazolyl)-1,1,-biphenyl (CBP) and molybdenum trioxide (MoO_3), were selected to be HTLs, and the blue OLED with TAPC HTL exhibited a maximum luminance of 2955 cd/m^2 and current efficiency (CE) of 5.75 cd/A at 50 mA/cm^2, which are 68% and 62% higher, respectively, than those of the minimum values found in the device with MoO_3 HTL. Fixing the hole transport material TAPC, the replacement of TPBi ETL with Bphen ETL can further improve the performance of the device, in which the maximum luminance can reach 3640 cd/m^2 at 50 mA/cm^2, which is 23% higher than that of the TPBi device. Furthermore, the lifetime of the device is also optimized by the change of ETL. These results indicate that the carrier mobility of transport materials and energy level alignment of different functional layers play important roles in the performance of the blue OLEDs. The findings suggest that selecting well-matched electron and hole transport materials is essential and beneficial for the device engineering of high-efficiency blue OLEDs.

Keywords: blue organic light emitting diodes; transport materials; host-dopant

1. Introduction

Organic light-emitting diodes (OLEDs) are constructed with several organic or inorganic layers between the anode and cathode, and have attracted great interest throughout the world owing to their advantages, such as fast response time, high contrast ratio, wide view angle, and low power consumption [1–10]. In addition, their remarkable ability to fabricate on flexible substrates is the most significant characteristic that facilitates the fabrication of OLED displays with various shapes and sizes, which is hardly implemented by using other existing technologies [5]. High-performance OLEDs with primary RGB colors are essential for developing high-quality full-color displays and white-light emission. However, state-of-the-art blue OLEDs have relatively poor emission performance compared with other red and green counterparts, in terms of luminous efficiency, color purity, and operational lifetime. Therefore, a lot of efforts have been made to improve the emission performance of blue OLEDs. It is widely accepted that a guest–host system is a useful method to improve emission characteristics of blue devices [4,6,11,12]. However, most of the research has been focused on the type and concentration of dopants, with only a few studies concentrated on the cooperation of hole and electron transport layers. Several papers about blue fluorescent OLEDs based on a host doped with BCzVBi have been reported, and a suitable host is expected to enable efficient energy transfer to the guest, giving rise to efficient blue luminescence [11,12].

Many efforts have been made to discuss the effects of hole and electron transport materials on the characteristics of OLED devices [13–20]. Jou et al. reported that exciton recombination and electric field distribution across the emission layer can be affected by charge transport materials, which play important roles in the luminance and turn-on voltage of devices [14]. Liu et al. [15] discussed the impact of electron transport materials on devices' stability, and Giebeler et al. [16] compared the effects of different hole transport materials on the emission characteristics. Although several groups have studied the effect of electron and hole transport materials on the performance of blue OLED devices, including efficiency and lifetime, quite a few of them systematically investigated the selection of appropriate hole transport material (HTM) and electron transport material (ETM) and analyzed the limiting factors to fabricate high-efficiency OLED devices with low turn-on voltage.

The energy levels and charge carrier mobility of charge transport materials are intimately related to the efficiency and operating voltage of devices. Since efficient injection of charges into emission layer (EML) leads to a high space charge limited current flow that reduces the operating voltage, it is suggested that HTM and ETM should have a high charge carrier mobility [14]. Meanwhile, only one kind of carrier with high mobility is usually unfavorable for efficiency. The mobility of holes is often higher than that of electrons, which will reduce the recombination of holes and electrons, and thus OLED devices introducing ETM and HTM with nearly equivalent carrier mobility should be employed. Therefore, it is necessary to make a proper medium to obtain the desired current efficiency and turn-on voltage, as well as other properties, via the selection of appropriate HTM and ETM, which can facilitate the injection and balanced transport of holes and electrons.

In this paper, the influence of charge transport materials on the performance of bis(9-carbazolyl)benzene (mCP)-BCzVBi-based OLEDs was studied in detail, especially in the aspect of efficiency and working voltage. The luminous characteristics of these devices were measured and compared with each other. In addition, the behavior of holes and electrons was discussed in detail, to re-examine the relationship between charge transport materials and devices' performance. From the above experiments and analysis, we hope that a more excellent performance of blue OLED devices can be obtained by selecting well-matched ETM and HTM materials.

2. Experiments

Blue OLEDs with a mCP:20wt.% BCzVBi host–guest system were fabricated on pre-patterned commercial indium-tin-oxide (ITO) coated glass substrates, whose sheet resistance was around 15 Ω/□. Typically, the substrates were ultrasonically cleaned in acetone, methanol, and deionized water for 5 min sequentially. Then, the substrates were dried with a nitrogen gun and exposed to oxygen plasma for 5 min, which can effectively modify the work function of ITO. After that, the substrates were transferred to a physical vapor thermal evaporation system with a base pressure of $\sim 3 \times 10^{-7}$ Torr. All functional layers, including organic, inorganic, and metal materials, were deposited on unheated substrates with different rates less than 0.1 nm/s, which guaranteed the film quality. The blue OLEDs had a 30 nm thick light-emitting layer, comprising an mCP host doped with 20 wt.% BCzVBi, sandwiched between a 40 nm thick hole transport layer (HTL) and a 40 nm thick electron transport layer (ETL). In this paper, four kinds of HTLs (1,1-Bis[(di-4-tolylamino)phenyl]cyclohexane (TAPC), N,N'-Di(1-naphthyl)-N,N'-diphenyl-(1,1'-biphenyl)-4'-diamine (NPB), 4,4;-Bis(N-carbazolyl)-1,1,-biphenyl (CBP) and molybdenum trioxide (MoO_3)) and two kinds of ETLs (2,2',2"-(1,3,5-

Benzinetriyl)-tris(1-phenyl-1-H-benzimidazole) (TPBi) and 4,7-Diphenyl-1,10-phenanthroline (Bphen)) were applied in our devices. Finally, a 0.5 nm thick lithium fluoride (LiF) and a 100 nm Al were used for the electron injection layer and cathode, respectively. Each substrate contained four devices, with an effective area of 0.1 cm^2. Figure 1a shows the schematic structure of the blue OLED devices studied in this work. All fabricated samples were encapsulated with a glass lid in a N_2-filled chamber and characterized at room temperature. Figure 1b illustrates the energy level diagram of blue OLEDs, showing the energy levels of the highest occupied molecular orbital (HOMO) and lowest unoccupied molecular orbital (LUMO) of organic materials, and the working functions of the two electrodes, which are referenced in the literature [4–11].

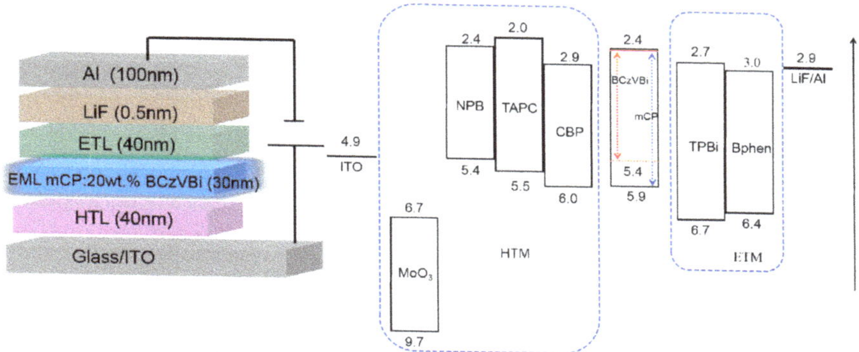

(a) (b)

Figure 1. (**a**) The schematic structure of our blue organic light-emitting diodes (OLEDs). (**b**) The energy level diagram of the blue OLEDs.

The current density–voltage (J–V) characteristics of OLEDs were tested using a Keithley 2400 source meter system and the electroluminescence (EL) spectra of the devices were measured by an Ocean Optics fiber-optic spectrometer. The photoluminescence (PL) spectra were measured by a HITACHI F-4600 luminescence spectrometer with an excitation wavelength of 320 nm. The luminance of the devices was measured by a Keithley 2000 multimeter, coupled with a calibrated silicon photodetector (1 cm in diameter), which was put directly onto the surface of an individual device, ensuring all photons emitted from the glass side were captured. Finally, in order to test the stability of the devices, as-fabricated OLEDs were stressed at a constant current density of 20 mA/cm^2, and the voltage and luminance were collected with a time interval of 5 s. The OLED lifetime was determined according to the luminance evolution profile recorded at 20 mA/cm^2.

3. Results and Discussion

Firstly, the blue OLEDs with the structure of ITO/HTL/mCP:20 wt.% BCzVBi/TPBi/LiF/Al were fabricated and characterized, in which TAPC, NPB, CBP, and MoO_3 were used for HTL. In the following, these four devices are labeled as TAPC device, NPB device, CBP device, and MoO_3 device, for clarity. Figure 2a displays the current density–voltage (J–V) characteristics of blue OLEDs based on different HTMs. As shown in Figure 2a, of these four kinds of devices, the MoO_3-based OLED had the lowest turn-on voltage of ~3.61 V, whereas the CBP-based OLED had the highest value of ~4.82 V. When the current density was set to be 40 mA/cm^2, the operating voltages of the blue OLEDs based on TAPC, NPB, CBP, and MoO_3 were 6.92 V, 7.96 V, 9.8 V, and 6.67 V, respectively. These characteristics can be explained by the hole injection from the ITO anode and the hole transport across the HTL. The former is usually related to the energy level alignment at the HTL/anode and HTL/EML interfaces, and the latter is mainly influenced by the hole mobility of HTMs. It can be seen in Figure 1b that holes injected from ITO to CBP must overcome a big energy barrier of ~1.1 eV at the ITO/HTL interface, while for the devices with NPB and TAPC HTLs, the energy barrier decreases to 0.5 eV and 0.6 eV, respectively. Thus, one can conclude that the NPB and TAPC devices have more efficient hole injection from ITO to HTL than the CBP device. As reported in other articles, the hole mobility of NPB, TAPC, and CBP are 8.8×10^{-4}, 1×10^{-2}, and 2×10^{-3} $cm^2 \cdot V^{-1} \cdot s^{-1}$, respectively [4,11,12,20]. Although the hole mobility of CBP is larger than that of NPB, the operating voltage of the CBP device is higher, suggesting that both hole mobility and energy level alignment play important roles in the J–V characteristics. For the CBP device, high hole mobility leads to high-efficiency hole transport, while the larger energy barrier impedes the hole injection, which finally causes the highest operating voltage among the NPB, TAPC, and CBP devices. In addition, the energy barrier of the TAPC device is 0.1 eV lager than that of the NPB device, while the hole mobility of TAPC is ~10 times higher than that of NPB, leading to the result that the operating voltage of the TAPC device is smaller than that of the NPB device. MoO_3 is widely used as hole injection material. It is also an excellent hole transport material due to its high electron conductivity, good stability, and deep-lying energy levels [19]. As seen in Figure 1b, the energy levels of MoO_3 lie below the HOMO levels of mCP and BCzVBi, which suggests that MoO_3 can accept electrons from mCP and BCzVBi, generating free holes in the EML. This process is equivalent to hole injection and transport from the MoO_3 layer to EML. Moreover, this hole injection and transport process are quicker and more efficient than other conventional devices using NPB, TAPC, and CBP HTLs. Thus, it is reasonable that the MoO_3 device shows the best performance seen in the J–V characteristics.

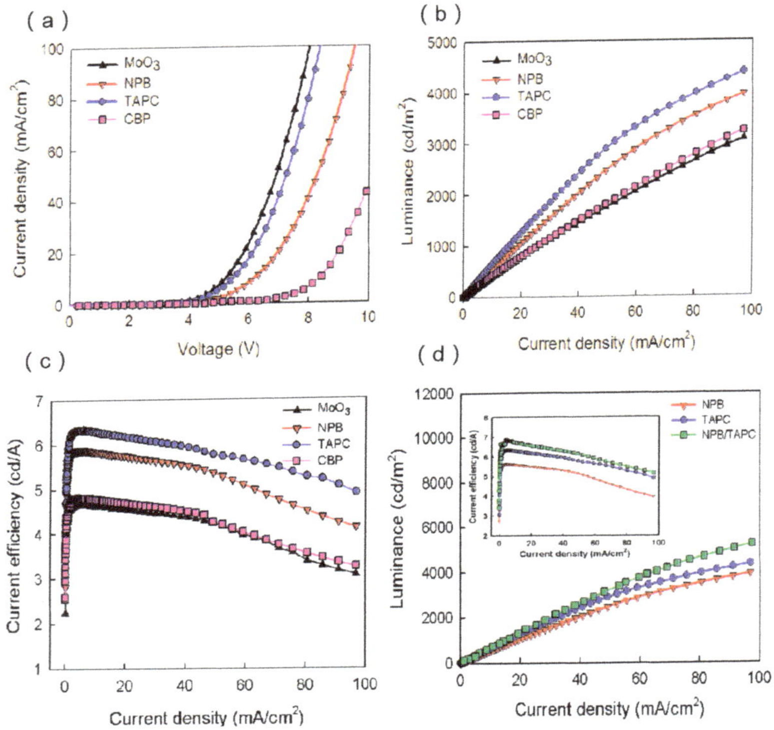

Figure 2. (**a**) Current density versus voltage curves, (**b**) luminance versus current density curves, (**c**) current efficiency versus current density curves of blue OLEDs with 2,2′,2″-(1,3,5-Benzinetriyl)-tris(1-phenyl-1-H-benzimidazole) (TPBi) electron transport layers (ETL) and four different hole transport layers (HTLs). (**d**) Luminance versus current density curves of blue OLEDs with TPBi ETL and N,N′-Di(1-naphthyl)-N,N′-diphenyl-(1,1′-biphenyl)-4′-diamine (NPB), 1,1-Bis[(di-4-tolylamino)phenyl]cyclohexane (TAPC), NPB/TAPC HTLs.

Figure 2b shows the luminance of blue OLEDs as functions of current density. As seen in Figure 2b, the luminance of the TAPC, NPB, CBP, and MoO_3 based OLEDs were 2955, 2558, 1875, and 1760 cd/m^2 at 50 mA/cm^2, respectively. The luminance is strongly related to the number of excitons generated in the EML, and thus the balance between electrons and holes in the EML is of vital importance [9]. The energy barrier between LiF/Al and TPBi is 0.2 eV and the electron mobility of TPBi is ~6.5×10^{-5} cm^2·V^{-1}·s^{-1} [19]. It can be seen that both of the MoO_3 and CBP devices had similar L-J characteristics, in which the luminance was lower than other two devices, since the charge balance was worse. As mentioned above, the hole injection and transport from MoO_3 to EML was very efficient, while the electron transport was relatively slow from TPBi to EML because of the low electron mobility. This may lead to the fact that the number of holes reaching EML is larger than that of the electrons, causing seriously unbalanced holes and electrons, and correspondingly the lower luminance and efficiency. Similarly, the low luminance of the CBP device can be explained by the unbalanced holes and electrons in EML, which originated from the gap between the carrier mobility of CBP (2×10^{-3} cm^2·V^{-1}·s^{-1}) and TPBi (~6.5×10^{-5} cm^2·V^{-1}·s^{-1}). In addition, the low-efficiency hole injection of the CBP device caused by the large energy levels also led to the small number of holes in EML, which was related to the low luminance. For the NPB device, since NPB and TPBi have the similar carrier mobility of ~10^{-5}–10^{-4} cm^2·V^{-1}·s^{-1}, the NPB device can achieve well-balanced holes and electrons in the EML, which could generate more excitons and lead to a higher luminance [15,19]. However, the TAPC device with the large gap between the carrier mobility of TAPC (1×10^{-2} cm^2·V^{-1}·s^{-1}) and TPBi (~6.5×10^{-5} cm^2·V^{-1}·s^{-1}) showed a higher luminance than the NPB device, which does not correlate well to the above reasoning. It is noted that, for the TAPC device, the LUMO levels of mCP and BCzVBi were both 2.4 eV, which is 0.4 eV lower than TAPC. Therefore, there was effective electron blocking at the EML/TAPC interface which is beneficial to the recombination of excitons and the balanced carriers in EML [4]. However, the LUMO level of NPB was well aligned with mCP and BCzVBi, and electrons in EML can escape from EML to NPB, which ultimately reduced the effective number of excitons.

Figure 2c illustrates the current efficiency–current density characteristics based on the four different HTMs. When the current density was 50 mA/cm^2, the current efficiencies of the TAPC, NPB, CBP, and MoO_3 devices were 5.75, 5.04, 3.78, and 3.56 cd/A, respectively. This result is consistent with the luminance discussed above, both indicating that the devices' efficiencies are proportional to the number of excitons generated in the EML region, which depend on the balance between the electrons and holes in the EML. On increasing the current density, the luminance (or the number of excitons) increased rapidly, leading to an increase in the current efficiency. Afterward, the luminance increased more slowly with a further increase in the current density, which caused the decrease in the current efficiency, namely the efficiency roll-off. As reported, the efficiency roll-off of fluorescent doped OLEDs is mainly attributed to charge imbalance, while quenching processes appear to have only a minor role [13,21–24]. Therefore, the higher current efficiency in the TAPC device can be attributed to the balance of the number of electrons and holes in the EML, due to excellent hole injection and transport into the EML.

In LEDs, the current efficiency of devices is proportional to the value of external quantum efficiency (EQE) defined by the equation: $EQE = \eta_{out} r q \gamma$, where η_{out} is the fraction of emitted photons that are coupled out of the device, r is the fraction of excitons that can potentially radiative decay, q is the photoluminescence quantum yield (PLQY) of emitters, and γ is the charge balance ($\gamma \leq 1$) [25,26]. Generally, the internal operation of LEDs does not influence η_{out}, while r and q would be set with emitters. Hence, from the perspective of device engineering, the change of current efficiency with different HTLs is most sensitive to γ. In terms of charge mobility, both of the MoO_3 and CBP devices exhibit the lowest efficiency since their charge balance is the worst among four devices, namely the hole mobility of MoO_3 and CBP HTLs is much higher than the electron mobility of TPBi ETL. While the NPB device exhibits the medium efficiency since NPB and TPBi have the similar carrier mobility of ~10^{-5}–10^{-4} cm^2·V^{-1}·s^{-1}. The TAPC device exhibits the highest efficiency, which does not correlate well to the above reason. It is suggested that the other two factors affecting the balance of carriers, including the low contact barrier energy and effective electron blocking at HTL/EML interface, lead to the high efficiency of the TAPC device. As reported in another literature, the hole and electron mobility of the mCP host is 1.2×10^{-4} and 4×10^{-5} $cm^2 \cdot V^{-1} \cdot s^{-1}$, respectively. [27] The different charge mobilitis of the host may mainly affect the location and width of the recombination zone, governing the behaviors of charge injection and transport in the EML. [28,29]

In order to examine that the effective electron blocking can extend the luminance, we inserted 10 nm TAPC to the NPB/EML interface in the original NPB device. The L-J characteristics of the NPB, TAPC, and NPB/TAPC devices are displayed in Figure 2d. As can be seen, the NPB/TAPC device showed the highest luminance of 3354 cd/m^2 at 50 mA/cm^2, which is higher than those of the NPB and TAPC devices, suggesting the favorable effect of electron blocking caused by TAPC. The inset in Figure 2d is the corresponding current efficiency as functions of current density. Higher current efficiency in the NPB/TAPC device can be attributed to the balance of the number of electrons and holes caused by the effective electron blocking [4].

The normalized electroluminescence (EL) spectra of the blue OLEDs with different HTLs are displayed in Figure 3a. All the devices have a main blue peak at ~485 nm and a shoulder peak at ~445 nm, which indicate the existence of the energy transfer from host material to the guest dopant molecule. The photoluminescence (PL) spectra of the mCP and BCZVBi films were also measured, as shown in Figure 3b. The PL peak of BCzVBi is at ~485 nm and the PL peaks of mCP are at ~445 nm and 433 nm, which is consistent with the EL results. In the EL spectra, the main peak at ~485 nm can be attributed to the radiation transition of BCzVBi excitons which are transferred from mCP, while the shoulder peak at ~445 nm originates from the mCP excitons. In addition, the shapes of the EL spectra do not change as we change the HTM, indicating that the behavior of holes in HTL does not influence the blue emission.

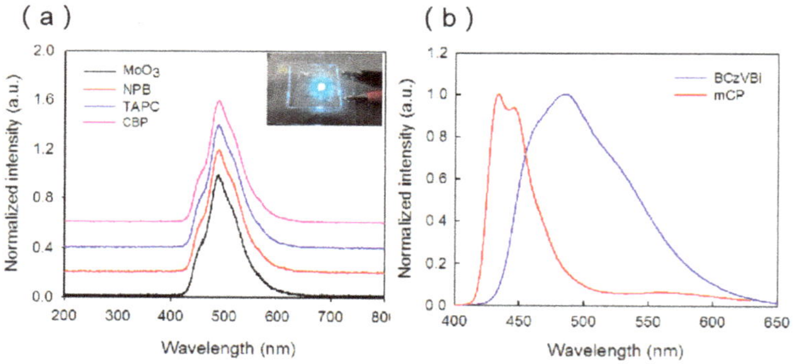

Figure 3. (**a**) Normalized electroluminescence spectra of OLEDs with TPBi ETL and four different HTLs. The inset is a luminescence image of the TAPC device at 50 mA/cm^2. (**b**) Photoluminescence (PL) spectra of 30 nm mCP and BCzVBi films deposited on ITO substrates.

In order to further improve the balance of carriers in EML, another ETL with higher electron mobility was introduced. As reported in another paper, the electron mobility of Bphen is 3.3×10^{-4} $cm^2 \cdot V^{-1} \cdot s^{-1}$, which is ~10 times higher than that of TPBi [19]. The Bphen devices were also fabricated and characterized in order to study the role of ETL on the balanced transport. As discussed above, one can find that the TAPC-based OLED shows a better luminous performance than the other three kinds of devices. So, fixing the hole transport material TAPC, the blue OLEDs with Bphen and TPBi ETLs were compared and discussed. The device structure was ITO/TAPC/mCP:20wt.% BCzVBi/ETL/LiF/Al, which are labeled as the Bphen device and TPBi device for clarity. Figure 4a displays the current density–voltage characteristics of the devices with different ETLs. It can be seen that the Bphen device had a turn-on voltage of ~3.4 V, which was smaller than that of the TPBi device. The LUMO level of Bphen was 3 eV, which is 0.3 eV lower than that of TPBi, and thus the electron injection from cathode to Bphen was much easier. Moreover, the transmission of electrons in Bphen will be faster than that in TPBi because of its high electron mobility. Figure 4b illustrates the luminance–current density characteristics of the Bphen device and the TPBi device, whose luminance were 3640 and 2955 cd/m^2, respectively. These results suggest that the carrier balance in EML was improved by substituting TPBi with Bphen. The hole mobility of TAPC (1×10^{-2} cm^2·V^{-1}·s^{-1}) was much larger than the electron mobility of TPBi (6.5×10^{-5} cm^2·V^{-1}·s^{-1}), resulting in more holes than electrons in EML. Bphen (3.3×10^{-4} cm^2·V^{-1}·s^{-1}) had a higher electron mobility than that of TPBi, increasing the number of electrons in EML, and finally improving the balance of holes and electrons in EML. This helps to generate more excitons, which increases the efficiency of the devices. Figure 4c illustrates the current efficiency–current density characteristics of the two devices, which are consistent with the luminance curves. Figure 4d shows the normalized EL spectra of the two devices. The main peak at ~485 nm and the shoulder peak at ~445 nm can be attributed to the BCzVBi excitons and mCP excitons, respectively. It is noted that the strength of the shoulder peak increases slightly, which may be caused by the shift of recombination zoom to the HTL/EML interface because of the high electron mobility.

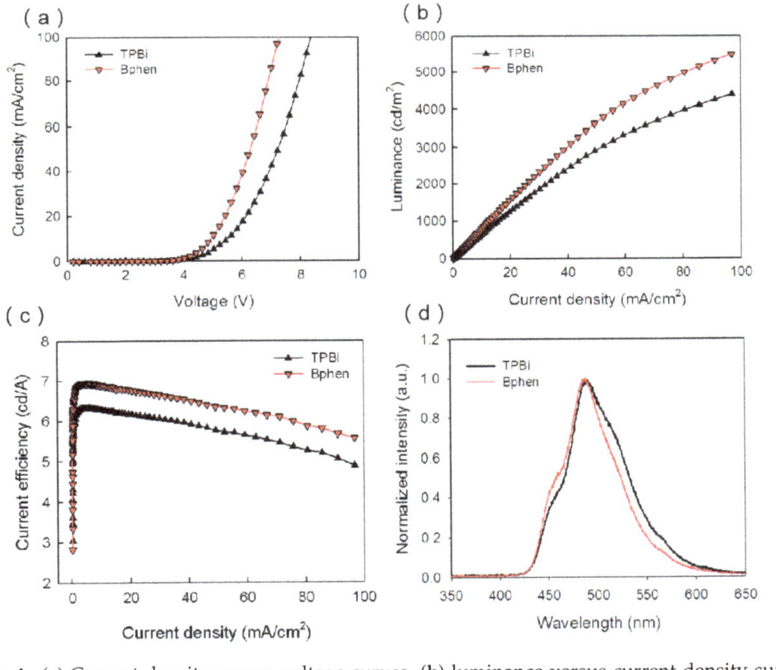

Figure 4. (**a**) Current density versus voltage curves, (**b**) luminance versus current density curves, (**c**) current efficiency versus current density curves, (**d**) normalized EL spectra of blue OLEDs with TAPC HTL and 4,7-Diphenyl-1,10-phenanthroline (Bphen), TPBi ETLs.

Figure 5 displays the evolution of the normalized luminance and operating voltage of the blue OLEDs with Bphen and TPBi ETLs at 20 mA/cm^2. The luminance of these two devices exhibited similar stretched exponential decay. The average values of half-life of the Bphen and TPBi devices were tested to be 3.22 h and 1.55 h, respectively. Clearly, the blue OLED with the Bphen ETL is relatively reliable under stressing, and has a lifetime nearly two times longer than that with TPBi ETL. With the decrease in luminance, the operating voltage increases, and both the rates of luminance decay and voltage rise in the Bphen device are much slower. These results illustrate that ETM has an important impact on the OLED reliability, including the intrinsic stability of electron transport materials, the LUMO level, and electron mobility [15,19]. Although TPBi (Tg = 127 °C) [18] has a higher glass transition temperature than Bphen (Tg = 66 °C) [12], the Bphen device exhibited a more stable performance than the TPBi device. There is a fact that Bphen has higher electron mobility than that of TPBi and better energy level alignment with the working function of the Al cathode, which facilitate the high-efficiency injection and transport of electrons. For the TPBi device, the misalignment of the TPBi LUMO level with the working function of the Al cathode resulted in an energy barrier as well as an additional voltage drop at the ETL/cathode interface, which would lead to localized joule heating and accelerate electrochemical reactions [19]. Moreover, lower electron mobility means that a higher voltage drop across the ETL is needed for charge transport. These reasons may be responsible for the faster degradation of the TPBi OLED relative to the Bphen device.

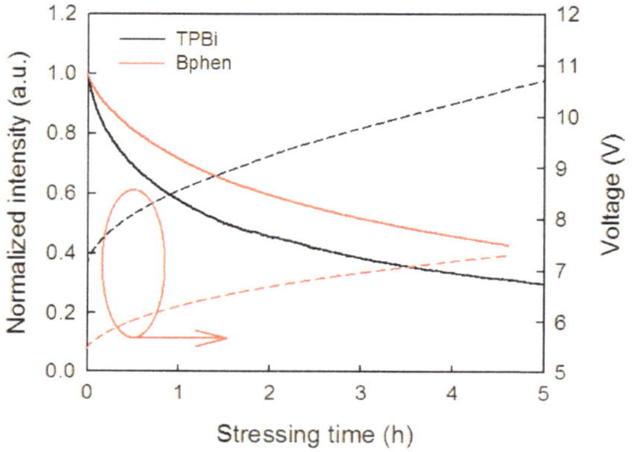

Figure 5. Evolution of the normalized luminance and voltage of blue OLEDs with Bphen and TPBi ETL stressed under 20 mA/cm^2.

The lifetime of our devices is higher than those of some phosphorescent OLEDs with the mCP host because of the better stability of fluorescent materials [18,29,30], which meets our initial expectation. However, it is noted that the lifetime of both the TPBi and Bphen devices is not very long and comparable with some results of fluorescent OLEDs reported by other groups [31–33]. Our results suggest that current stressing may induce joule heating, causing accelerated device degradation [19,29,30]. Besides, concentration quenching, often referred to as singlet exciton-charge annihilation, may also exist in our devices with high concentration (20 wt.%) BCzVBi doped in the host [34]. At high concentration, dopant aggregation occurs, facilitating charge injection and transport, while also enhancing exciton interaction and nonradiative decay. What's more, the stability of host and fluorescent materials used in this study is also an important factor. Further improvement of devices' lifetime will be conducted in our future research work using other methods such as pulsed current stressing and device structure engineering.

4. Conclusions

We have investigated the luminous characteristics of thermal evaporation blue OLEDs with various HTL and ETL materials. Fixing the electron transport material TPBi, four hole transport materials, including NPB, TAPC, CBP, and MoO_3, were selected to be HTLs. Of these four kinds of devices, the blue OLED with TAPC HTL exhibited a maximum luminance of 2955 cd/m^2 and current efficiency (CE) of 5.75 cd/A at 50 mA/cm^2, which are 68% and 62% higher, respectively, than those of the minimum values in the device with the MoO_3 ETL. This can be explained by the charge balance in the EML. In order to improve the injection and transport of electrons, Bphen was introduced and a maximum luminance of 3640 cd/m^2 at 50 mA/cm^2 was obtained, which is 23% higher than that of the TPBi device. Furthermore, the Bphen device showed a half-period of ~3.22 h, which is two times longer than that of the TPBi device stressed under 20 mA/cm^2. These results indicate that the carrier mobility of transport materials and energy level alignment of different functional layers play important roles in the performance of OLEDs. The findings present a rational strategy for the device engineering of high-efficiency blue OLEDs.

Author Contributions: Conceptualization, Y.Z. and F.M.; Funding acquisition, Y.Z. and F.M.; Investigation, N.L., W.S. and J.F.; Methodology, N.L., S.M., D.S., J.X., Y.J. and X.C.; Writing—original draft, N.L.; Writing—review and editing, Y.Z. and F.M.

Funding: This research was funded by the Foundation of Hubei Provincial Science and Technology Department (Grant No. 2016BKJ005), the Leading Plan of Green Industry of Hubei University of Technology (Grant No. YXQN2016003), the International Science and Technology Cooperation Program of China (Grant No. 2016YFE0124300), the Open Foundation of Hubei Key Laboratory for High-efficiency Utilization of Solar Energy and Operation Control of Energy Storage System (Grant Nos. HBSEES201801, HBSEES201705), and the National Natural Science Foundation of China (Grant Nos. 51602099, 51371079, 11304092, and 11305056).

Conflicts of Interest: The authors declare no conflict of interest.

References

1. Tang, C.W.; VanSlyke, S.A. Organic electroluminescent diodes. *Appl. Phys. Lett.* **1987**, *51*, 913–915. [CrossRef]
2. McCarthy, M.A.; Liu, B.; Donoghue, E.P.; Kravchenko, I.; Kim, D.Y.; So, F.; Rinzler, A.G. Low-voltage, low-power, organic light-emitting transistors for active matrix displays. *Science* **2011**, *332*, 570–573. [CrossRef] [PubMed]
3. Chen, S.; Kwok, H.S. Full color organic electroluminescent display with shared blue light-emitting layer for reducing one fine metal shadow mask. *Org. Electron.* **2012**, *13*, 31–35. [CrossRef]
4. Liu, N.; Shi, W.X.; Zhou, Y.M.; Cao, X.A. Impact of dopant aggregation on the EL of blue fluorescent host-dopant emitters. *IEEE Electron Device Lett.* **2019**, *40*, 750–753. [CrossRef]
5. Jung, J.H.; Ha, M.Y.; Park, D.Y.; Lee, M.J.; Choi, S.J.; Moon, D.G. Effect of hole transporting materials on the emission characteristics of soluble processed organic light-emitting devices on the plastic substrate. *Mol. Cryst. Liq. Cryst.* **2017**, *644*, 214–220. [CrossRef]
6. Takita, Y.; Takeda, K.; Hashimoto, N.; Nomura, S.; Suzuki, T.; Nakashima, H.; Uesaka, S.; Seo, S.; Yamazaki, S. Highly efficient deep-blue fluorescent dopant for achieving low-power OLED display satisfying BT. 2020 chromaticity. *J. Soc. Inf. Disp.* **2018**, *26*, 55–63. [CrossRef]
7. Abdurahman, A.; Obolda, A.; Peng, Q.M.; Li, F. Efficient deep blue fluorescent oleds with ultra-low efficiency roll-off based on 4h-1, 2, 4-triazole cored DA and DAD type emitters. *Dye. Pigment.* **2018**, *153*, 10–17. [CrossRef]
8. Bae, H.W.; Kim, G.W.; Lampande, R.; Park, J.H.; Ko, I.J.; Yu, H.J.; Lee, C.Y.; Kwon, J.H. Efficiency enhancement in fluorescent deep-blue OLEDs by boosting singlet exciton generation through triplet fusion and charge recombination rate. *Org. Electron.* **2019**, *70*, 1–6. [CrossRef]
9. Liu, Y.D.; Zhao, Y.; Zhang, H.H.; Song, X.C.; Zhou, J.H.; Wu, Z.S.; Zhang, J.; Guo, W.J.; Mi, Y.H. The application of the nanostructure aluminum in the blue organic light-emitting devices. *Org. Electron.* **2018**, *57*, 1–6. [CrossRef]
10. Wang, X.; Zhong, Z.M.; Zhao, S.; Guo, T.; Jiang, J.X.; Zou, J.H.; Ying, L.; Peng, J.B.; Cao, Y. An efficient blue emitter based on a naphthalene indenofluorene core. *Org. Electron.* **2018**, *55*, 157–164. [CrossRef]
11. Du, Q.Q.; Wang, W.J.; Li, S.H.; Zhang, D.; Li, W.L.; Zheng, W.Q. Study on blue organic light-emitting diodes doped with 4, 4′-bis (9-ethyl-3carbazovinylene)-1, 1′-biphenyl in various host materials. *Opt. Commun.* **2016**, *366*, 253–259. [CrossRef]
12. He, S.J.; Wang, Z.B.; Wang, D.K.; Jiang, N.; Lu, Z.H. Highly efficient blue fluorescent organic light-emitting diodes with a high emitter/host ratio. *Appl. Phys. Lett.* **2013**, *103*, 083301. [CrossRef]
13. Meng, M.; Song, W.; Kim, Y.H.; Lee, S.Y.; Jhun, C.G.; Zhu, F.R.; Ryu, D.H.; Kim, W.Y. Effects of Electron Transport Material on Blue Organic Light-Emitting Diode with Fluorescent Dopant of BCzVBi. *J. Nanosci. Nanotechnol.* **2013**, *13*, 294–299. [CrossRef] [PubMed]
14. Jou, J.H.; Weng, J.W.; Chavhan, S.D.; Yadav, R.A.K.; Liang, T.W. Investigation of charge-transporting layers for high-efficiency organic light-emitting diode. *J. Phys. D Appl. Phys.* **2018**, *51*, 454002. [CrossRef]

15. Liu, L.; Li, S.; Zhou, Y.M.; Liu, L.Y.; Cao, X.A. High-current stressing of organic light-emitting diodes with different electron-transport materials. *Microelectron. Reliab.* **2017**, *71*, 106–110. [CrossRef]
16. Giebeler, C.; Antoniadis, H.; Bradley, D.D.C.; Shirota, Y. Influence of the hole transport layer on the performance of organic light-emitting diodes. *J. Appl. Phys.* **1999**, *85*, 608–615. [CrossRef]
17. Qin, P.L.; He, Q.; Ouyang, D.; Fang, G.J.; Choy, W.C.H.; Li, G. Transition metal oxides as hole-transporting materials in organic semiconductor and hybrid perovskite based solar cells. *Sci. China Chem.* **2017**, *60*, 472–489. [CrossRef]
18. Lee, H.; Ahn, H.; Lee, C. Device characteristics of blue phosphorescent organic light-emitting diodes depending on the electron transport materials. *J. Inf. Disp.* **2011**, *12*, 219–222. [CrossRef]
19. Shi, W.X.; Liu, N.; Zhou, Y.M.; Cao, X.A. Effects of Postannealing on the Characteristics and Reliability of Polyfluorene Organic Light-Emitting Diodes. *IEEE Trans. Electron Devices* **2019**, *66*, 1057–1062. [CrossRef]
20. Huh, D.H.; Kim, G.W.; Kim, G.H.; Kulshreshtha, C.; Kwon, J.H. High hole mobility hole transport material for organic light-emitting devices. *Synth. Met.* **2013**, *180*, 79–84. [CrossRef]
21. Kwak, J.; Chae, J.; Son, S.; Jung, B.J. Highly efficient blue fluorescent organic light-emitting diodes by engineering hole-transporting/exciton-blocking layer. *ECS Solid State Lett.* **2015**, *4*, R5–R9. [CrossRef]
22. Giebink, N.C.; Forrest, S.R. Quantum efficiency roll-off at high brightness in fluorescent and phosphorescent organic light emitting diodes. *Phys. Rev. B* **2008**, *77*, 235215. [CrossRef]
23. Anikeeva, P.O.; Madigan, C.F.; Halpert, J.E.; Bawendi, M.G.; Bulović, V. Electronic and excitonic processes in light-emitting devices based on organic materials and colloidal quantum dots. *Phys. Rev. B* **2008**, *78*, 085434. [CrossRef]
24. Murawski, C.; Leo, K.; Gather, M.C. Efficiency roll-off in organic light-emitting diodes. *Adv. Mater.* **2013**, *25*, 6801–6827. [CrossRef] [PubMed]
25. Shirasaki, Y.; Supran, G.J.; Bawendi, M.G.; Bulović, V. Emergence of colloidal quantum-dot light-emitting technologies. *Nat. Photonics* **2013**, *7*, 13–23. [CrossRef]
26. Liu, B.Q.; Wang, L.; Gu, H.S.; Sun, H.D.; Demir, H.K. Highly Efficient Green Light-Emitting Diodes from All-Inorganic Perovskite Nanocrystals Enabled by a New Electron Transport Layer. *Adv. Opt.Mater.* **2018**, *6*, 1800220. [CrossRef]
27. Hsiao, C.H.; Liu, S.W.; Chen, C.T.; Lee, J.H. Emitting layer thickness dependence of color stability in phosphorescent organic light-emitting devices. *Org. Electron.* **2010**, *11*, 1500–1506. [CrossRef]
28. Lee, S.; Tang, C.W.; Rothberg, L.J. Effects of mixed host spatial distribution on the efficiency of blue phosphorescent organic light-emitting diodes. *Appl. Phys. Lett.* **2012**, *101*, 043303. [CrossRef]
29. Yang, R.Y.; Li, X.M.; Cao, X.A. Role of wide bandgap host in the degradation of blue phosphorescent organic light-emitting diodes. *J. Appl. Phys.* **2017**, *122*, 075501. [CrossRef]
30. Yang, R.Y.; Cao, X.A. Thermal and Nonthermal Factors Affecting the Lifetime of Blue Phosphorescent Organic Light-Emitting Diodes. *IEEE Trans. Electron Devices* **2018**, *65*, 3300. [CrossRef]
31. Kim, H.G.; Kim, K.H.; Kim, J.J. Highly Efficient, Conventional, Fluorescent Organic Light-Emitting Diodes with Extended Lifetime. *Adv. Mater.* **2017**, *29*, 1702159. [CrossRef] [PubMed]
32. Song, W.; Kim, T.; Lee, J.Y.; Lee, Y.; Jeong, H. Investigation of degradation mechanism of phosphorescent and thermally activated delayed fluorescent organic light-emitting diodes through doping concentration dependence of lifetime. *J. Ind. Eng. Chem.* **2018**, *68*, 350–354. [CrossRef]
33. Kim, S.H.; Cho, I.; Sim, M.K.; Park, S.; Park, S.Y. Highly efficient deep-blue emitting organic light emitting diode based on the multifunctional fluorescent molecule comprising covalently bonded carbazole and anthracene moieties. *J. Mater. Chem.* **2011**, *21*, 9139. [CrossRef]
34. Lee, B.M.; Kim, N.H.; Yoon, J.A.; Lee, S.E.; Kim, Y.K.; Kim, W.Y.; Mascher, P. Luminescence characteristics of hybrid dual emitting layers in blue organic light-emitting diodes by controlling the fluorescent doping concentration. *J. Lumin.* **2014**, *148*, 72–78. [CrossRef]

micromachines

MDPI

Article

Enhancement of Light Extraction Efficiency for InGaN/GaN Light-Emitting Diodes Using Silver Nanoparticle Embedded ZnO Thin Films

Po-Hsun Lei [1,*], Chyi-Da Yang [2], Po-Chun Huang [1] and Sheng-Jhan Yeh [1]

1 Institute of Electro-Optical and Material Science, National Formosa University, No. 64, Wunhua Rd., Huwei, Yunlin County 632, Taiwan; d917205@oz.nthu.edu.tw (P.-C.H.); gy31999@gmail.com (S.-J.Y.)

2 Department of Microelectronics Engineering, National Kaohsiung University of Science and Technology, Kaohsiung 811, Taiwan; cdyang@stu.nkmu.edu.tw

* Correspondence: pohsunlei@gmail.com; Tel.: +886-5-6315-668

Received: 19 March 2019; Accepted: 9 April 2019; Published: 10 April 2019

Abstract: In this study, we propose a liquid-phase-deposited silver nanoparticle embedded ZnO (LPD-Ag NP/ZnO) thin film at room temperature to improve the light extraction efficiency (LEE) for InGaN/GaN light-emitting diodes (LEDs). The treatment solution for the deposition of the LPD-Ag/NP ZnO thin film comprised a ZnO-powder-saturated HCl and a silver nitrate ($AgNO_3$) aqueous solution. The enhanced LEE of an InGaN/GaN LED with the LPD-Ag NP/ZnO window layer can be attributed to the surface texture and localized surface plasmon (LSP) coupling effect. The surface texture of the LPD-Ag/NP ZnO window layer relies on the $AgNO_3$ concentration, which decides the root-mean-square (RMS) roughness of the thin film. The LSP resonance or extinction wavelength also depends on the concentration of $AgNO_3$, which determines the Ag NP size and content of Ag atoms in the LPD-Ag NP/ZnO thin film. The $AgNO_3$ concentration for the optimal LEE of an InGaN/GaN LED with an LPD-Ag NP/ZnO window layer occurs at 0.05 M, which demonstrates an increased light output intensity that is approximately 1.52 times that of a conventional InGaN/GaN LED under a 20-mA driving current.

Keywords: Liquid phase deposition method; InGaN/GaN light-emitting diode; silver nanoparticle; zinc oxide; localized surface plasmon

---

1. Introduction

Because gallium-nitride (GaN)-based blue light-emitting diodes (LEDs) have the inherent advantages of a wide bandgap in the green to ultraviolet range, a relatively long lifetime, and low energy consumption, they have received intensive and extensive investigation. InGaN/GaN LEDs with a high light output intensity and low power consumption have been widely applied to products such as solid-state lighting, backlight units for liquid crystal displays, car headlights, traffic lights, and full-color displays [1–4]. More recently, GaN-based LEDs and solid-state lighting were widely used as transmitting devices in the visible light communication (VLC) system, which has the characteristics of high-speed light communication with low transmission loss, the absence of electromagnetic interference, and license and high security [5–7]. A key issue for GaN-based LEDs and solid-state lighting in these applications is a high external quantum efficiency. The external quantum efficiency (EQE), which is the product of the internal quantum efficiency (IQE) and light extraction efficiency (LEE), determines the light output intensity of InGaN/GaN LEDs. The IQE is affected by the concentration of defects, amount of overflow carriers, and spreading area of injection current in the active region. To obtain a modern InGaN/GaN LED with a high IQE, various device structures, such as a designed active region including double-heterostructures and multiple quantum wells (MQWs), the insertion of an electron-blocking layer, and an expanded current-spreading layer, have been investigated [2]. To date, IQEs of higher than 80% have been achieved for InGaN/GaN LEDs because of the rapid development of growth methods and technologies [8–10]. The LEE of InGaN/GaN LEDs is also a crucial factor in determining the EQE. The LEE for InGaN/GaN LEDs is very low because of the high index contrast between GaN (n = 2.5) and free space. Several studies have reported that the IQE and LEE of InGaN/GaN LEDs can be enhanced using a patterned sapphire substrate to reflect and adjust the trajectory of light emitting to the sapphire substrate [11–13] or flip-chip structured-LEDs to increase the probability of light escaping from the sapphire [14,15]. However, the mechanically and chemically strong nature of sapphire makes patterning a challenging task. In addition, achieving the small dimensions of scattering objects through photolithography technologies is impossible because of the short wavelength of nitride-based LEDs. Some studies [16–21] have proposed the fabrication of textured GaN or a window layer to improve the LEE of InGaN/GaN LEDs through the changed trajectory of incident light beyond the escape cone. However, the textured InGaN/GaN LEDs demonstrate deteriorated electrical and temperature properties because of the inhomogeneous spatial field distribution under driving. Moreover, controlling the process and achieving a high repeatability of the rough window layer are difficult tasks. Some studies have reported that photonic crystals [22] and nanopyramids [23] can increase the LEE of InGaN/GaN LEDs through either diffraction or resonant coupling. However, these approaches are not practical because they require technologies that can accurately control the dimension, such as e-beam lithography, which is not suitable for obtaining a high throughput.

Recently, the surface plasmon (SP) coupling effect, which can improve the extraction efficiency of InGaN/GaN LEDs, has received considerable research attention [24–26]. SP coupling effects are the collective oscillations of electrons at the interface of a metal and a dielectric, and they can be classified as surface plasmon polaritons (SPPs) at metal surfaces and localized surface plasmons (LSPs) of local oscillation among isolated metallic nanostructures (NPs) with a resonant frequency. The SPP-enhanced LEE of InGaN/GaN LED can be realized by placing a metal layer near InGaN/GaN MQWs [27,28]. However, the inherently high reflection of the metal may block light emission from the InGaN/GaN active region, thus degrading the light output intensity. In contrast, several reports [23,24] have indicated that the LEE of InGaN/GaN blue LEDs can be improved by LSPs. Because the emission wavelength of InGaN/GaN blue LEDs matches the resonance wavelength of LSPs, the photons emitted from the InGaN/GaN active region can be closely coupled with the LSP modes to enhance the light output intensity of InGaN/GaN LED [29–31]. The resonant wavelength for the LSPs depends on the metal, such as silver (Ag) for blue, gold (Au) for green, and aluminum (Al) for UV [32,33]. In addition, because the interface is sufficiently rough to scatter the SPs, the coupling energy can be transferred into free space photons.

A well-known method to obtain Ag NPs is the solid-state dewetting process, which transforms an Ag thin film into Ag NPs under a high annealing temperature. However, a high-temperature process may result in the redistribution of the dopant and thermal stress, degrading the performance of devices. We are the first to synthesize a liquid-phase-deposited Ag nanoparticle embedded zinc oxide (LPD-Ag NP/ZnO) thin film through a chemical reaction in an aqueous solution at a low deposition temperature (i.e., even at room temperature) [34]. This method offers several advantages, including a low growth temperature, low cost, large area growth, good step coverage, and simple process and deposition equipment. In this study, the textured surface and LSP coupling effect, which result from the Ag NPs and ZnO in LPD-Ag NP/ZnO thin films, can enhance the LEE of InGaN/GaN LEDs with the LPD-Ag NP/ZnO window layer. The surface texture and LSP coupling effect of an LPD-Ag NP/ZnO thin film depend on the concentration of the silver nitrate ($AgNO_3$) aqueous solution, which determines the root-mean-square (RMS) roughness, Ag content, and Ag NP size of the LPD-Ag NP/ZnO thin film. The measured results indicate that the main factor for enhancing the LEE of InGaN/GaN LEDs with the LPD-Ag NP/ZnO window layer is the LSP coupling effect.

2. Materials and Methods

2.1. Fabrication of InGaN/GaN LEDs with LPD-Ag NPs/ZnO Thin Films

GaN epi-wafers were grown on a c-face (0001) sapphire substrate using a metal-organic chemical vapor deposition (MOCVD) system [29]. The InGaN/GaN LED structure comprises a GaN buffer layer grown at a low temperature, a heavily Si-doped n-type GaN layer, an InGaN/GaN MQW active region, and an Mg-doped p-type GaN layer. Indium-tin-oxide (ITO; Tyntek Co., Hsinchu, Taiwan) was deposited on the p-type GaN layer to form a transparent conductive layer (TCL). The epi-wafers were then patterned using standard photolithographic and partial etching to define the emitting regions. A Ti/Pt/Au alloy was used as ohmic contact metal in the p- and n-GaN contact regions, and the wafer was alloyed in an N_2 atmosphere for 5 min at 450 °C. The finished wafer was placed in the treatment solution to deposit the LPD-Ag NPs/ZnO window layer. The size of the emission window for the InGaN/GaN LEDs with ITO TCL and an LPD-Ag NPs/ZnO window layer was 300 × 300 μm^2. The schematics of the InGaN/GaN LED with an LPD-Ag NP/ZnO window layer are shown in Figure 1.

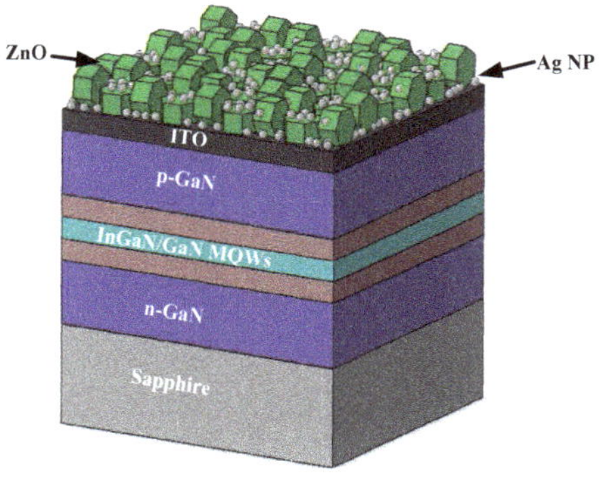

Figure 1. The schematics of InGaN/GaN LED with an LPD-Ag NP/ZnO window layer.

2.2. Preparation of Treatment Solution and Growth of LPD-Ag NP/ZnO Thin Films

The treatment solution for LPD-Ag NP/ZnO thin films grown on the sapphire substrate was prepared according to the following steps. First, ZnO powder (J. T. Baker) was added into deionized water-diluted HCl (12 M, Taimax, Taiwan), and the mixed solution was then stirred for 1 h at 25 °C to ensure that the HCl solution was saturated with the ZnO powder. Second, a aqueous solution of $AgNO_3$ was added to the ZnO-saturated HCl solution to form the treatment solution. The pH scale of the treatment solution was dominated by the concentration of HCl. The optimum pH scale to grow the LPD-Ag NP/ZnO thin film on the sapphire substrate was approximately 4–5.5. The LPD system for the deposition of the LPD-Ag NP/ZnO thin films comprised a Teflon vessel immersed in a temperature-controlled water bath system. The temperature-controlled water bath system can provide a uniform temperature distribution with an accuracy of 0.1 °C. The following process was implemented to deposit the LPD-Ag NP/ZnO thin film on the InGaN/GaN LED. First, a Teflon vessel containing the treatment solution was placed in the temperature-controlled water bath system at 25 °C for 3–5 min. This preheating step ensured a uniform deposition of the LPD-Ag NP/ZnO thin film on the inserted substrate. Second, the InGaN/GaN LED was inserted in the treatment solution for LPD-Ag NP/ZnO thin film deposition. The surface morphology of LPD-Ag NPs/ZnO on the sapphire substrate was investigated through scanning electron microscopy (SEM), and the Ag content in the LPD-Ag NP/ZnO thin film was detected through energy-dispersive X-ray spectroscopy (EDS).

3. Results and Discussion

To obtain a fair deposition rate, which determines the RMS roughness of the LPD-ZnO thin film, the pH scale of treatment solution and diluted HCl concentration should be maintained at 4–5.5 and 6 M, respectively, during the deposition process and the deposition temperature needs a temperature distribution with an accuracy of 0.1 °C. Figure 2 shows the SEM images for the LPD-ZnO thin film (Figure 2a) and those for the LPD-Ag NP/ZnO thin films deposited at $AgNO_3$ concentrations of 0.01, 0.05, and 0.1 M (Figure 2b–d). The estimated percentages of Ag NP size in the LPD-Ag NP/ZnO thin films deposited at $AgNO_3$ concentrations of 0.05 and 0.1 M are shown in Figure 2e,f. The deposition temperature of the all thin films was 25 °C. The LPD-ZnO thin film exhibited a hexagonal and flake-shaped structure on the sapphire substrate, as shown in Figure 2a. Figure 2b–d show that the structure of the LPD-Ag NP/ZnO thin films reflected the randomly distributed Ag NPs on the hexagonal and flake-shaped ZnO. In addition, Figure 2b–d show that the Ag NP distribution of the LPD-Ag NP/ZnO thin film deposited at a high $AgNO_3$ concentration is denser than that of the thin film deposited at a low $AgNO_3$ concentration. The average size of the randomly distributed Ag NP on the flake-shaped ZnO was approximately 76.8 and 142.6 nm for $AgNO_3$ concentrations of 0.05 and 0.1 M, respectively. The chemical reaction for the LPD-deposited ZnO thin film can be expressed as follows [34,35]:

$$2ZnO + HCl + 2H_2O \leftrightarrows Zn(OH)_4{}^{2-} + Zn^{2+} + H^+ + Cl^-, \quad (1)$$

$$Zn(OH)_4{}^{2} \leftrightarrows ZnO_2{}^{2-} + H_2O, \quad (2)$$

The concentration of zinc species including Zn^{2+} and $Zn(OH)_4{}^{2-}$ was used to determine the growth rate of the LPD-ZnO thin film in our previous study [35]. $AgNO_3$ powder can dissolve in H_2O and form an Ag-complex, namely $Ag(OH)_2{}^-$, in $AgNO_3$ aqueous solution through the following chemical reactions [36]:

$$AgNO_3 \leftrightarrows Ag^+ + NO_3{}^-, \quad (3)$$

$$Ag^+ + 2H_2O \leftrightarrows Ag(OH)_2{}^- + 2H^+, \quad (4)$$

The Ag-complex reacts with Zn^{2+} and $ZnO_2{}^{2-}$ to form the LPD-Ag NP/ZnO thin film according to the following equation [35,36]:

$$ZnO_2{}^{2-} + Zn^{2+} + 2Ag(OH)_2{}^- \leftrightarrows 2ZnO + 2Ag + 2H_2O + 2OH^-, \quad (5)$$

The formation of a flake-shaped ZnO thin film depends on the pH scale of the treatment solution [31]. To obtain a similar flake-shaped ZnO thin film, the pH scale of the treatment solution for LPD-Ag NP/ZnO thin films grown on the sapphire substrate was maintained at 4–5.5. The concentration of Ag^+ and $Ag(OH)_2{}^-$ in treatment solution depends on the concentration of $AgNO_3$. A high $AgNO_3$ concentration can increase the concentration of $Ag(OH)_2{}^-$, resulting in large and dense Ag NPs embedded on the flake-shaped ZnO.

Table 1 lists the Ag contents of LPD-Ag NP/ZnO thin films measured using EDS for various $AgNO_3$ concentrations. The Ag content of LPD-Ag NP/ZnO thin films increased with the $AgNO_3$ concentration, resulting from a high Ag^+ concentration, as shown in Equation (3). Moreover, the density of Ag NPs distributed on the flake-shaped ZnO for LPD-Ag NP/ZnO thin films observed from the SEM images exhibited a similar trend to the Ag content of the LPD-Ag NP/ZnO thin films shown in Table 1. We quantitatively determined the density of the Ag NP distributed on the flake-shaped ZnO according to the Ag content of the LPD-Ag NP/ZnO thin films, and we used the varying $AgNO_3$ concentrations of the treatment solution to obtain the specific distribution and size of Ag NPs on the flake-shaped ZnO. However, adding $AgNO_3$ to the treatment solution might cause Ag atoms to dope into the flake-shaped ZnO, rather than form an Ag NP embedded on the flake-shaped ZnO during LPD-Ag NP/ZnO thin film deposition. The extinction-peak position depends on the Ag NP diameter and distance between Ag NPs [32]. Increasing the diameter of Ag NP or decreasing the distance between Ag NPs will shift the extinction-peak wavelength to a long value. Therefore, Ag NPs with a large size or high content under a high $AgNO_3$ concentration result in a red-shift for the extinction-peak, as shown in Table 1. The insets of Figure 2a,c show the EDS spectra for the LPD-ZnO and LPD-Ag NPs/ZnO with an $AgNO_3$ concentration of 0.05 M. The O and Zn atoms contents of the LPD-ZnO thin film were 72.77 and 27.23 at %, respectively, and the compositions of O, Zn, and Ag atoms in the LPD-Ag NPs/ZnO were 70.59, 25.64, and 3.76 at %, respectively. The ratio of O to Zn for the LPD-ZnO thin film (2.67) was similar to that in the LPD-Ag NPs/ZnO thin film (2.75), implying that the Ag NPs were randomly distributed on the flake-shaped ZnO rather than doped during the deposition process.

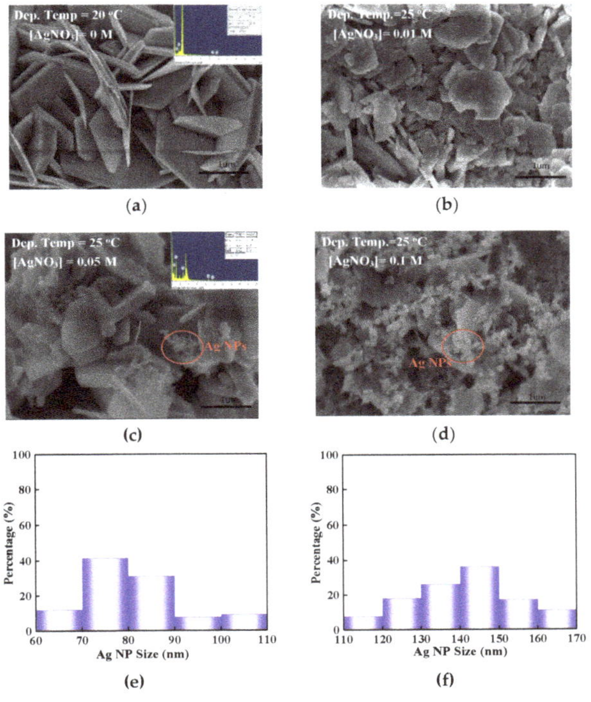

Figure 2. SEM images for (**a**) LPD-ZnO and LPD-Ag NP/ZnO thin films under $AgNO_3$ concentrations of (**b**) 0.01, (**c**) 0.05, and (**d**) 0.1 M. The insets of (**a**) show the content of Zn and O of LPD-ZnO and (**c**) represent the content of Zn, O, and Ag and Ag NP size of LPD-Ag NP/ZnO under an $AgNO_3$ concentration of 0.05 M. The estimated percentage of Ag NP size in LPD-Ag NP/ZnO thin films deposited at (**e**) 0.05 and (**f**) 0.1 M

Table 1. Relationship between measured Ag NP diameter, Ag content, and extinction-peak position under different $AgNO_3$ concentrations.

$AgNO_3$ Concentration (M)	Measured Ag NP Diameter (Average) (nm)	Ag Content (at %)	Extinction-Peak Position (nm)
0.01	-	0.21	-
0.03	60.3	1.64	442
0.05	76.8	3.76	463
0.07	82.3	5.89	489
0.1	142.6	9.46	492

The photons emitted from the InGaN/GaN active region to the LPD-Ag NP/ZnO window layer were localized along the interface of the randomly-distributed Ag NPs and ZnO. The near-field strength at this interface was enhanced and the incident light from the InGaN/GaN active region was coupled with the LSPs in the Ag NPs. The enhanced light coupled with the LSPs was then decoupled by the Ag NPs and re-emitted into the free space [34]. The wavelength of incident light from the InGaN/GaN active region should be close to the LSP resonant wavelength to enhance the LEE. To determine the LSP resonance wavelength related to the Ag NPs, the measured extinction spectra of the LPD-Ag NP/ZnO thin film are shown in Figure 3 as functions of the $AgNO_3$ concentration. In addition, the diameter of the Ag NPs, the Ag content of the LPD-Ag NP/ZnO thin film, and the peak position of the aforementioned extinction spectra are summarized in Table 1. The peak position of the extinction spectra in Figure 3 reveals an obvious red-shift as the $AgNO_3$ concentration increases from 0.03 to 0.1 M. This is attributed to the following reasons. First, the resonant electromagnetic behavior of the Ag NPs depends on the confinement of conduction electrons to the small particle volume [37]. Because the extinction of light with the wavelength is smaller than that of the diameter of Ag NPs, the electrons move and resonate at a specific frequency (or wavelength), which is called the particle dipole plasmon frequency (or wavelength). As the size of the Ag NPs increases, the conduction electron cannot move in phase, leading to a reduced depolarization field for the red-shift phenomenon. The size of the Ag NPs increases with increasing $AgNO_3$ concentration, resulting in a reduced depolarization field, as indicated in Table 1. Second, the peak position of the absorption or extinction spectrum for Ag NP/ZnO thin films, which depends on the LSP resonant wavelength, can be represented as follows [38,39]:

$$\lambda_P = \sqrt{\frac{4\pi^2 c^2 m_{eff} \varepsilon_0}{Ne^2}}, \tag{6}$$

where λ_P is the LSP-resonant wavelength, m_{eff} is the effective mass of the free electron of the metal, and N is the electron density of metal. The LSP resonant wavelength relates to the electron density of metal. The Ag content increases from 1.64 to 9.46 at % as the $AgNO_3$ concentration increases from 0.03 to 0.1 M, as shown in Table 1. The electron density of the Ag NPs decreases with increasing Ag content ($AgNO_3$ concentration) because electrons move from the Ag NPs to the conduction band of the flake-shaped ZnO, as shown in the inset of Figure 3. According to Equation (6), the reduced electron density of Ag NPs leads to a red-shift phenomenon [40].

To obtain the optimal $AgNO_3$ concentration, which determines the Ag NP size, to maximize LEE, we used the OptiFDTD computational software, which is based on the finite-difference time-domain, to calculate the optical intensity for InGaN/GaN LEDs with or without the Ag NP/ZnO window layer. A random distribution of Ag NPs with various diameters on the flake-shaped ZnO window layer in InGaN/GaN LED was applied to determine the Ag NP distribution in the LPD-Ag NP/ZnO thin film, as shown in Figure 2. Figure 4a shows the calculated optical intensity of conventional InGaN/GaN LEDs, InGaN/GaN LEDs with the flake-shaped ZnO window layer, and InGaN/GaN LEDs with Ag NPs distributed on the flake-shaped ZnO window layer for Ag NPs with various diameters. The emitting wavelength was set at 460 nm. Figure 4 indicates that the InGaN/GaN LED with Ag NPs distributed on the flake-shaped ZnO thin film achieves the maximum optical intensity when the diameter of the Ag NPs is 75 nm. Figure 4b,c show the electric field distribution of InGaN/GaN LEDs with LPD-ZnO and 75-nm-sized LPD-Ag NP/ZnO window layers. Although the color distribution in Figure 4c is similar to Figure 4b, the scale of the color bar in Figure 4c is larger than that in Figure 4b. These results indicate that an LSP coupling effect is found in InGaN/GaN LED with an LPD-Ag NP/ZnO window layer. Consequently, the enhanced light intensity or LEE might be attributed to the LSP coupling effect of the Ag NPs and the textured LPD-Ag NPs/ZnO thin films, which increase the probability of photons escaping from p-GaN.

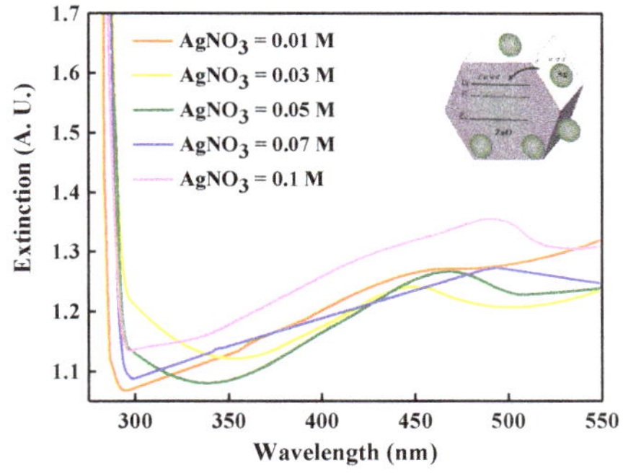

Figure 3. Measured extinction spectra of the LPD-Ag NP/ZnO thin films for different $AgNO_3$ concentrations. The inset shows a schematic of the movement of electrons between Ag and ZnO.

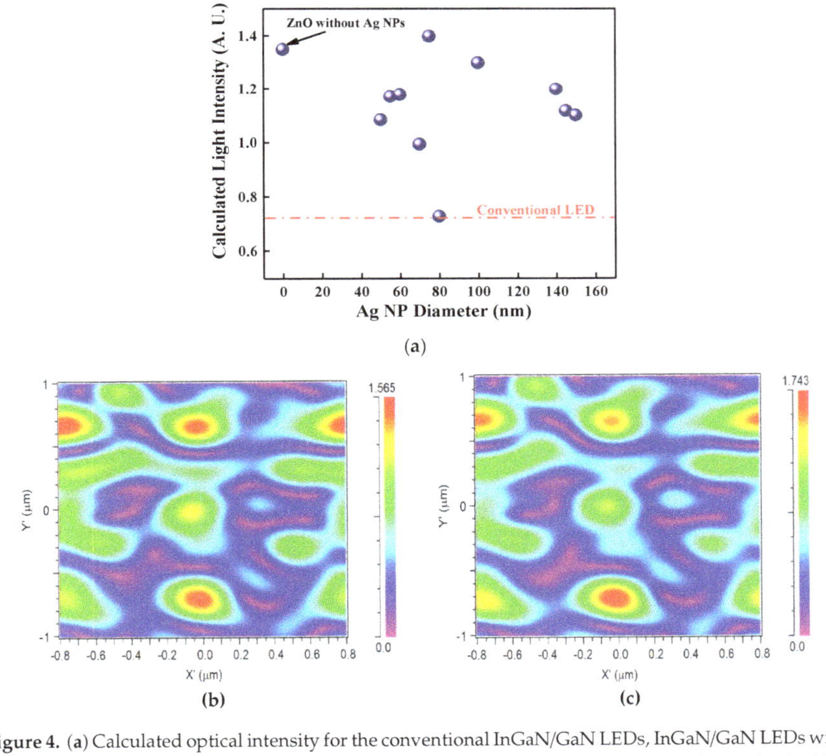

Figure 4. (**a**) Calculated optical intensity for the conventional InGaN/GaN LEDs, InGaN/GaN LEDs with flake-shaped-ZnO window layers, and InGaN/GaN LEDs with Ag NP/ZnO window layers for Ag NPs with various diameters. The electric field distribution of InGaN/GaN LEDs with (**b**) flake-shaped-ZnO window layers, and with (**c**) 75-nm-sized Ag NP/ZnO window layers.

The current–voltage (I–V) and light output intensity–current (L–I) characteristics for the conventional, LPD-ZnO-enhanced, and LPD-Ag NP/ZnO-enhanced InGaN/GaN LEDs are shown in Figure 5a. The forward voltages for the InGaN/GaN LEDs with an LPD-ZnO window layer and LPD-Ag NP/ZnO window layers grown at $AgNO_3$ concentrations of 0.01, 0.03, 0.05, 0.07, and 0.1 M were 3.64, 3.62, 3.65, 3.67, 3.66, and 3.66 V, respectively, at 20 mA, whereas that for the conventional

InGaN/GaN LED was 3.58 V at the same driving current level. The forward voltage of the InGaN/GaN LEDs with LPD-ZnO and LPD-Ag NP/ZnO window layers was higher than that for the conventional InGaN/GaN LED because of the deterioration of the ohmic contact resulting from the LPD-ZnO and LPD-Ag NP/ZnO growth on the periphery of the contact. In addition, the series resistance of the InGaN/GaN LEDs with LPD-ZnO and LPD-Ag NP/ZnO window layers was slightly higher than that of conventional InGaN/GaN LED. The series resistance of the InGaN/GaN LEDs depends on the contact metal, metal/ITO interface, ITO, p-GaN, InGaN/GaN active region, and n-GaN, and the increased resistance is caused by the deterioration in the contact during LPD-Ag NP/ZnO deposition. The light output intensities for InGaN/GaN LEDs with an LPD-ZnO window layer and LPD-Ag NP/ZnO window layers grown at $AgNO_3$ concentrations of 0.01, 0.03, 0.05, 0.07, and 0.1 M were 132.9, 133.3, 133.8, 143.5, 136.2, and 134.2 mcd, respectively, for a 20-mA driving current, whereas that for the conventional InGaN/GaN LED was 95.0 mcd at the same driving current. The insets of Figure 5a show the microscope images of light emission for InGaN/GaN LEDs with and without an LPD-Ag NP/ZnO window layer. The light output intensity of the InGaN/GaN LED with an LPD-Ag NP/ZnO window layer was higher than that without an LPD-Ag NP/ZnO window layer because of the improved LEE. Figure 5b represents the calculated light intensity (from Figure 4a) and measured light output intensity (from Figure 5a) under the driving current of 100 mA as a function of Ag NP size. Figure 5c performs the transmittance spectra of LPD-Ag NP/ZnO thin films as a function of $AgNO_3$ concentration over the visible wavelength and the inset of Figure 5c shows the transmittance of LPD-ZnO and LPD-Ag NP/ZnO thin films with the changed $AgNO_3$ concentration at the wavelength of 460 nm. The calculated and measured results indicate that the highest light intensity occurred at the Ag NP size of 75 and 76.8 nm ($AgNO_3$ concentration of 0.05 M) because of the highest transmittance at the 460-nm wavelength shown in the inset of Figure 5c. The highest transmittance can possibly be attributed to the LSP effect [41]. However, the calculated light intensity decreases with Ag NP size increasing form 0 to 60 nm, but the measured light output intensity increases slightly at the same range. This may be attributed to the close transmittance for LPD-ZnO and LPD-Ag NP/ZnO with the Ag NP size of 60 nm, as shown in the inset of Figure 5c. A low Ag NP density on fake-shaped ZnO and a weak LSP effect lead to a slight increase in light output intensity for InGaN/GaN LED with the LPD-Ag NP/ZnO window layer (60-nm AgNP) compared with that with the LPD-ZnO window layer.

The increased light output intensity may be attributed to the textured surface and LSP coupling effect. It is important to confirm the dominated factor of enhanced LEE for InGaN/GaN LED with an LPD-Ag NP/ZnO window layer. Figure 6 represents the RMS roughness for LPD-Ag NP/ZnO thin films as a function of $AgNO_3$ concentration. The insets of Figure 6 show the AFM images for LPD-Ag NP/ZnO thin films grown at the $AgNO_3$ concentration of 0.01, 0.05, and 0.1 M. In our previous study [31], the LEE of InGaN/GaN LED with an LPD-ZnO window layer depended on the surface texture rather than the ratio of Zn/O. The surface texture depends on the deposition rate of ZnO flake, which was determined by the HCl concentration, and the ratio of Zn/O is almost the same under the varied HCl concentration. Therefore, the enhanced LEE of InGaN/GaN LED caused by the textured LPD-Ag NP/ZnO window layer can be investigated by the RMS roughness. The LEE of an LED gives the ratio of the number of useful emitting-photons to the number of injection charge particles and is defined as [3]

$$\eta_{LEE} = \frac{dP}{dI}\frac{e\lambda}{hc}, \tag{7}$$

where P is the light output power, I is the injection current, h is the Plank constant, and λ is the emitting wavelength. A textured window layer applied to InGaN/GaN LED can improve the escaped probability of the photons emitted from the InGaN/GaN active region. The RMS roughness of the LPD-Ag NP/ZnO thin film decreases with an increase in the $AgNO_3$ concentration. This is attributed to the following reasons. First, increasing the $AgNO_3$ concentration leads to a large number of Ag NPs aggregated at the interface between ZnO flakes to reduce the RMS roughness of the LPD-Ag NP/ZnO thin film. Secondly, the ZnO flakes in LPD-Ag NP/ZnO thin films tend to bend under a high $AgNO_3$

concentration, resulting in a low RMS roughness. This is attributed to the fact that the concentration of the Ag precursor has a great effect on the morphology of ZnO [42]. LED with a weak textured window layer (with low RMS roughness) shows a lower LEE than that with a strong one (with large RMS roughness) [3]. However, the light output intensity of InGaN/GaN LED increases as the $AgNO_3$ concentration rises from 0 to 0.05 M, as shown in Figure 5. Consequently, the increasing light output intensity of InGaN/GaN LED with an LPD-Ag NP ZnO window layer is attributed to the LSP coupling effect, which occurred because the extinction wavelength was similar to the wavelength of the photon emitted from the InGaN/GaN active region, as shown in Figure 3 [41]. However, further increasing the $AgNO_3$ concentration above 0.07 M reduced the light output intensity because of the weak LSP coupling effect caused by the difference between the resonant wavelength and the emission wavelength from the InGaN/GaN active region. In addition, a strong LSP coupling effect can be obtained by choosing an extinction wavelength similar to and smaller than the emission wavelength from the LED active region.

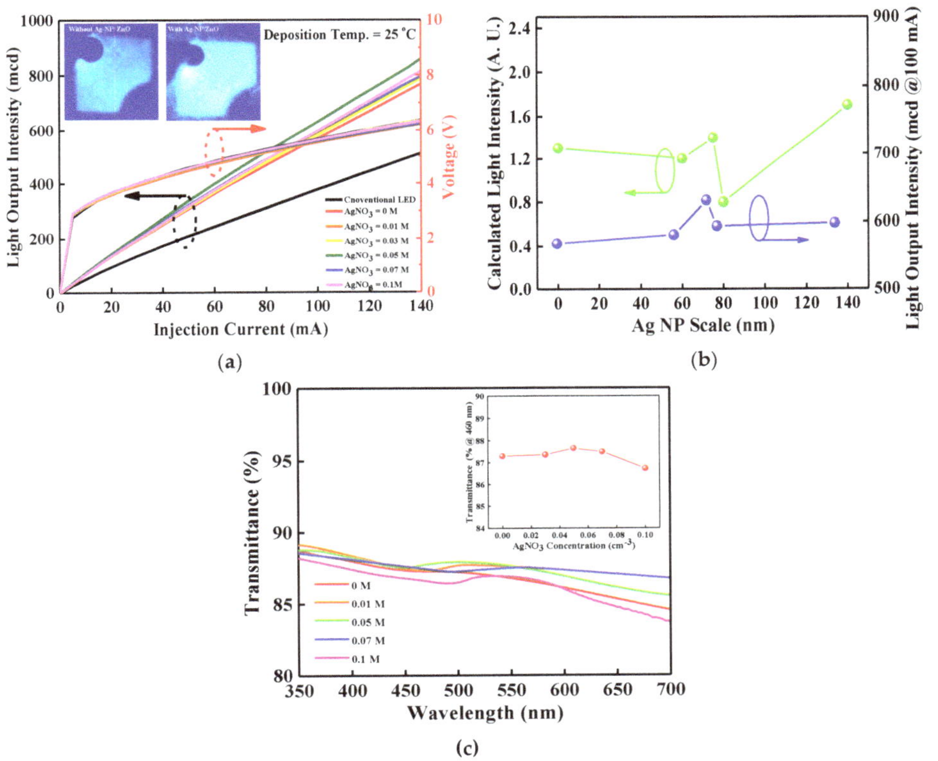

Figure 5. (**a**) I–V and L–I plots for the conventional, LPD-ZnO-enhanced, and LPD-Ag NP/ZnO-enhanced InGaN/GaN LEDs. The micrographs of light emission for InGaN/GaN LEDs with and without the LPD-Ag NP/ZnO window layer are shown in the inset, (**b**) calculated light intensity and measured light output intensity at a 100-mA driving current as a function of Ag NP size, and (**c**) transmittance spectra of LPD-Ag NP/ZnO thin films as a function of $AgNO_3$ concentration over the visible range. The inset of Figure 5c shows the transmittance at a 460-nm wavelength for LPD-ZnO and LPD-Ag NP/ZnO thin films deposited at varied $AgNO_3$ concentrations.

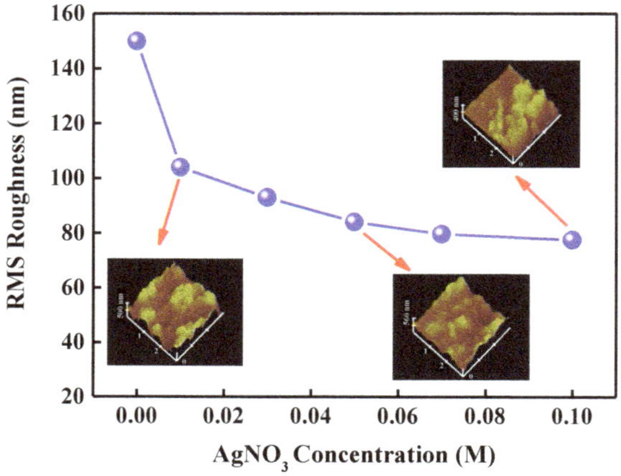

Figure 6. Root mean square (RMS) roughness for LPD-Ag NP/ZnO thin films as a function of $AgNO_3$ concentration. The insets are AFM images for LPD-Ag NP/ZnO thin films grown at 0.01, 0.05, and 0.1 M.

Figure 7 shows the average forward voltage and light output intensity under an injection current of 20 mA for the chosen chips from the InGaN/GaN wafer with the optimal LPD-Ag NP/ZnO and LPD-ZnO window layer for varied runs. The uniformity of the LPD-Ag NP/ZnO thin film on the InGaN/GaN wafer is noteworthy because it is crucial when determining the performance of the InGaN/GaN LEDs. Because the size and distribution of LPD-Ag NPs on the flake-shaped ZnO were similar, the chosen chips from the InGaN/GaN wafer with the optimal LPD-Ag NP/ZnO window layer revealed variations of approximately 5% in the forward voltage and 4.3% in the light output intensity under the same driving current conditions. In addition, the standard deviation of the measured enhancement of emission intensity of the InGaN/GaN wafer with LPD-Ag NP/ZnO to that with LPD-ZnO was about 1.4%.

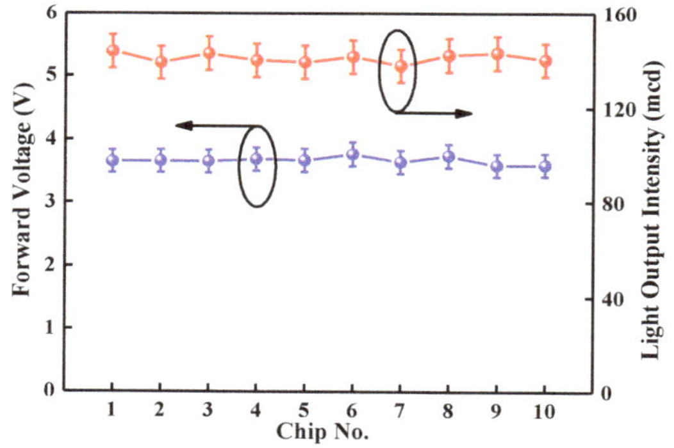

Figure 7. Average forward voltages and light output intensities at an injection current of 20 mA for the selected chips on the InGaN/GaN wafers with the optimal LPD-Ag NP/ZnO and LPD-ZnO window layer for varied runs.

Figure 8 shows the electroluminescence (EL) spectra of the InGaN/GaN LEDs with and without an LPD-Ag NP/ZnO window layer deposited at various $AgNO_3$ concentrations under a 20-mA driving current. The LPD-Ag NP/ZnO thin film deposited at an $AgNO_3$ concentration of 0.05 M exhibited the highest light output intensity because of the LSP coupling effect, and it was approximately 1.52 times higher than that of the conventional InGaN/GaN LEDs. The inset of Figure 8 depicts the full width at half maximum (FWHM) of the EL spectra for the LPD-Ag NP/ZnO-enhanced and conventional InGN/GaN LEDs. The FWHM of the InGaN/GaN LED with an LPD-Ag NP ZnO window layer is narrower than that of a conventional InGaN/GaN LED because of the LSP coupling effect, which can concentrate the emitting spectrum at the resonant wavelength.

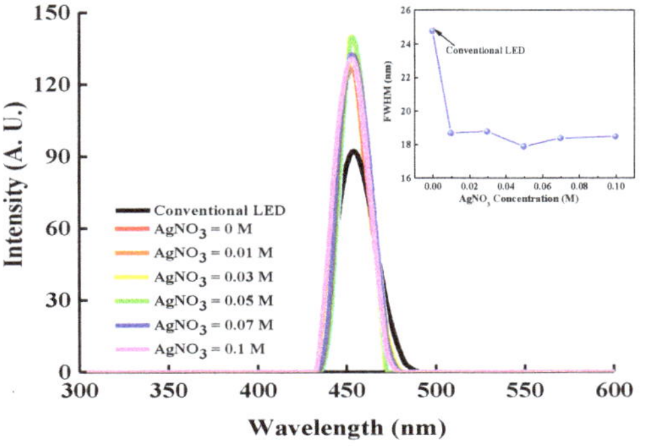

Figure 8. Electroluminescence (EL) spectra of InGaN/GaN LEDs with and without LPD-Ag NP/ZnO thin films deposited at different $AgNO_3$ concentrations under a 20-mA driving current. The inset shows the FWHMs of the EL spectra.

4. Conclusions

An LPD-Ag NP/ZnO window layer can enhance the light output intensity of InGaN/GaN LEDs. The improved LEE was attributed to the LSP coupling effect, which was caused by the Ag NPs distributed on the flake-shaped ZnO. The size and distribution of the Ag NPs on the flake-shaped ZnO were adjusted according to the concentration of the $AgNO_3$ aqueous solution that was added to the ZnO-powder-saturated HCl aqueous solution and the pH scale of treatment solution. We quantitatively demonstrated the density of the Ag NPs distributed on the flake-shaped ZnO by determining the Ag content of the LPD-Ag NP/ZnO thin films, because they had similar variations in $AgNO_3$ concentration. The experimental result indicated that the maximum light output intensity occurred when Ag NP with the diameter of approximately 76.8 nm, which is close to the calculated result of Ag NP with the diameter of 75 nm, and with the content of 3.76 at %, were distributed on the flake-shaped ZnO, because the LSP resonance wavelength for this condition was similar to the emission wavelength of the InGaN/GaN LEDs. The light output intensity of the InGaN/GaN LEDs with the optimal LPD-Ag NP/ZnO window layer was 1.52 times higher than that of the conventional InGaN/GaN LED.

Author Contributions: This research was designed and analyzed by L.-P.H. The experimental process and data measurement were performed by Y.-C.D. and Y.-S.J. The data calculation and collection steps were performed by H.-P.C.

Acknowledgments: The authors gratefully acknowledge the financial support of the Ministry of Science and Technology, R. O. C. (107-2221-E-150-033).

Conflicts of Interest: The authors declare no conflict of interest.

References

1. Dupuis, R.D.; Krames, M.R. History, Development, and applications of high-brightness visible light-emitting diodes. *J. Lightwave Technol.* **2008**, *26*, 1154–1171. [CrossRef]
2. Li, L.P.; Zhang, Y.H.; Bi, W.G.; Zhang, Z.H.; Kuo, H.C. On the hole injection for III-Nitride based deep ultraviolet light-emitting diodes. *Materials* **2017**, *10*, 1221. [CrossRef] [PubMed]
3. Schubert, E.F. *Light–Emitting Diodes*, 2nd ed.; Cambridge University Press: Cambridge, UK, 2003.

4. Koike, M.; Shibata, N.; Kato, H.; Takahashi, Y. Development of high efficiency InGaN/GaN multiquantum-well light-emitting diodes and their application. *IEEE J. Sel. Top. Quantum Electron.* **2002**, *8*, 271–277. [CrossRef]
5. Nikolaevsky, L.; Shchori, T.; Malka, D. Modeling a 1 × 8 MMI Green Light Power Splitter Based on Gallium-Nitride Slot Waveguide Structure. *IEEE Photon. Technol. Lett.* **2018**, *30*, 720–723. [CrossRef]
6. Shoresh, T.; Katanov, N.; Malka, D. 1 × 4 MMI visible light wavelength demultiplexer based on a GaN slot-waveguide structure. *Photonic Nanostruct. Fundam. Appl.* **2018**, *30*, 45–49. [CrossRef]
7. Dadabayev, R.; Shabairou, N.; Zalevsky, Z.; Malka, D. A visible light RGB wavelength demultiplexer based on silicon-nitride multicore PCF. *Opt. Laser Technol.* **2019**, *111*, 411–416. [CrossRef]
8. Ngo, T.H.; Chery, N.; Valvin, P.; Courville, A.; Mierry, P.; Damilano, B.; Ruterana, P.; Gil, B. Internal quantum efficiency in polar and semipolar (11–22) InxGa1-xN/InyGa1-yN quantum wells emitting from blue to red. *Superlattices Microstruct.* **2018**, *113*, 129–134. [CrossRef]
9. Ryu, H.Y.; Ryu, G.H.; Choi, Y.H.; Ma, B.J. Modeling and simulation of efficiency droop in GaN-based blue lightemitting diodes incorporating the effect of reduced active volume of InGaN quantum wells. *Curr. Appl. Phys.* **2017**, *17*, 1298–1302. [CrossRef]
10. Koronski, K.; Strak, P.; Wierzbicka, A.; Grzanka, E.; Borysiuk, J.; Sobczak, K.; Jakiela, R.; Sobanska, M.; Klosek, K.; Monroy, E.; et al. Experimental and theoretical analysis of influence of barrier composition on optical properties of GaN/AlGaN multi-quantum wells: Temperature- and pressure-dependent photoluminescence studies. *J. Alloys Compd.* **2018**, *769*, 1064–1071. [CrossRef]
11. Zhou, Q.B.; Xu, M.S.; Li, Q.X.; Wang, H. Improved efficiency of GaN-based green led by a nano-micro complex patterned sapphire substrate. *IEEE Photonic Technol. Lett.* **2017**, *29*, 983–986. [CrossRef]
12. Liou, J.K.; Chen, W.C.; Chang, C.H.; Chang, Y.C.; Tsai, J.H.; Liu, W.C. Enhanced light extraction of a high-power GaN-based light-emitting diode with a nanohemispherical hybrid backside reflector. *IEEE Trans. Electron Dev.* **2015**, *62*, 3296–3301. [CrossRef]
13. Huang, S.W.; Chang, C.C.; Lin, H.Y.; Li, X.F.; Lin, Y.C.; Liu, C.Y. Fabrication of nano-cavity patterned sapphire substrate using self-assembly meshed Pt thin film on c-plane sapphire substrate. *Thin Solid Film* **2017**, *628*, 127–131. [CrossRef]
14. Zhou, S.J.; Zheng, C.J.; Lv, J.J.; Gao, Y.L.; Wang, R.Q.; Liu, S. GaN-based flip-chip LEDs with highly reflective ITO/DBR p-type and via hole-based n-type contacts for enhanced current spreading and light extraction. *Opt. Laser Technol.* **2017**, *92*, 95–100. [CrossRef]
15. Chang, S.J.; Lin, N.M.; Shei, S.C. GaN-based power flip-chip LEDs with SILAR and hydrothermal ZnO nanorods. *IEEE J. Sel. Top. Quantum Electron.* **2015**, *21*, 9100405.
16. Lin, T.S.; Wang, S.J.; Hung, C.H. Enhanced light extraction of GaN-based vertical LEDs with patterned trenches and nanostructures. *Mater. Sci. Semicond. Process.* **2017**, *57*, 77–82. [CrossRef]
17. Uthirakumar, P.; Kang, J.H.; Ryu, B.D.; Kim, H.G.; Hong, C.H. Nanoscale ITO/ZnO layer-texturing for high-efficiency InGaN/GaN light emitting diodes. *Mater. Sci. Eng. B* **2010**, *166*, 230–234. [CrossRef]
18. Lee, T.X.; Chou, C.C. Scale-dependent light scattering analysis of textured structures on LED light extraction enhancement using hybrid full-wave finite-difference time-domain and ray-tracing methods. *Energy* **2017**, *10*, 424. [CrossRef]
19. Xu, K.; Xie, Y.Y.; Ma, H.L.; Du, Y.X.; Zeng, F.G.; Ding, P.; Gao, Z.Y.; Xu, C.; Sun, J. ZnO nanorods/graphene/Ni/Au hybrid structures as transparent conductive layer in GaN LED for low work voltage and high light extraction. *Solid-State Electron.* **2016**, *126*, 5–9. [CrossRef]
20. Yang, Y.B.; Ren, Y.; Chen, Y.X.; Liu, M.G.; Chen, W.J.; Han, X.B.; Lin, X.Q.; Liao, Q.; Zang, W.J.; Lou, H.; et al. Light output enhancement of GaN-based light-emitting diodes by maskless surface roughening. *Microelectron. Eng.* **2015**, *139*, 39–42. [CrossRef]
21. Byeon, K.J.; Cho, J.Y.; Jo, H.B.; Lee, H. Fabrication of high-brightness GaN-based light-emitting diodes viathermal nanoimprinting of ZnO-nanoparticle-dispersed resin. *Appl. Surf. Sci.* **2015**, *346*, 354–360. [CrossRef]
22. Lei, P.H.; Yang, C.D.; Yang, Y.S.; Lin, J.H. Preparation of a periodic polystyrene nanosphere array using the dip-drop method with post-deposition etching and its application of improving light extraction efficiency of InGaN/GaN LEDs. *Nanoscal. Res. Letts.* **2018**, *13*, 180. [CrossRef] [PubMed]
23. Mikulics, M.; Winden, A.; Marso, M.; Moonshiram, A.; Luth, H.; Grutzmacher, D.; Hardtdegen, H. Nano-light-emitting-diodes based on InGaN mesoscopic structures for energy saving optoelectronics. *Appl. Phys. Letts.* **2016**, *109*, 041103. [CrossRef]

24. Fadil, A.; Iida, D.; Chen, Y.T.; Ou, Y.Y.; Kamiyama, S.; Ou, H.Y. Influence of near-field coupling from Ag surface plasmons on InGaN/GaN quantum-well photoluminescence. *J. Lumin.* **2016**, *175*, 213–216. [CrossRef]
25. Li, Y.F.; Wang, S.A.; Su, X.L.; Tang, W.H.; Li, Q.A.; Guo, M.F.; Zhang, Y.; Zhang, M.Y.; Yun, F.; Hou, X. Efficiency droop suppression of distance-engineered surface plasmon-coupled photoluminescence in GaN-based quantum well LEDs. *AIP Adv.* **2017**, *7*, 115118. [CrossRef]
26. Lu, C.H.; Wu, S.E.; Lai, Y.L.; Li, Y.L.; Liu, C.P. Improved light emission of GaN-based light-emitting diodes by efficient localized surface plasmon coupling with silver nanoparticles. *J. Alloys Compd.* **2014**, *585*, 460–464. [CrossRef]
27. Zhao, H.P.; Zhang, J.; Liu, G.Y.; Tansu, N. Surface plasmon dispersion engineering via double-metallic Au/Ag layers for III-V nitride based light-emitting diodes. *Appl. Phys. Lett.* **2011**, *98*, 151115. [CrossRef]
28. Zhang, H.S.; Zhu, J.; Zhu, Z.D.; Li, Q.Q.; Jin, G.F. Surface-plasmon-enhanced GaN-LED based on the quasi-symmetrical planar waveguide structure. *Opt. Commun.* **2013**, *311*, 311–316. [CrossRef]
29. Hsieh, C.; Yao, Y.F.; Chen, C.F.; Shih, P.Y.; Lin, C.H.; Su, C.Y.; Chen, H.S.; Chen, C.H.; Yu, C.K.; Kiang, Y.W.; et al. Localized surface plasmon coupled light-emitting diodes with buried and surface Ag nanoparticles. *IEEE Photonics Technol. Lett.* **2014**, *26*, 1699–1702. [CrossRef]
30. Niu, B.J.; Wu, L.L.; Tang, W.; Zhang, X.T.; Meng, Q.G. Enhancement of near-band edge emission of Au/ZnO composite nanobelts by surface plasmon resonance. *Cryst. Eng. Commun.* **2011**, *13*, 3678–3681. [CrossRef]
31. Cho, C.Y.; Hong, S.H.; Park, S.J. Improvement of optical and electrical properties of indiumtin oxide layer of GaN-based light-emitting diode by surface plasmon in silver nanoparticles. *Thin Solid Films* **2015**, *590*, 76–79. [CrossRef]
32. Lee, I.H.; Jang, L.W.; Polyakov, A.Y. Performance enhancement of GaN-based light emitting diodes by the interaction with localized surface plasmons. *Nano Energy* **2015**, *13*, 140–173. [CrossRef]
33. Tateishi, K.; Funato, M.; Kawaka, Y.; Okamoto, K.; Tamada, K. Highly enhanced green emission from InGaN quantum wells due to surface plasmon resonance on aluminum films. *Appl. Phys. Lett.* **2015**, *106*, 121112. [CrossRef]
34. Lei, P.H.; Lee, Y.C.; Lin, C.T. Enhanced extraction efficiency of InGaN/GaN light-emitting diodes by liquid phase-deposited zinc-oxide-textured window layer. *IEEE Trans. Electron. Dev.* **2012**, *59*, 1672–1678.
35. Yao, T.; Hong, S.K. *Advance in Material Research 12 Oxide and Nitride Semiconductors Processing, Properties, and Applications*; Springer: New York, NY, USA, 2009.
36. Xie, J.S.; Wu, Q.S. One-pot synthesis of ZnO/Ag nanospheres with enhanced photocatalytic activity. *Mater. Lett.* **2010**, *64*, 389–392. [CrossRef]
37. Zhang, S.G.; Zhang, X.W.; Yin, Z.G.; Si, F.T.; Gao, H.L.; Dong, J.J.; Liu, X. Optimization of electroluminscence from n-ZnO/AlN/p-GaN light-emitting diodes by tailoring Ag localized surface plasmon. *J. Appl. Phys.* **2012**, *112*, 013112. [CrossRef]
38. Shan, G.; Xu, L.H.; Wang, G.R.; Liu, Y. Enhanced Raman scattering of ZnO quantum dots on silver colloids. *J. Phys. Chem.* **2007**, *C111*, 3290–3293. [CrossRef]
39. Xie, W.; Li, Y.Z.; Sun, W.; Huang, J.C.; Xie, H.; Zhao, X.J. Surface modification of ZnO with Ag improves its photocatalytic efficiency and photostability. *J. Photochem. Photobiol. A* **2010**, *216*, 149–155. [CrossRef]
40. Han, Z.Z.; Ren, L.L.; Cui, Z.H.; Chen, C.Q.; Pan, H.B.; Chen, J.Z. Ag/ZnO flower heterostructures as a visile-light driven photocatalyst via surface plasmon resonance. *Appl. Catal. B* **2012**, *126*, 298–305. [CrossRef]
41. Yu, Z.Z.; Li, Q.; Fan, Q.G.; Zhu, Y.X. Investigation on surface-plasmon-enhanced light emission of InGaN/GaN multiple quantum wells. *Superlattices Microstruct.* **2018**, *117*, 200–206. [CrossRef]
42. Shan, G.Y.; Zheng, S.J.; Chen, S.P.; Chen, Y.W.; Liu, Y.C. Multifunctional ZnO/Ag nanorod array as highly sensitive substrate for surface enhanced Raman detection. *Colloid Surf. B Biointerface* **2012**, *94*, 157–162. [CrossRef]

MDPI
St. Alban-Anlage 66
4052 Basel
Switzerland
Tel. +41 61 683 77 34
Fax +41 61 302 89 18
www.mdpi.com

Micromachines Editorial Office
E-mail: micromachines@mdpi.com
www.mdpi.com/journal/micromachines

www.ingramcontent.com/pod-product-compliance
Lightning Source LLC
Chambersburg PA
CBHW061501210926
51850CB00008B/1779
9783039369041